Hydrosystem Restoration Handbook
Streamflow Recharge (SFR) and Lake Rehabilitation (LR)

Hydrosystem Restoration Handbook

Streamflow Recharge (SFR) and Lake Rehabilitation (LR)

Volume 1

Edited by

Saeid Eslamian
Department of Water Sciences and Engineering, College of Agriculture, Isfahan University of Technology, Isfahan, Iran

Faezeh Eslamian
GHD, Montreal, Canada

ELSEVIER

Elsevier
Radarweg 29, PO Box 211, 1000 AE Amsterdam, Netherlands
125 London Wall, London EC2Y 5AS, United Kingdom
50 Hampshire Street, 5th Floor, Cambridge, MA 02139, United States

Notices

Knowledge and best practice in this field are constantly changing. As new research and experience broaden our understanding, changes in research methods, professional practices, or medical treatment may become necessary.

Practitioners and researchers must always rely on their own experience and knowledge in evaluating and using any information, methods, compounds, or experiments described herein. In using such information or methods they should be mindful of their own safety and the safety of others, including parties for whom they have a professional responsibility.

To the fullest extent of the law, neither the Publisher nor the authors, contributors, or editors, assume any liability for any injury and/or damage to persons or property as a matter of products liability, negligence or otherwise, or from any use or operation of any methods, products, instructions, or ideas contained in the material herein.

ISBN: 978-0-443-29802-8

For Information on all Elsevier publications
visit our website at https://www.elsevier.com/books-and-journals

Publisher: Candice Janco
Acquisitions Editor: Maria Elekidou
Editorial Project Manager: Rupinder Heron
Production Project Manager: Paul Prasad Chandramohan
Cover Designer: Mark Rogers

Typeset by MPS Limited, Chennai, India

Dedication

To my dearest wife, Jacklin

Your love is the ink that writes the story of my life. Every moment with you is a chapter filled with joy, laughter, and endless support. Your presence is my inspiration, and this book is a testament to the beautiful journey we share. Thank you for being my everything.

With all my love,
Saeid

Contents

Part II
Best management

4. Managing streams through restored floodplains: a case of Ganga River in the middle Ganga plain

Ankit Modi, Saeid Eslamian and Vishal Kapoor

5. Best management practices in stream: debris and runoff reduction, riparian buffers and plantings, and stabilizing stream banks

Sagarika Patowary, Mridusmita Debnath and Arup K. Sarma

6. River restoration involves using riverside vegetation to enhance ecological health while implementing effective management strategies

Meenakshi Pawar and Pushkar Pawar

Part III
River flow modeling

7. Evaluating the reliability of open-source hydrodynamic models in flood inundation mapping: an exhaustive approach over a sensitive coastal catchment

Dev Anand Thakur, Vijay Suryawanshi, H. Ramesh and Mohit Prakash Mohanty

8. Simulation of river flow (as a primary component for aquifer recharge) using deep learning approach

Mohammad Javad Zareian and Fatemeh Salem

Part IV
Climate changes impacts and adaptation

9. Accuracy of climate and weather early warnings for sustainable crop water and river basin management

Punnoli Dhanya, Vellingiri Geethalakshmi, Subbiah Ramanathan, Kandasamy Senthilraja, Manickam Dhasarathan, Punnoli Sreeraj, Ganesan Dheebakaran, Chinnasamy Pradipa, Kulanthaisamy Bhuvaneshwari, N.S. Vidhya Priya, Sasirekha Sivasubramaniam, Prasad Arul and S. Vigneswaran

10. Flood risk assessment for the integrated disaster risk management and climate change adaptation action plan: La Mojana Region, Colombia

Omar Dario Cardona, Gabriel Andres Bernal and Mabel Cristina Marulanda

11. Impact of climate changes on stream flow discharge of some African rivers

A. Adediji, M.O. Ibitoye, J.J. Idolor and Saeid Eslamian

12. Stream morphology changes under climate changes

Bhawana Nigam

Part V
Case studies

13. Stream flow restoration case studies in the United States and Canada

Saeid Eslamian and Mousa Maleki

14. PATRICOVA viewer for teaching the risk of flooding: a resource to improve resilience to natural hazards

*Álvaro-Francisco Morote, Jorge Olcina and
Saeid Eslamian*

15. Lake rehabilitation case studies in China

*Jihui Fan, Majid Galoie, Artemis Motamedi,
Ning Niu and Saeid Eslamian*

16. Analysis of surface flows of Urmia Lake Basin: a review

*Mina Mahdizadeh, Saeid Eslamian and
Yaser Sabzevari*

20. Rejuvenation of streams and rivers through decentralized and community-driven rainwater capture for dignified livelihoods, climate resilience, and peace: living examples from Rajasthan, India

Indira Khurana

List of contributors

A. Adediji Department of Geography, Obafemi Awolowo University, Ile-Ife, Nigeria

Prasad Arul Krishi Vigyan Kendra, Tamil Nadu Agriculture University, Tirur, Tamil Nadu, India; District Agrometeorology Unit, Chennai, Tamil Nadu, India

Gabriel Andres Bernal INGENIAR: Risk Intelligence, Bogota, Colombia; Universidad Nacional de Colombia, Ingenieria Civil y Agricola, Bogota, Colombia

Kulanthaisamy Bhuvaneshwari Agro Climatic Research Centre, Tamil Nadu Agricultural University, Coimbatore, Tamil Nadu, India

Omar Dario Cardona Universidad Nacional de Colombia, IDEA, Manizales, Colombia; INGENIAR: Risk Intelligence, Bogota, Colombia

Fatemeh Dadvand Department of Water Sciences and Engineering, College of Agriculture, Isfahan University of Technology, Isfahan, Iran

Mridusmita Debnath Scientist, Indian Council of Agricultural Research- Research Complex for Eastern Region, Patna, Bihar, India

Punnoli Dhanya Agro Climatic Research Centre, Tamil Nadu Agricultural University, Coimbatore, Tamil Nadu, India

Manickam Dhasarathan Tamil Nadu Agricultural University, Coimbatore, Tamil Nadu, India

Ganesan Dheebakaran Agro Climatic Research Centre, Tamil Nadu Agricultural University, Coimbatore, Tamil Nadu, India

Saeid Eslamian Department of Water Sciences and Engineering, College of Agriculture, Isfahan University of Technology, Isfahan, Iran

Jihui Fan Institute of Mountain Hazards and Environment, Chinese Academy of Sciences, Chengdu, Sichuan, P.R. China

Majid Galoie Civil Engineering Department, Imam Khomeini International University, Qazvin, Iran

Vellingiri Geethalakshmi Tamil Nadu Agriculture University, Coimbatore, Tamil Nadu, India

M.O. Ibitoye Department of Remote Sensing and Geoinformation Science, Federal University of Technology, Akure, Nigeria

J.J. Idolor Institute of Ecology & Environmental Management, Obafemi Awolowo University, Ile-Ife, Nigeria

Dejana Jakovljević Geographical Institute "Jovan Cvijić", Serbian Academy of Sciences and Arts, Belgrade, Serbia

Vishal Kapoor Indian Institute of Technology Kanpur, Kanpur, Uttar Pradesh, India

Indira Khurana Indian Himalayan River Basins Council, New Delhi, India

Hemraj Ramdas Kumavat Department of Civil Engineering, R C Patel Institute of Technology, Shirpur, Maharashtra, India

Purushottam Kumar Mahato Floodkon Consultants LLP., Sector 132, Noida, Uttar Pradesh, India

Mina Mahdizadeh Department of Soil Science and Engineering, College of Agriculture, Isfahan University of Technology, Isfahan, Iran

Mousa Maleki Department of Water Sciences and Engineering, College of Agriculture, Isfahan University of Technology, Isfahan, Iran

S. Manasi Centre for Research in Urban Affairs (CRUA), Institute for Social and Economic Change (ISEC), Bengaluru, Karnataka, India

Dipak Mandal Centre for Research in Urban Affairs (CRUA), Institute for Social and Economic Change (ISEC), Bengaluru, Karnataka, India

Mabel Cristina Marulanda INGENIAR: Risk Intelligence, Bogota, Colombia

Ana Milanović Pešić Geographical Institute "Jovan Cvijić", Serbian Academy of Sciences and Arts, Belgrade, Serbia

Ankit Modi Department of Civil Engineering, Indian Institute of Technology Kanpur, Kanpur, Uttar Pradesh, India

Mohit Prakash Mohanty Department of Water Resources Development and Management, Indian Institute of Technology Roorkee, Roorkee, Uttarakhand, India

Álvaro-Francisco Morote Department of Experimental and Social Sciences Education, Faculty of Teaching Training, University of Valencia, Valencia, Spain

Artemis Motamedi Civil Engineering Department, Buein Zahra Technical University, Qazvin, Iran

Bhawana Nigam Department of Geography, Anugrah Narayan College, Patliputra University, Patna, Bihar, India

Ning Niu Institute of Mountain Hazards and Environment, Chinese Academy of Sciences, Chengdu, Sichuan, P.R. China; University of Chinese Academy of Sciences, Beijing, P.R. China

Jorge Olcina Department of Regional Geographical Analysis and Physical Geography, University of Alicante, Alicante, Spain

Padam Jee Omar Department of Civil Engineering, Babasaheb Bhimrao Ambedkar University, Lucknow, Uttar Pradesh, India

Sagarika Patowary North Eastern Regional Institute of Water and Land Management, Tezpur, Assam, India

Meenakshi Pawar Department of Architecture and Planning, Indira Gandhi Delhi Technical University for Women, Government of Delhi, New Delhi, India

Pushkar Pawar Department of Environmental Planning, School of Planning & Architecture, New Delhi (SPAD), Ministry of Education, Government of India, New Delhi, Delhi, India

Chinnasamy Pradipa Agro Climatic Research Centre, Tamil Nadu Agricultural University, Coimbatore, Tamil Nadu, India

Subash Prasad Rai Floodkon Consultants LLP., Sector 132, Noida, Uttar Pradesh, India

Subbiah Ramanathan Agro Climatic Research Centre, Tamil Nadu Agricultural University, Coimbatore, Tamil Nadu, India

H. Ramesh Department of Water Resources and Ocean Engineering, National Institute of Technology Karnataka, Mangaluru, Karnataka, India

Yaser Sabzevari Department of Water Sciences and Engineering, College of Agriculture, Isfahan University of Technology, Isfahan, Iran

Fatemeh Salem Faculty of Computer Science and Engineering, Shahid Beheshtei University, Tehran, Iran

Arup K. Sarma Indian Institute of Technology Guwahati, Guwahati, Assam, India

Kandasamy Senthilraja Tamil Nadu Agricultural University, Coimbatore, Tamil Nadu, India

Shashank Singh Floodkon Consultants LLP., Sector 132, Noida, Uttar Pradesh, India

Sasirekha Sivasubramaniam ECE, Coimbatore, Tamil Nadu, India

Punnoli Sreeraj Thangal Kunju Musaliar College of Engineering, Kollam, Kerala, India

Zoran Stevanović Center for Karst Hydrogeology of the Department of Hydrogeology, Faculty of Mining & Geology, University of Belgrade, Belgrade, Serbia

Vijay Suryawanshi Department of Water Resources and Ocean Engineering, National Institute of Technology Karnataka, Mangaluru, Karnataka, India

Dev Anand Thakur Department of Water Resources Development and Management, Indian Institute of Technology Roorkee, Roorkee, Uttarakhand, India

Harinarayan Tiwari Floodkon Consultants LLP., Sector 132, Noida, Uttar Pradesh, India

Ravi Prakash Tripathi Department of Civil Engineering, Rajkiya Engineering College, Sonbhadra, Uttar Pradesh, India

N.S. Vidhya Priya ECE, Coimbatore, Tamil Nadu, India

S. Vigneswaran Institute of Forest Genetics and Tree Breeding, Coimbatore, Tamil Nadu, India

Mohammad Javad Zareian Department of Water Resources Study and Research, Water Research Institute (WRI), Tehran, Iran

About the editors

Saeid Eslamian received his PhD in civil and environmental engineering from the University of New South Wales, Australia, in 1998. Saeid was a visiting professor at Princeton University and ETH Zurich in 2005 and 2008, respectively. He has contributed to more than 1000 publications in journals, conferences, and books. Eslamian has been honored as a 2% Top Researcher by Stanford University and Elsevier for total career. Currently, he is a Full Professor of resilient climate and water systems and the Director of the Excellence Center in Risk Management and Natural Hazards, Isfahan University of Technology. His scientific interests are floods, droughts, water reuse, climate change adaptation, sustainability and resilience.

Faezeh Eslamian is a PhD holder of bioresource engineering from McGill University. Her research focuses on the development of a novel lime-based product to mitigate phosphorus loss from agricultural fields. Faezeh completed her bachelor's and master's degrees in civil and environmental engineering from Isfahan University of Technology, Iran, where she evaluated natural and low-cost absorbents for the removal of pollutants such as textile dyes and heavy metals. Furthermore, she has conducted research on worldwide water quality standards and wastewater reuse guidelines. Faezeh is an experienced multidisciplinary researcher with research interests in soil and water quality, environmental remediation, water reuse, and drought management. She is now working in GHD, Montreal, Canada.

Preface

This preface introduces the book's comprehensive approach to sustainable water resources management, highlighting its relevance in the context of current environmental challenges. It provides a clear roadmap of the book's structure and content, setting the stage for readers to engage with the detailed information and insights provided in each part and chapter.

In the face of increasing environmental challenges, the necessity for sustainable water resource management has never been more critical. As populations grow and climate change impacts intensify, the degradation of our vital water bodies becomes an urgent issue demanding comprehensive and innovative solutions. The *Hydrosystem Restoration Handbook: Streamflow Recharge (SFR) and Lake Rehabilitation (LR)*, is a timely contribution to the field, providing a thorough examination of the latest methods and practices in stream and lake restoration. Other volumes are as follows:

Vol. 2: Groundwater Natural Recharge (GNR)
Vol. 3: Groundwater Artificial Recharge with Conventional Water (GARC)
Vol. 4: Groundwater Artificial Recharge with Unconventional Water (GARU)

This four-volume handbook explains the techniques and experiences that entail with rehabilitation/restoration of surface water and groundwater resources using both natural and artificial recharge of conventional and unconventional water.

"Hydrosystem Restoration: Streamflow Recharge (SFR) and Lake Rehabilitation (LR)" (Vol. 1) includes five parts:

Part I, **Introduction**, sets the stage with foundational concepts in stream and lake rehabilitation. We begin with "An Introduction to Stream Rehabilitation Planning," which offers an essential overview of the principles and objectives guiding effective restoration projects. This is followed by "Assessment of Site Suitability for Surface Water Retention and Springshed Rejuvenation," presenting methods for evaluating potential sites for water retention initiatives. The final chapter in this section, "Catchment Areas in Karst: Advantages and Disadvantages," explores the unique characteristics of karst landscapes and their implications for water management.

Part II, **Best Management**, delves into practical strategies for managing stream environments. "Managing Streams through Restored Floodplains: A Case of Ganga River in the Middle Ganga Plains" provides a detailed case study highlighting the benefits of floodplain restoration. The subsequent chapters discuss various best management practices, such as debris and runoff reduction, riparian buffers, and stream bank stabilization. "Riverside Vegetation Planting" examines the ecological benefits and management strategies essential for successful river restoration efforts.

Part III, **River Flow Modeling**, addresses the technical aspects of hydrodynamic modeling and its applications. "Evaluating the Reliability of Open-Source Hydrodynamic Models in Flood Inundation Mapping" presents a critical assessment of model reliability in coastal catchments. This is complemented by "Simulation of River Flow as a Primary Component for Aquifer Recharge using Deep Learning Network," which explores the integration of advanced computational techniques in water management.

Part IV, **Climate Change Impacts and Adaptation**, focuses on the intersection of climate science and water resources management. "Accuracy of Climate and Weather Early Warnings for Sustainable Crop Water and River Basin Management" underscores the importance of reliable forecasting in mitigating climate impacts. The subsequent chapters explore flood risk assessment, the effects of climate change on streamflow discharge, and morphological changes in streams due to climate variations.

Part V, **Case Studies**, brings theory and practice together through a series of real-world examples. These case studies span diverse geographical regions and contexts, from streamflow restoration projects in the United States and Canada to community-based conservation initiatives in India. Each chapter provides valuable insights and lessons learned, highlighting the successes and challenges faced in various restoration efforts.

This book aims to serve as a comprehensive resource for researchers, practitioners, policymakers, and students engaged in the field of water resources management. By combining theoretical foundations with practical applications and case studies, we hope to inspire innovative approaches and foster a deeper understanding of the complexities involved in stream and river rehabilitation. Ultimately, our goal is to contribute to the sustainable management and preservation of our vital water resources for future generations.

This handbook will benefit a range of individuals, including undergraduate and postgraduate students, school instructors, university professors, and members of scientific research centers, by providing valuable resources for teaching, researching, experimenting, assessing, and applying updated procedures.

The primary audience of this work would be in disciplines of water science and engineering, soil sciences, and watershed management.

The secondary readers of this work will include professionals from various sectors such as water board staff, State Department of Agriculture officials, Ministry of Energy representatives, consultant engineers, nongovernmental organizations (NGOs), and government policymakers.

This book finally helps in solving the water scarcity crisis, particularly in arid and semiarid regions in the world's changing climate.

We extend our gratitude to all contributors and anonymous reviewers whose expertise and dedication have made this book possible. May this collective effort guide and inform the ongoing journey toward resilient and sustainable water ecosystems.

Saeid Eslamian
Faezeh Eslamian

Part I

Introduction

Chapter 1

An introduction to stream rehabilitation planning

Dejana Jakovljević and Ana Milanović Pešić

Geographical Institute "Jovan Cvijić", Serbian Academy of Sciences and Arts, Belgrade, Serbia

1.1 Introduction

Water resources present one of the most important natural resources for humanity, and many consider it the most important strategic resource of the 21st century (United Nations, 2023). The population growth, as well as accelerated urbanization, have caused river regime fluctuation and water quality degradation in many watercourses. River flow decrease and deficit often cause hydrological drought (Blauhut et al., 2022; Herbert & Döll, 2023; Kreibich et al., 2019), water supply issues, and reduces crop production and hydroelectricity production (Souza & Reis, 2022; Tzanakakis et al., 2020). On the other hand, extremely high flow often causes severe floods in many urban areas worldwide (Arrighi et al., 2018; Gavrilović et al., 2012; Kundzewicz et al., 2013; Marchi et al., 2010; Jakovljević, 2020). In addition, the urbanization effect, along with flow variability, leads to degraded water quality (Chen, 2017; Milanović et al., 2011; Milanović Pešić et al., 2020; Radu et al., 2020; Tang et al., 2019; Uddin & Jeong, 2021; Walker et al., 2015; Xu et al., 2019) and ecosystems (Adeyemo, 2003; Rajaram & Das, 2008; Wright et al., 2015; Zahoor & Mushtaq, 2023; Zhu et al., 2002). Considering that the world is facing numerous issues related to water, the United Nations declared two important international decades for action. The first one was titled "Water for Life" 2005–15 (including International Year of Water Cooperation in 2013), and the second one is dedicated to Water for Sustainable Development, 2018–28. The goal of all these global actions was to highlight the necessity of pure water resources for health and human and economic development (United Nations, 2023) and to undertake numerous activities in order to restore degraded water resources. Therefore, river restoration (Fernandes, 2013; Macedo et al., 2022; Petts et al., 2000) and waterfront development (Aouissi et al., 2023; Hein & Hillmann, 2016; Hoyle et al., 1988) are in the mode and increasingly present worldwide. In order to upgrade treatment types of running water, rehabilitation and restoration are used as the most common concepts. All river and stream rehabilitation projects aim to determine the reintroduction of various habitats and to revitalize damaged ecosystems (King et al., 2003). The first river rehabilitation intervention started with rapid urbanization in the middle of the 20th century. In period from 1950s to 1970s, the river management approach included various technical measures for flood prevention, directed toward the individual river; from the 1980s to 2000s, it additionally included techniques for improvement of ecological state of the river. In the 2000s, the rehabilitation concept was enriched, and besides technical measures, it included the simultaneous restructuration of the river and the city (Şimşek, 2014).

The term *river and stream rehabilitation* is defined in various ways in international literature. However, the common is that all definitions indicate measures by which watercourses can be returned to their original natural state or in as good condition as possible. In defining this term, some authors emphasized concerns for the natural values of riverine sites. Wade et al. (1998) defined rehabilitation as a process of partially returning to former natural state or good working condition. Shields et al. (2003) expressed the opinion that river rehabilitation and river restoration are the same thing and represent the process of restoring a degraded river as close as possible to its remaining natural potential. According to the some authors, rehabilitation is a process of returning the rivers to a former or predegradation state by applying various measures (Roni et al., 2008), which is dedicated to ecological condition (Schanze et al., 2005); activities of artificial stimulation of natural processes or structures (Muhar et al., 2018) or the process of ecological river state improvement in order to bring them closer to their predisturbed state (Fernandes, 2013; Macedo et al., 2022; Petts et al., 2000). However, some authors point out that river rehabilitation does not lead to the complete recovery of rivers

Hydrosystem Restoration Handbook. DOI: https://doi.org/10.1016/B978-0-443-29802-8.00001-7

and their ecosystems (Brierley & Fryirs, 2005). Other studies have pointed out that rehabilitation is more a social than a technical process and should be seen in a sociocultural context (Folke, 2006; Fryirs & Brierley, 2009; Higgs, 2003). It often involves the improvement of natural resources of economic importance and cultural or spiritual significance (Roni et al., 2008) and must take place together with defining the design of public space on the riverbank Fernandes (2013). In addition, river rehabilitation implies stabilizing a degraded river while respecting the land use that is already represented in the river basin (King et al., 2003).

Over time, meaning of the term *river and stream restoration*, as well as its goals, has been transformed from improvement of the water quality (in the 1950s and 1960s) through application of modern techniques (in the 1970s), management of landscapes (in the 1980s), ecosystems, and nature (in the 1990s) to the relation improvement and balance between natural (river) and anthropogenic values (in the 2000s) (Wada, 2010). There is no international consensus or clearly denoted difference in the use and definition of the terms *river rehabilitation* and *river restoration*. Based on a brief overview of scientific literature dealing with ecological issues, it is interesting to emphasize that these terms are used alternately, as they refer to the similar concepts or the activities that are interconnected. Both of these concepts are directed toward ecosystem improvement in degraded rivers or streams. They are often packaged as a set of activities to achieve a self-sustaining condition of rivers and to promote multiple-function river systems (*European Centre for River Restoration*, 2023). Although similarities of these concepts are most often emphasized (King et al., 2003), some scientists pointed out that there are specific differences between them (Shields et al., 2003; Simsek, 2012).

The term *restoration* is denoted differently, with a focus on various approaches and aspects. Meier (1998) considered river and stream restoration as an opportunity to return the river to its original state as much as possible, depending on the prevailing anthropogenic influence to maintain a healthy ecosystem that can be used by society in a sustainable way. Furthermore, restoration is defined as any level of intervention that enables the process of returning the ecosystem to the pre-existing natural state and functions (Roni et al., 2008; Sear, 1994) and creating a self-sustaining ecosystem that would not require additional interventions to maintain the restored state (Fogg & Wells, 1998; Petts et al., 2000). Numerous scientists (Brierley & Fryirs, 2000; Fogg & Wells, 1998; Meier, 1998; Petts et al., 2000) shared the opinion that restoration aims to return, to some degree, a degraded ecosystem to its previous state and that complete recovery is not possible due to the lack of historical sources that would serve as a guide for this process or due to significant modification of the watercourses during the time. They considered restoration as a main term that includes the different levels of intervention (King et al., 2003). According to Rutherfurd et al. (2000), restoration involves activities aimed at restoring the disturbed ecosystem to its previous state of water quality, natural sediments, river regime, morphology, and channel stability, as well as restoration of co-indigenous aquatic plants and animals and plant communities on the coast. The term restoration is often used to describe the processes directed toward recreating rivers' former natural physical, chemical, and biological state (Shields et al., 2003; Wade et al., 1998) or to describe the rehabilitative management of freshwater ecosystems (King et al., 2003). Also, restoration is described as a process of returning a site to its former natural state by applying measures and interventions that contribute to the reconstruction of its structure, functionality, diversity, and dynamics (Bernhardt & Palmer, 2007; Findlay & Taylor, 2006; Muhar et al., 1995; Wohl et al., 2005). According to Simsek (2012), river restoration presents the process of improving the degraded river with the goal of returning to its original condition and includes all aspects, from riverbed morphology to water quality and ecosystems. Shahady (2022) indicated that river restoration presents a methodology for improving degraded water resources and damaged river ecosystems with the aim of returning them to a close to natural state.

The term *river and stream reclamation* is usually in use in the United Kingdom instead of the term *river restoration*, which is common in the United States (Simsek, 2012). It relates to the set of activities directed toward river and stream environmental state improvement. The improved status of the river is indicated by expanded habitat, decreased erosion process of the river banks and beds, improved water quality (through decreased pollutant levels), and achieved self-sustainability of river regime in watercourses without the need for additional periodic human interventions (European Environment Agency, 2023b). Success in river restoration depends on the implementation of a water management plan, which includes several activities such as flood protection measures, regulation of rainfall water, control of sediment inflow from urban areas, among others (Kondolf & Keller, 1991). River restoration enables a future-proof river ecosystem with healthy and safe water (*European Centre for River Restoration*, 2023).

The term *river and stream remediation*, unlike restoration and rehabilitation, implies measures that contribute to the improvement of the ecological condition of watercourses. However, it does not necessarily have to be up to the level of the original state of the river (Rutherfurd et al., 2000). This term is similar to the term *enhancement*, which refers to improving of the ecosystem current state in order to repair or mitigate disturbance effects (Petts et al., 2000), but without reaching its initial state (Calow & Petts, 1994). Therefore, it can be emphasized that remediation is characterized by the goal absence that is related to the state before the degradation. This distinguishes it from restoration and rehabilitation (King et al., 2003).

In general, it can be emphasized that all marked terms rehabilitation, restoration, reclamation, and remediation aim to improve ecosystems in degraded river and stream.

This chapter aims to present an overview of the planning and implementation of river and stream rehabilitation measures, to give examples of successful project implementations worldwide, and to address limitations in applying these measures.

1.2   Historical background

River and stream rehabilitation is closely connected with urbanization, economic development, and population growth. Depending on these processes, the river and stream rehabilitation process has various phases. According to Şimşek (2014), few phases of river rehabilitation practice could be identified during history. Before the 1850s, most rivers were in a natural state. From the 1860s to the 1960s, a few artificial constructions were built. At the end of this period, from the 1950s to 1960s and during the 1970s, with rapid urbanization, the first urban rehabilitation measures were undertaken in flood prevention (Şimşek, 2014). The measures for their prevention included building straightened concrete channels, deepening and enlargement of river beds, and increasing banks for flood prevention. However, during this phase, the environmental functions of rivers, including habitats, self-purification, and riparian function were lost (Şimşek, 2014). This engineering paradigm is called the "river degradation era" (Hillman & Brierley, 2005). Degradation of the aquatic ecosystems was caused by physical disturbance, hydrological manipulation, and chemical disturbance. Physical disturbance, such as channelization, including construction of new channels or modification of existing ones leads to changes in river regime (such as changes in speed and sediment deposition) and loss of ecological biodiversity. Hydrological modification, including dams and reservoirs building, interbasin transfer, and diversion, direct abstraction of surface and groundwater, agricultural activities caused changes in river regime (changes in high and low flows) and sediment regime, and further changes in channel morphology, threatened water quality and aquatic life. Chemical disturbance includes point sources of pollution (industry, households) and nonpoint sources of pollution (agriculture, forestry, traffic), as well as the use of rivers as wastewater recipients, causing water quality decline and ecosystem degradation in terms of abundance and biodiversity (King et al., 2003).

In the next phase, during the late 1980s, the river condition improvement, such as water quality and aesthetics, was in the focus. In this period, parks were constructed, mainly in the floodplains. In the next phase, during the 1990s, ecological river techniques were introduced to enhance the ecological function of the river (Şimşec, 2014). This approach is also called "the age of repair" (Hillman & Brierley, 2005). In the last phase, from the 2000s, the focus is on the connection between river and human. In this concept, so-called "spatial integration," rivers should be the path of the social and cultural assets of the city (Şimşek, 2014).

1.3   Legislation

The long-term and widespread degradation of rivers and freshwater ecosystems has increased interest in their rehabilitation worldwide. The United Nations has also recognized these water issues. It is the goal of the United Nations Decade on Ecosystem Restoration 2021−2030 to support activities that will stop and reverse further degradation of terrestrial, coastal, and marine ecosystems worldwide (United Nations, 2020). This is a global call to action for massively scaled up restoration (including riverine ecosystems) that includes scientific research and financial and political support.

Legislation regarding water management is an important segment in the implementation of the rehabilitation process of rivers and freshwater ecosystems. Framework legislation at the international or national level represents a valuable guideline for implementing river rehabilitation. Additionally, specific policies, strategies, and regulations for water management of international river basins adopted by various international associations and commissions are important.

In Europe, European Water Framework Directive (WFD) is the key directive for establishing environmental assessment programs in the European Union (EU) member states. It was declared by the European Parliament in 2000 in order to serve as a guideline for water management in EU countries and the harmonization of water legislations (European Commission, 2000). WFD had set targets for achieving good ecological status in surface water in these countries by 2015, and now this deadline has been extended to 2027. This Directive includes documents on the characterization of basins, categorization of surface water bodies, ecological classification of water bodies (in five classes, from high to poor), program of measures for the protection and recovery of water ecosystems, recommendations for ecological monitoring of surface water, and guidelines for the evaluation of the success of river rehabilitation projects. In EU countries and countries with the status of EU candidates, WFD has been incorporated into the national Water Laws to establish various environmental targets that should be achieved. In addition, the WISE Water Framework Directive database (European Environment Agency, 2023a) was established, which collects data on river basins of first and second category within the EU countries, along with Great Britain, Norway, and Iceland. In addition, in 2023, Nature Restoration Law was established by the Council of the European Union and the European Parliament. According to this law, EU states are committed to establishing recovery measures with a global goal of restoring 30% of degraded ecosystems by 2030 (European Commission, 2023).

At the international level, numerous international associations and commissions are significant for river rehabilitation activities and projects. Among them is the European Centre for River Restoration (ECRR), which collects data on river rehabilitation projects and provides recommendations for measures and activities that lead to successful rehabilitation of river ecosystems. Also, the International Commission for the Protection of the Danube River deals with regulation related to the sustainable use of water resources and the Danube River Basin, as well as activities related to river restoration in line with EU WFD and national and regional water management policies. The work of this commission is based on the principles of the international Convention on the Protection of the Danube River (1994) (DRPC, 1994). In Asia, the Mekong River Commission was established in 1995 among the governments of Cambodia, Lao, Thailand, and Viet Nam. This agreement includes international cooperation in water resources management in the Mekong River Basin, from water use to the protection measures. Besides that, in 2002, the Basin Development Plan was initiated in order to determine priority projects for implementation at the basin level (Dudgeon, 2005).

Recommendations for ecological monitoring of rivers and freshwater ecosystems and guidelines for their rehabilitation are often incorporated in national Water Laws. However, it is important to emphasize that many countries do not have legislation related to the rehabilitation of rivers and river ecosystems. Water laws of several countries in Asia can be listed as the examples of including of river rehabilitation measures in water legislation. In China, 1984 Law on the Prevention and Control of Water Pollution (revised in 1996 and 2008) and 1988 Water Law (revised in 2010) include water quality rehabilitation in line with national standards (Chinalawinfo Database, 2023). However, these laws do not contain recommendations for technical measures, so the Guidelines for Aquatic Ecological Protection and Restoration Planning was adopted in 2015. It includes technical requirements for rehabilitating the riverine ecosystem. In addition, China has drafted national environmental protection strategies for the Yangtze River (2017) and for the Yellow River (2019). Also, Yangtze River Protection Law was declared in 2021, which is China's first law for one river basin management (Ministry of Ecology and Environment of the Peoples Republic of China, 2023). In Japan, river rehabilitation was incorporated in the River Law in 1997, and the river, stream, and ecosystem conservation and improvement were established as the primary goal (Infrastructure Development Institute, 1998). According to the Water Environment Conservation Act (2018), government in the Republic of Korea has developed national programs for the rehabilitation of rivers and river ecosystems (Korea Legislation Research Institute, 2020).

In the United States, the basic guide for ecosystem monitoring and assessment is Clean Water Act from 1972 (United States Environmental Protection Agency, 2023). Due to a lack of standard protocols for monitoring, analysis, and reporting, reports between states in the United States were diverse and incomparable for all rivers (Paulsen et al., 1998), so the Environmental Monitoring and Assessment Program (EMAP) was initiated in 1997 (Lazorchak et al., 1998). In Canada, the water legislation of some provinces prescribes rehabilitation measures in cases when water quality or the river ecosystem is degraded. At state level, measures regarding river rehabilitation are included in Canadian Fisheries Act 1985 (Justice Laws Website, 2023).

In Brazil, CONAMA Resolution 357 of 2005 also includes the biological assessment of water resources. It also emphasizes the importance of maintaining and protecting coastal vegetation and aquatic biodiversity (CONAMA, 2005). Also, Colombia has guidelines for standardized ecological monitoring of water resources and ecological quality index (Ríos-Touma & Ramírez, 2019). Furthermore, Costa Rica has recently adopted a policy for the protection and preservation of streams and coastal areas, titled National Policy on Protection Areas for Quebradas, Arroios and Nacientes Rivers 2020 (Ministro de Ambiente y Energía Costa Rica, 2020).

In South Africa, National Water Act of South Africa also incorporates resource quality objectives for water objects (Department of Water Affairs & Forestry, 2008). This law can serve as a guideline for river rehabilitation efforts. In Australia, ecological monitoring and assessments of rivers are in the jurisdiction of the regional governments, and particularly important is the Water Resources Act 2007 (Australian Capital Territory, 2019).

Despite legislation, it is important to emphasize that the implementation of activities related to river rehabilitation can be complex due to absence of accompanying ecosystem monitoring, so it remains unknown whether and to what extent there is an improvement in ecological quality. This also makes it challenging to gain new knowledge for river restoration. Furthermore, countries that share the same river basins often have different legislations related to environmental protection, which disables global improvement in the basin.

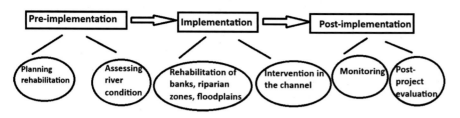

FIGURE 1.1 Phases in the rehabilitation process (based on Gilvear et al., 2013; King et al., 2003) phases in the rehabilitation process (based on Gilvear et al., 2013; King et al., 2003).

1.4 Planning and implementation

Fig. 1.1 Phases in the rehabilitation process (based on Gilvear et al., 2013; King et al., 2003).

The preimplementation phase includes planning rehabilitation and assessing river conditions. Planning rehabilitation includes the collection of specialist data (input) from relevant organizations and individuals. Assessing river condition includes a comparison of the river current state with its former state or natural features. This referent state can be used for setting rehabilitation goals (King et al., 2003).

The implementation phase includes rehabilitation of banks, riparian zones, floodplains, and interventions in the channel. These steps involve different techniques and measures for improving flood protection, water quality rehabilitation of riparian zones, and recreation of ecosystem diversity.

The postimplementation phase covers monitoring and postproject evaluation. Monitoring includes continuous observation of selected elements of the ecosystem, while evaluation presents an assessment of the efficacy of the river rehabilitation process by comparing collected data on certain elements with predefined values for those elements (values that represent the targets of rehabilitation).

1.5 Restoration measures

Restoration measures consist of various tools and techniques, and their application differs from case to case. Frissell and Ralph (1998) concluded that the choice of the restoration measures depends on the anthropogenic influences and characteristics of the river (channel type and its geomorphological and hydrological setting). Besides that, the selection of measures depends on the aspect that should be improved (water quality, river regime, and biodiversity).

Many authors have tried to classify these measures into various categories. King et al. (2003) listed following measures: environmentally sensitive river maintenance activities (modify one bank, reduce or eliminate dredging operations, control disposal of spoil, maintain or establish riparian and aquatic vegetation, control aquatic weeds, and create sediment traps), soft bioengineering approaches (stream bank protection and revegetation), and mitigation of hard engineering practices (in bank zone: structural control, and protection or retaining walls or barriers; full-width structures, partial-width structures, manipulation of substrata and wood debris, providing cover or shelter; in riparian zone and floodplains: reconnecting the floodplain with the river, reinstating meanders, stabilizing streambanks, riparian zone rehabilitation, riparian buffer strips, and constructing wetlands for water quality management). Roni et al. (2008) reviewed various stream rehabilitation techniques throughout the world including road improvements, riparian rehabilitation, floodplain connectivity, and rehabilitation (reconnection of lakes; remeandering a straightened stream; increase flow; restoration of the natural regime, establishing new floodplain habitats), instream habitat improvement (placement of log or boulder structures, spawning gravels, brush, or other covers), and nutrient enrichment. Similarly, Muhar et al. (2018) summarize restoration strategies and applications in integrated river basin management. These measures include re-establishing morphological river type (removing bed and bank stabilization, reconnecting or creating side arms, reconnecting oxbow or meanders, and restructuration of riparian zone); re-establishing lateral connectivity or floodplain habitat restoration (removing or replacing or lowering dams and lowering the floodplain area); flow management (increasing and adapting residual and dynamic environmental flow); hydropeaking mitigation (hydropower production without hydropeaking, hydropeaking mitigation through compensatory reservoirs, and avoiding overlapping hydropeaking); temperature mitigation; sediment management; flushing management; re-establishing longitudinal continuum (deconstructing barrier, rebuilding a fish passes); and modifying land use (encouraging extensive sustainable agriculture and creation of buffer zones). Schanze et al. (2004) classified rehabilitation techniques into the following groups for improvement: hydrology and hydrodynamics, stream morphology and connectivity, water quality, biodiversity, and features of public health and safety. Based on the level of intervention, Matthews et al. (2010) classified rehabilitation interventions into in-channel habitats (small-scale interventions to improve the condition in river channel), morphological (larger-scale interventions in channel morphology), and lateral connectivity (connections between river channel and floodplain, such as removing embankments). Jakovljević et al. (2019) have recommended an integrated system of measures consisting of two main groups: regular and interventional measures. Regular measures involve technical measures (measures for pollution mitigation, emission decrease, an increase of watercourses capacity, and control), legal-organizational measures (establishment of the legal framework for standard effluents and control of point sources of pollution), urban planning measures (the integral plan of regulation, protection, and use of watercourses on the specific territory), economic measures (principles "user pays" and "polluter pays"), and development of information database (plants cadaster, cadaster of water pollutants, and cadaster of limit values for toxic and hazardous substances). Interventional measures include measures for the prevention of accident pollution (identification and removal of pollution sources, continuous monitoring, and preventing the pollution spread), restrictive measures (prohibiting the construction of new facilities and closing existing ones), and notifying the users and prohibition of water use in case of accidental pollution.

Other authors have studied specific aspects of the restoration process. Schmutz and Moog (2018) emphasized mitigation measures for dams: re-establishing longitudinal continuity through fish passes and downstream mitigation measures involving flow regulation and sediment transport. Seliger and Zeiringer (2018) also suggested measures for restoring longitudinal continuity, which involves facilities for upstream and downstream fish migration. The selection of measures for upstream continuity restoration depends on barriers, local conditions, and financial resources, including the removal of barriers, natural-like fish passes, technical fish passes, and special constructions. Measures for downstream continuity restoration involve facilities for improving safe passage and prohibiting transit from harmful hydropower plant components. Greimel et al. (2018) recommended a group of measures in order to reduce the hydrological impact caused by hydropeaking including power plant operational measures (increasing the minimal base flow and reducing flow fluctuation rates, amplitudes, and frequencies), constructional measures (retention basins, hydropeaking drainage via side channels, hydropeaking diversion to new hydropower plants, and side channels with more stable flow), creation of refugial habitats (channel widening, reconnection of tributaries, and construction of side channel with stable flow), and habitat improvement (channel reconstruction and increase of the permanent wetted surface). Biabanaki et al. (2012) addressed following drainage management measures: limiting development of land exposed to flood, guiding development away from locations exposed to flood, contributing in reducing individual losses in flood, land use planning including recognition of flood hazards and damage and value of riparian zone, planning of extension of public facilities in case of flood hazards, and using available public programs and administrative devices for implementing flood management.

With regard to sediment management options, Hauer et al. (2018) suggest structural and nonstructural measures. Structural measures include the installation of boulders and deadwood, river widening, changing the bed (energy) slope, artificial gravel dumping, and removing sediments from the reservoir. Nonstructural measures could be land use change, which should be associated with changes in agricultural and forestry practices leading to the erosion decrease.

Considering that anthropogenic impacts and natural processes degrade water quality, last years, various studies have been conducted in terms of recommendations of measures for water quality improvement. Weigelhofer et al. (2018) addressed measures for managing diffuse nutrient emission to streams and rivers, so-called "end-of-pipe" measures, including riparian areas, vegetated buffer strips, floodplains, and denitrifying bioreactors. Schäfer and Bundschuh (2018) divided "end-of-pipe" measures and technologies into measures for point and nonpoint sources of pollution. The wastewater treatment plant is the most common measure that is applied to prevent point source of pollution. Besides vegetated buffer strips, natural and constructed wetlands are often applied in the reduction of toxicant concentration as a measure for nonpoint sources of pollution.

Nowadays, naturalness is considered the authentic approach in river rehabilitation (Fryirs & Brierley, 2009). Therefore, nature-based solutions, such as constructed wetlands and buffer strips, have essential and multiple roles in this process.

Constructed wetlands are systems projected to use natural processes between vegetation, soil, and microorganisms (Tang et al., 2009; Zemanová et al., 2010). They have multifunctional purposes such as water quality improvement, flood storage and mitigation, and maintenance of surface water during droughts, nutrient cycling, and habitats for wildlife, fish, and plants (Jakovljević et al., 2021; Nikolić et al., 2010). The primary function of constructed wetlands is the reduction of nutrients, pesticides, and metals (Austin & Yu, 2016; Mander et al., 2017; Schäfer & Bundschuh, 2018; Tournebize et al., 2017; Vymazal & Březinová, 2015; Wang et al., 2018; Weigelhofer et al., 2018). Constructed wetlands could be designed to include infiltration, which would increase the flow of rivers and streams during the dry season (Austin & Yu, 2016). Newly constructed wetlands could serve as habitats for aquatic and terrestrial species instead of some former natural wetlands that were drained for agricultural purposes or during the urbanization process (Austin & Yu, 2016; Bae & Lee, 2018).

Buffer strips consist of different types of vegetation, such as trees, shrubs, and grass, and they are also multipurpose. Their functions are to stabilize streambanks, slow and capture runoff, regulate stream temperature, filter pollutants, provide habitats and food, and act as wildlife corridors (Cunningham et al., 2009; Jakovljević et al., 2021). The leading role of buffer strips is in nutrient removal and reduction of toxicants (pesticides, metals, and salts) from agriculture, urban areas, and roads (Mander et al., 2017; Schäfer & Bundschuh, 2018; Walton et al., 2020; Weigelhofer et al., 2018), provide bank stabilization, shading, and increased habitat biodiversity, and improve microclimate conditions (Weigelhofer et al., 2018). Grass buffer strips are used to stabilize soil against erosion and to reduce flow velocity and peak flow rate (Vymazal & Březinová, 2015).

1.6   Examples of river rehabilitation projects

River and stream rehabilitation projects have been implemented worldwide. According to Macedo et al. (2022), rehabilitation projects of urban streams have been implemented since 1970 in Global North countries, while this practice has recently come into existence in Global South countries. Feio et al. (2021) concluded that the greatest implementation has been recorded in Japan, South Korea, Singapore, Europe, North America, and Australia. According to the same authors, most implemented measures have been targeted to improve water quality and river connectivity for fish and riparian vegetation, while the measures for improving biological conditions are limited (Table 1.1).

TABLE 1.1 Examples of river rehabilitation.

Location	Measure and effect	Source
Kushiro and Kamisaigo Rivers (Japan)	Widening, increased biodiversity	Feio et al. (2021)
Yangjaecheon, Anyangcheon and Cheonggyecheon River (South Korea)	Flood control, water supply, water quality	Feio et al. (2021)
Kallang River (Singapore)	Naturalization, increased biodiversity	Feio et al. (2021)
Yangtze River (China)	Suspension of dams project, annual fish moratorium, fish breeding and restocking, improved river health, increased biodiversity	(Dudgeon, 2005)
Mekong River (Cambodia, Lao, Thailand, Viet Nam)	Reforestation, fish breeding and restocking, establishment of fish spawning, rearing and feeding areas	(Dudgeon, 2005), Roni et al. (2008)
Danube River (Austrian and Romanian section)	Levee breaching and modification, reconnection of floodplains, increased habitats and biodiversity	Buijse et al. (2002), Roni et al. (2008)
Rhine River (Netherland section)	Reconnection of floodplains, increased fish species	Buijse et al. (2002), Roni et al. (2008)
Gelså River (Denmark)	Remeandering, small increase in biodiversity	Roni et al. (2008)
Drava River (Austria)	Increased natural flood retention, spawning habitats for amphibians and fish, achieved good ecological status and additional ecosystem services (recreation)	Muhar et al. (2018)
Traisen River (Austria)	Restored river section instead of artificial canal, lowering floodplain terrain, established of native willow, fish migration and spawning sites	Muhar et al. (2018)
Montego River (Portugal)	Rehabilitation of longitudinal connectivity (fish passes)	Feio et al. (2021)
Clearwater River (Idaho, USA)	Dam removal, improved habitat and salmon runs	Roni et al. (2008)
Cosumnes River (California, USA)	Levee breaching, improved floodplain	Roni et al. (2008)
Three urban streams in Belo Horizonte (Brazil)	Improvement of water quality, stabilization of river banks, revegetation of riparian zones, riverbed naturalization	Macedo et al. (2022)
Martín stream (Argentina)	Transplanting macrophytes, habitat restoration	Altieri et al. (2021)
Williams River (Australia)	Using large wood, halted river degradation, increased sediment storage, and improved fish habitat	Brooks et al. (2006)
Kuils River (South Africa)	Channelization, uniformed channel shape, upstream banh erosion, massive bank collapse, reduced aquatic habitat and invertebrates, increased alien weeds	(King et al., 2003)

1.6.1   Asia

In Japan, river rehabilitation has been conducted in more than 23,000 cases since 2004. These measures included only nature-friendly river engineering. Some examples are the Kushiro and Kamisaigo Rivers, where the rivers are widening and the species number is increased. Recently, river rehabilitation projects have included green infrastructure and recovery from disasters. In South Korea, flood control, water supply, and water quality have been the most important results. The first urban rehabilitated river was the Yangjaecheon River in Seoul. After that, rehabilitation projects have been conducted in the Anyangcheon River in Anyang and the Cheonggyecheon River in Seoul. River naturalization with the emphasis on aquatic ecology have become the main concept in river management. The most important project in Singapore has been the naturalization of the 3.2 km of the Kallang River, which was a concrete canal before that. After rehabilitation, an increase in the number of bird species, dragonfly species, and freshwater otters has been recorded (Feio et al., 2021). In other parts of Asia, such as monsoonal Asia, restoration to the original state of the largest rivers, including the Mekong River, Yangtze River, and Gang River, is practically impossible due to the previous activities, which included flow regulation, construction of large dams, fishery, irrigation, and hydropower production. However, some rehabilitation efforts, such as the cancellation of major dam projects, the annual fishing moratorium along the Yangtze River, and breeding and restocking endangered fish species in the Mekong River and Yangtze River, have enhanced river health and preserved biodiversity (Dudgeon, 2005). In Cambodia, reforestation is applied to establish areas for fish spawning, rearing, and feeding in seasonally flooded forests near the Mekong River. In India and Bangladesh, the reconnection of secondary channels and floodplain lakes has resulted in increased fish catch (Roni et al., 2008).

1.6.2   Europe

River Basin Management Plans in Europe should be revised every 6 years to contain Programs of Measures for all water bodies to achieve good ecological status. These measures should include reduction of diffuse pollution sources and the reduction of hydromorphological pressures by installing fish passages or removing barriers. The ECRR is focused on ecological, physical, spatial, and management measures. Other European projects, such as RESTORE, REFORM, and FORECAST, include 1325 projects in 31 countries. Most projects are realized in the United Kingdom, but they also exist in the Netherlands, France, Spain, Denmark, Austria, and Italy. These projects involve measures related to bank stabilization, installation of fish passes, dam removals, levee breaching, secondary channel construction, reconnection of floodplain habitats, and riparian vegetation plantings, while some projects address multiple problems at the same time (Feio et al., 2021; Roni et al., 2008). Levee breaching and modification on the Danube River have positively affected fish, amphibians, and dragonflies. Reconnecting floodplain habitats has resulted in an additional 50 km of habitats and increased fish species number.

Similarly, the reconnection of floodplain lakes and channels in the lower Rhine River led to an increase in the abundance of cyprinids fish (Buijse et al., 2002; Roni et al., 2008). Remeandering is a measure that has been applied in stream rehabilitation and Gelså River in Denmark and has resulted in a small increase in macroinvertebrates, fish fauna, and aquatic vegetation (Roni et al., 2008). Positive examples of morphological river restoration have been recorded in Austria: the Austrian section of the Drava River and the Traisen River. Restoration of Drava River has been performed in the scope of different projects: EU LIFE projects from 1999 to 2011 and REFORM from 2011 to 2015. These projects have increased natural flood retention, achieved good ecological status, increased spawning habitats for amphibians and fish, and provided additional ecosystem services such as recreation. Traisen River restoration has been conducted in the scope of the project "LIFE + Project Traisen. Restoration measures have included a 9.5 km restored river section, which has replaced artificial canal and lowered floodplain for specific riparian plants, enabling the establishment of native willow and fish migration and spawning sites (Muhar et al., 2018).

A positive example is also the rehabilitation of the longitudinal continuity of the Montego River in Portugal, where fish passes have been constructed. This project enabled migratory aids and 45 km of habitats for diadromous fish species (Feio et al., 2021). In Hungarian streams, the preservation of near-natural hydromorphological and riparian conditions significantly mitigated the effects of urbanization, leading to good biological conditions (Zerega et al., 2021). River rehabilitation in 46 projects in Switzerland could be divided into three types related to improving riparian zones: culvert removal, one riverside widening, and both riversides widening (Langhans et al., 2014). River widening projects contributed to the establishment of riparian plants (Roni et al., 2008).

1.6.3   North and South America

Successful examples of rehabilitation in North America included instream structures, channel reconstruction, naturalized stream flows, fish passes, and restricted livestock grazing (Feio et al., 2021). In Oregon, streams, excluding livestock grazing, have improved vegetation and stream morphology (Roni et al., 2008). Reconnecting isolated habitats, rehabilitation of floodplains, and improvement of stream habitats usually increase local fish abundance (Feio et al., 2021). For example, 70% of fish production increases in Idaho streams have been recorded as a result of barrier removal and other rehabilitation techniques. Dam removal on the Clearwater River has improved habitat quality and salmon runs. Levee breaching on the Cosumnes River in California has resulted in successful floodplain restoration. Besides these measures, road abandonment or complete removal has reduced sediment delivery to streams in Montana and Redwood National Park in California (Roni et al., 2008). In Florida, hydromorphological measures such as stream reclamation, flow changes, bank stabilization, channel reconstruction, and floodplain reconnection are addressed in most rehabilitation projects (Zerega et al., 2021). In Canada, rehabilitation is focused on channel design, fish habitats, riparian vegetation, removal of small dams and barriers, and reduction of diffuse pollution loading to streams (Feio et al., 2021).

In Belo Horizonte in Brazil, three urban streams have been rehabilitated in the period 2006–2007 and results have been assessed 10 years later. The rehabilitation included improving water quality through waste management, stabilization of river banks, revegetation of riparian zones, and riverbed naturalization (Macedo et al., 2022). In Argentina, transplanting macrophytes has been used as a rehabilitation technique in the lowland Martín stream. The results showed that this technique has been suitable for restoring habitat heterogeneity, which was previously deteriorated by engineering works (Altieri et al., 2021).

1.6.4 Australia and Oceania

River rehabilitation has been conducted in south-eastern Australia in Williams River using large wood to stimulate the natural system, degraded due to human disturbance over two centuries. Results have shown that further river degradation has been halted, sediment storage has been increased, and the quality of fish habitats has been improved (Brooks et al., 2006). In New Zealand, riparian planting is a dominant measure in rehabilitation projects (Feio et al., 2021). Fencing and planting of riparian buffers have improved water quality and channel stability (Roni et al., 2008).

1.6.5 Africa

Rehabilitation projects have been conducted in Kuils River in South Africa but with negative consequences. Engineering works (channelization) in the middle sector of Kuils River have geomorphological and ecological impacts. Geomorphological changes such as widening and straightening have caused uniform channel shape, upstream bank erosion, and further massive bank collapse. Ecological changes included reduction in aquatic habitat and, consequently, abundance and diversity of aquatic invertebrates; on the other side, soil disturbance has increased the diversity of alien weeds (King et al., 2003).

Many rehabilitation projects have limitations related to monitoring (before and after implementation) (Feio et al., 2021; Roni et al., 2008), financial sources, technical possibilities, knowledge of flora and fauna (especially in Africa and Latin America), poor awareness of decision makers (Feio et al., 2021), inadequate assessment of historical conditions, and limiting factors (Roni et al., 2008).

1.7 Conclusions

The history of river and stream rehabilitation is closely connected with population growth, economic development, and urbanization. Due to anthropogenic pressures, many watercourses have been degraded in terms of water quality and quantity, changes of regime, and biodiversity. Depending on these processes, river rehabilitation measures could be divided into the following phases: 1860–1960 (few artificial structures were built); 1970s (flood prevention measures such as channelization); 1980s (construction of parks in floodplains); 1990s (ecological river improvement techniques); and 2000s (the establishment of a relationship between river and humans).

Numerous rehabilitation measures could be implemented depending on river conditions, previous human pressures, and intervention, as well as the aspects that should be improved. These measures could be grouped into various categories, such as environmentally sensitive river maintenance activities, soft bioengineering approaches, and mitigation of hard engineering practices. In the scope of these measures, some restoration techniques could be identified, such as road improvements, riparian and floodplain rehabilitation and connectivity, stream habitat improvement, nutrient enrichment, hydropeaking mitigation, and temperature mitigation. Other strategies could include flow management, flushing management, sediment management, and land use modification. Recently, nature-based solutions have been implemented worldwide. Among the various measures, constructed wetlands and buffer strips have essential and multifunctional roles. Constructed wetlands have multipurpose functions in water quality improvement, flood storage and mitigation, maintenance of surface water during droughts, nutrient cycling, and habitats for wildlife, fish, and plants. Buffer strips stabilize streambanks, slow and capture runoff, regulate stream temperature, filter pollutants, provide habitats and food, and act as wildlife corridors.

The implementation and effectiveness of applied measures varies from region to region. Generally, the greatest successful implementation has been recorded in Japan, South Korea, Singapore, Europe, North America, and Australia. In other regions (especially in Africa, South America, and Mexico), many limitation factors have been recorded, including lack of monitoring, financial sources, technical possibilities, knowledge of flora and fauna, and historical conditions and factors, and awareness of decision makers.

Acknowledgement

This research was supported by the by the Ministry of Science, Technological Development and Innovation of the Republic of Serbia (Contract No. 451-03-66/2024-03/200172

References

Adeyemo, O. K. (2003). Consequences of pollution and degradation of Nigerian aquatic environment on fisheries resources. *Environmentalist, 23*(4), 297−306. Available from https://doi.org/10.1023/B:ENVR.0000031357.89548.fb.

Altieri, P., Paz, L. E., Jensen, R. F., Donadelli, J., & Capítulo, A. R. (2021). Transplanting macrophytes as a rehabilitation technique for lowland streams and their influence on macroinvertebrate assemblages. *Anais da Academia Brasileira de Ciências, 93*(3). Available from https://doi.org/10.1590/0001-3765202120191029.

Aouissi, K. B., Madani, S., Hein, C., & Benacer, H. (2023). Morphological approach for the typological classification of waterfront revitalization. *Journal of the Geographical Institute Jovan Cvijic SASA, 73*(1), 109−122. Available from https://doi.org/10.2298/IJGI2301109A. https://ojs.gi.sanu.ac.rs/index.php/zbornik/article/download/521/354.

Arrighi, C., Brugioni, M., Castelli, F., Franceschini, S., & Mazzanti, B. (2018). Flood risk assessment in art cities: the exemplary case of Florence (Italy). *Journal of Flood Risk Management, 11*, S616−S631. Available from https://doi.org/10.1111/jfr3.12226.

Austin, G., & Yu, K. (2016). *Constructed Wetlands and Sustainable Development* (pp. 1−286). United States: Taylor and Francis http://www.tandfebooks.com/doi/book/10.4324/9781315694221, Available from https://doi.org/10.4324/9781315694221.

Australian Capital Territory, (2019). Water Resources Environmental Flow Guidelines 2019; Disallowable Instrument DI2019—37 Made under the Water Resources Act 2007, s12 (Environmental Flow Guidelines), Canberra, Australia. http://www.legislation.act.gov.au

Bae, S.-H., & Lee, S.-D. (2018). Construction and management plan of constructed wetland for promoting biodiversity. *Journal of People, Plants, and Environment, 21*(3), 185−202. Available from https://doi.org/10.11628/ksppe.2018.21.3.185.

Bernhardt, E. S., & Palmer, M. A. (2007). Restoring streams in an urbanizing world. *Freshwater Biology, 52*(4), 738−751. Available from https://doi.org/10.1111/j.1365-2427.2006.01718.x.

Biabanaki, M., Naeini, A. T., & Eslamian, S. S. (2012). Effects of urbanization on stream channel. *Journal of Civil Engineering and Urbanization, 2*(4), 136−142.

Blauhut, V., Stoelzle, M., Ahopelto, L., Brunner, M. I., Teutschbein, C., Wendt, D. E., Akstinas, V., Bakke, S. J., Barker, L. J., Bartošová, L., Briede, A., Cammalleri, C., Kalin, K. C., De Stefano, L., Fendeková, M., Finger, D. C., Huysmans, M., Ivanov, M., Jaagus, J., Jakubínský, J., Krakovska, S., Laaha, G., Lakatos, M., Manevski, K., Neumann Andersen, M., Nikolova, N., Osuch, M., van Oel, P., Radeva, K., Romanowicz, R. J., Toth, E., Trnka, M., Urošev, M., Urquijo Reguera, J., Sauquet, E., Stevkov, A., Tallaksen, L. M., Trofimova, I., Van Loon, A. F., van Vliet, M. T. H., Vidal, J.-P., Wanders, N., Werner, M., Willems, P., & Živković, N. (2022). Lessons from the 2018−2019 European droughts: a collective need for unifying drought risk management. *Natural Hazards and Earth System Sciences, 22*(6), 2201−2217. Available from https://doi.org/10.5194/nhess-22-2201-2022.

Brierley, G. J., & Fryirs, K. (2000). River styles, a geomorphic approach to catchment characterization: Implications for river rehabilitation in Bega catchment, New South Wales, Australia. *Environmental Management, 25*(6), 661−679. Available from https://doi.org/10.1007/s002670010052.

Brierley, G. J., Fryirs, K. A. (2005). Geomorphology and River Management: Applications of the River Styles Framework.

Brooks, A. P., Howell, T., Abbe, T. B., & Arthington, A. H. (2006). Confronting hysteresis: Wood based river rehabilitation in highly altered riverine landscapes of south-eastern Australia. *Geomorphology, 79*(3-4), 395−422. Available from https://doi.org/10.1016/j.geomorph.2006.06.035.

Buijse, A. D., Coops, H., Staras, M., Jans, L. H., Van Geest, G. J., Grift, R. E., Ibelings, B. W., Oosterberg, W., & Roozen, F. C. J. M. (2002). Restoration strategies for river floodplains along large lowland rivers in Europe. *Freshwater Biology, 47*(4), 889−907. Available from https://doi.org/10.1046/j.1365-2427.2002.00915.x.

Calow, P., & Petts, G. (1994). The Rivers Handbook: Hydrological and Ecological Principles. Oxford, UK: Blackwell Scientific Publications.

Chen, W. Y. (2017). Environmental externalities of urban river pollution and restoration: A hedonic analysis in Guangzhou (China). *Landscape and Urban Planning, 157*, 170−179. Available from https://doi.org/10.1016/j.landurbplan.2016.06.010, http://www.elsevier.com/inca/publications/store/5/0/3/3/4/7.

Chinalawinfo Database, (2023). 1984 Law of the People's Republic of China on Prevention and Control of Water Pollution. http://www.lawinfochina.com

CONAMA, (2005). Resolução 357. Dispõe Sobre a Classificação dos Corpos de Agua e Diretrizes Ambientais Para o Seu Enquadramento, Bem Como Estabelece as Condições e Padrões de Lançamento de Efluentes, e da Outras Providências; Conselho Nacional do Meio Ambiente: Brasília, Brazil.

Cunningham, K., Stuhlinger, C., Liechty, H., (2009). Riparian buffers: functions and values. University of Arkansas, Division of Agriculture [Cooperative Extension Service], US Department of Agriculture and county government cooperating, USA.

Department of Water Affairs and Forestry, 2008. Draft Regulations for the Establishment of a Water Resource Classification System. Government Gazette No. 31417, Pretoria, South Africa, pp. 129−144.

DRPC, (1994). Convention on cooperation for the protection and sustainable use of the river Danube (Convention for the protection of the Danube) (OJ L 342, 12.12.1997, pp. 19−43). https://eur-lex.europa.eu/EN/legal-content/summary/convention-for-the-protection-of-the-danube.html

Dudgeon, D. (2005). River rehabilitation for conservation of fish biodiversity in monsoonal Asia. *Ecology and Society*, *10*(2). Available from https://doi.org/10.5751/ES-01469-100215.

European Centre for River Restoration (2023).

European Commission, 2000. European Water Framework Directive. https://environment.ec.europa.eu/topics/water/water-framework-directive_en

European Commission, 2023. Nature Restoration Law https://environment.ec.europa.eu/topics/nature-and-biodiversity/nature-restoration-law_en

European Environment Agency (2023a). WISE Water Framework Directive Database. https://vvv.eea.europa.EU/data-and-maps/data/vise-vfd-4

European Environment Agency (2023b). Restoring European rivers and lakes in cities improves quality of life. Available: https://www.eea.europa.eu/highlights/restoring-european-rivers-and-lakes [2023, October 20].

Feio, M. J., Hughes, R. M., Callisto, M., Nichols, S. J., Odume, O. N., Quintella, B. R., Kuemmerlen, M., Aguiar, F. C., Almeida, S. F. P., Alonso-Eguíalis, P., Arimoro, F. O., Dyer, F. J., Harding, J. S., Jang, S., Kaufmann, P. R., Lee, S., Li, J., Macedo, D. R., Mendes, A., Mercado-Silva, N., Monk, W., Nakamura, K., Ndiritu, G. G., Ogden, R., Peat, M., Reynoldson, T. B., Rios-Touma, B., Segurado, P., & Yates, A. G. (2021). The biological assessment and rehabilitation of the world's rivers: An overview. *Water (Switzerland)*, *13*(3). Available from https://doi.org/10.3390/w13030371, http://www.mdpi.com/journal/water.

Fernandes D. T. (2013), 2013 Proceedings of the Fábos Conference on Landscape and Greenway Planning An Integrated Approach of Landscape Design in the Rehabilitation of an Urban River Corridor: River Tinto 4

Findlay, S. J., & Taylor, M. P. (2006). Why rehabilitate urban river systems? *Area*, *38*(3), 312−325. Available from https://doi.org/10.1111/j.1475-4762.2006.00696.x.

Fogg, J., Wells, G. (1998). Stream Corridor Restoration: Principles, Processes, and Practices. Federal Interagency Stream Restoration Working Group.

Folke, C. (2006). Resilience: The emergence of a perspective for social−ecological systems analyses. *Global Environmental Change*, *16*(3), 253−267. Available from https://doi.org/10.1016/j.gloenvcha.2006.04.002.

Frissell, C. A., & Ralph, S. C. (1998). *Stream and Watershed Restoration* (pp. 599−624). Springer Science and Business Media LLC. Available from doi:10.1007/978-1-4612-1652-0_24.

Fryirs, K., & Brierley, G. J. (2009). Naturalness and place in river rehabilitation. Resilience Alliance, Australia. *Ecology and Society*, *14*(1). Available from https://doi.org/10.5751/ES-02789-140120, http://www.ecologyandsociety.org/vol14/iss1/art20/.

Gavrilović, L. J., Milanović Pešić, A., & Urošev, M. (2012). A hydrological analysis of the greatest floods in Serbia in the 1960−2010 period. *Carpathian Journal of Earth and Environmental Sciences*, *7*, 107−116.

Gilvear, D. J., Spray, C. J., & Casas-Mulet, R. (2013). River rehabilitation for the delivery of multiple ecosystem services at the river network scale. *Journal of Environmental Management*, *126*, 30−43. Available from https://doi.org/10.1016/j.jenvman.2013.03.026.

Greimel, F., Schülting, L., Graf, W., Bondar-Kunze, E., Auer, S., Zeiringer, B., & Hauer, C. (2018). *Hydropeaking Impacts and Mitigation* (pp. 91−110). Springer Science and Business Media LLC. Available from https://doi.org/10.1007/978-3-319-73250-3_5.

Hauer, C., Leitner, P., Unfer, G., Pulg, U., Habersack, H., & Graf, W. (2018). *The Role of Sediment and Sediment Dynamics in the Aquatic Environment* (pp. 151−169). Springer Science and Business Media LLC. Available from https://doi.org/10.1007/978-3-319-73250-3_8.

Hein, C., & Hillmann, F. (2016). *The missing link: redevelopment of the urban waterfront as a function of cruise ship tourism. Waterfronts Revisited*. Routledge. Available from http://doi.org/10.4324/9781315637815, 222-238.

Herbert, C., & Döll, P. (2023). Analyzing the informative value of alternative hazard indicators for monitoring drought hazard for human water supply and river ecosystems at the global scale. *Natural Hazards and Earth System Sciences*, *23*(6), 2111−2131. Available from https://doi.org/10.5194/nhess-23-2111-2023.

Higgs, E. (2003). *Nature by Fesign: People, Natural Process and Ecological Restoration*. MIT Press.

Hillman, M., & Brierley, G. (2005). A critical review of catchment-scale stream rehabilitation programmes. *Progress in Physical Geography: Earth and Environment*, *29*(1), 50−76. Available from https://doi.org/10.1191/0309133305pp434ra.

Hoyle, B. S., Pinder, D. A., & Husain, M. S. (1988). *Revitalizing the Waterfront: International Dimensions of Dockland Redevelopment*. London, UK: Belhaven Press.

Infrastructure Development Institute, (1998). The River Law, with Commentary by Article: Legal Framework for River and Water Management in Japan. http://www.idi.or.jp/wp/wp-content/uploads/2018/05/RIVERE.pdf

Jakovljević, D. (2020). Assessment of water quality during the floods in May, 2014, Serbia. *Journal of Geographical Institute of Jovan Cvijic, 70*, 215−226. Available from https://doi.org/10.2298/IJGI2003215J.

Jakovljević, D., Milanović Pešić, A., & Milijašević Joksimović, D. (2019). Water Resources Management Issues in the Danube River Basin District—Examples from Serbia. *Water Resources*, *46*, 286−295. Available from https://doi.org/10.1134/S0097807819020039.

Jakovljević, D., Milanović Pešić, A., & Milijašević Joksimović, D. (2021). Protection from harmful effects of water—Examples from Serbia. In P. Samui, H. Bonakdari, & R. Deo (Eds.), *Water Engineering Modeling and Mathematic Tools* (pp. 157−175). Amsterdam, Netherlands; Oxford, UK, Cambridge, MA, USA: Elsevier. Available from https://doi.org/10.1016/B978-0-12-820644-7.00026-8.

Justice Laws Website, (2023). Fisheries Act (R.S.C., 1985, c. F-14). https://laws-lois.justice.gc.ca/PDF/F-14.pdf

King, J. M., Scheepers, A. C. T., Fisher, R. C., Reinecke, M. K., & Smith, L. B. (2003). River rehabilitation: literature review, case studies and emerging principles. *Water Research Commission*, 2003.

Kondolf, G. M., Keller, E. A., (1991). Management of urbanizing watersheds. California Watersheds at the Urban Interface, in Proceedings of the Third Biennial Watershed Conference, University of California, USA, pp. 27–40.

Korea Legislation Research Institute, (2020). Water Environment Conservation Act. https://faolex.fao.org/docs/pdf/kor195044.pdf

Kreibich, H., Blauhut, V., Aerts, J. C. J. H., Bouwer, L. M., Van Lanen, H. A. J., Mejia, A., Mens, M., & Van Loon, A. F. (2019). How to improve attribution of changes in drought and flood impacts. *Hydrological Sciences Journal*, *64*(1), 1–18. Available from https://doi.org/10.1080/02626667.2018.1558367, http://www.tandfonline.com/loi/thsj20.

Kundzewicz, Z. W., Pińskwar, I., & Brakenridge, G. R. (2013). Large floods in Europe, 1985–2009. *Hydrological Sciences Journal*, *58*, 1–7. Available from https://doi.org/10.1080/02626667.2012.745082.

Langhans, S. D., Hermoso, V., Linke, S., Bunn, S. E., & Possingham, H. P. (2014). Cost-effective river rehabilitation planning: Optimizing for morphological benefits at large spatial scales. *Journal of Environmental Management*, *132*, 296–303. Available from https://doi.org/10.1016/j.jenvman.2013.11.021, https://www.sciencedirect.com/journal/journal-of-environmental-management.

Lazorchak, J. M., Klemm, D. J., Peck, D. V. (1998). Environmental Monitoring and Assessment Program- Surface Waters: Field Operations and Methods for Measuring the Ecological Condition of Wadeable Streams. EPA/620/R-94/004F. Environmental Protection Agency. https://archive.epa.gov/emap/archive-emap/web/pdf/sec01-2.pdf

Macedo, D. R., Callisto, M., Linares, M. S., Hughes, R. M., Romano, B. M. L., Rothe-Neves, M., & Silveira, J. S. (2022). Urban stream rehabilitation in a densely populated Brazilian metropolis. Frontiers Media S.A., Brazil. *Frontiers in Environmental Science*, *10*. Available from https://doi.org/10.3389/fenvs.2022.921934, journal.frontiersin.org/journal/environmental-science.

Mander, Ü., Tournebize, J., Tonderski, K., Verhoeven, J. T. A., & Mitsch, W. J. (2017). Planning and establishment principles for constructed wetlands and riparian buffer zones in agricultural catchments. *Ecological Engineering*, *103*, 296–300. Available from https://doi.org/10.1016/j.ecoleng.2016.12.006, http://www.elsevier.com/inca/publications/store/5/2/2/7/5/1.

Marchi, L., Borga, M., Preciso, E., & Gaume, E. (2010). Characterisation of selected extreme flash floods in Europe and implications for flood risk management. *Journal of Hydrology*, *394*(1-2), 118–133. Available from https://doi.org/10.1016/j.jhydrol.2010.07.017.

Matthews, J., Reeze, B., Feld, C. K., & Hendriks, A. J. (2010). Lessons from practice: Assessing early progress and success in river rehabilitation. *Hydrobiologia*, *655*(1), 1–14. Available from https://doi.org/10.1007/s10750-010-0389-2.

Meier C. I. (1998). The ecological basis of river restoration: 2. Defining restoration from an ecological perspective, in Engineering Approaches to Ecosystem Restoration, American Society of Civil Engineers (ASCE) Chile, pp. 392–397, Available from https://doi.org/10.1061/40382(1998)70.

Milanović, A, Milijašević, D., & Brankov, J. (2011). Assessment of polluting effects and surface water quality using water pollution index: A case study of hydro-system Danube-Tisa-Danube, Serbia. *Carpathian Journal of Earth & Environmental Science*, *6*, 269–277.

Milanović Pešić, A., Brankov, J., & Milijašević Joksimović, D. (2020). Water quality assessment and populations' perceptions in the National park Djerdap (Serbia): Key factors affecting the environment. *Environment, Development and Sustainability*, *22*, 2365–2383. Available from https://doi.org/10.1007/s10668-018-0295-8.

Ministro de Ambiente y Energía Costa Rica (2020). Politica Nacional de Areas de Proteccion de Rios Quebradas, Arroios and Nacientes. https://da.go.cr/wp-content/uploads/2018/05/Politica_Nacional_Areas_Proteccion_2020.pdf.

Ministry of Ecology and Environment of the Peoples Republic of China. (2023). *The Yangtze River Protection Law, 2017*. Available from https://www.mee.gov.cn/.

Muhar, S., Schmutz, S., & Jungwirth, M. (1995). River restoration concepts: Goals and perspectives. In F. Schiemer, M. Zalewski, & J. E. Thorpe (Eds.), *The Importance of Aquatic-Terrestrial Ecotones for Freshwater Fish. Developments in Hydrobiology* (105, pp. 183–194). Dordrecht: Springer. Available from https://doi.org/10.1007/978-94-017-3360-1_1.

Muhar, S., Sendzimir, J., Jungwirth, M., & Hohensinner, S. (2018). *Restoration in Integrated River Basin Management* (pp. 273–299). Springer Science and Business Media LLC. Available from https://doi.org/10.1007/978-3-319-73250-3_15.

Nikolić, V., Milićević, D., Milenković, Milivojević, S. (2010). Constructed wetland application in waste water treatment processes in Serbia. BALWOIS, BALWOIS.

Paulsen, S. G., Hughes, R. M., & Larsen, D. P. (1998). Critical elements in describing and understanding our nation's aquatic resources. *Journal of American Water Resources Association*, *34*, 995–1005. Available from https://doi.org/10.1111/j.1752-1688.1998.tb04148.x.

Petts, G., Sparks, R., & Campbell, I. (2000). River restoration in developed economies. In P. Boon, B. R. Davies, & G. E. Petts (Eds.), *Global Perspectives on River Conservation*. Chichester, UK: Wiley.

Radu, V. M., Ionescu, P., Deak, G., Diacu, E., Ivanov, A. A., Zamfir, S., & Marcus, M. I. (2020). Overall assessment of surface water quality in the Lower Danube River. *Environmental Monitoring and Assessment*, *192*(2). Available from https://doi.org/10.1007/s10661-020-8086-8, https://link.springer.com/journal/10661.

Rajaram, T., & Das, A. (2008). Water pollution by industrial effluents in India: Discharge scenarios and case for participatory ecosystem specific local regulation. *Futures*, *40*(1), 56–69. Available from https://doi.org/10.1016/j.futures.2007.06.002.

Roni, P., Hanson, K., & Beechie, T. (2008). Global review of the physical and biological effectiveness of stream habitat rehabilitation techniques. *North American Journal of Fisheries Management*, *28*, 856–890. Available from https://doi.org/10.1577/M06-169.1.

Rutherfurd, I. D., Jerie, K., Marsh, N. (2000). A rehabilitation manual for Australian streams. Cooperative Research Centre for Catchment Hydrology, Land and Water Resources Research and Development Corporation. Canberra, Australia.

Ríos-Touma, B., & Ramírez, A. (2019). Multiple stressors in the neotropical region: Environmental Impacts in biodiversity hotspots. In S. Sabater, A. Elosegi, & R. Ludwig (Eds.), *In Multiple Stressors in River Ecosystems: Status, Impacts and Prospects for the Future* (pp. 205−220). Cambridge, MA, USA: Elsevier. Available from https://doi.org/10.1016/B978-0-12-811713-2.00012-1.

Schanze, J., Olfert, A., Tourbier, J. T., Gersdorf, I., & Schwager, T. (2004). *Existing urban river rehabilitation schemes. Final report of WP2 from Urban River Basin Enhancement Methods (URBEM).* European Commission.

Schanze, J., Olfert, A., Tourbier, J. T., & Gersdorf, I. (2005). Urban river rehabilitation schemes. In J. T. Tourbier, & J. Schanze (Eds.), *International Conference on Urban River Rehabilitation URRC 05, Proceedings* (pp. 5−13). Germany: Dresden.

Schmutz, S., & Moog, O. (2018). *Dams: Ecological Impacts and Management* (pp. 111−127). Springer Science and Business Media LLC. Available from https://doi.org/10.1007/978-3-319-73250-3_6.

Schäfer, R. B., & M.Bundschuh. (2018). *Ecotoxicology* (pp. 225−239). Springer Science and Business Media LLC. Available from https://doi.org/10.1007/978-3-319-73250-3_12.

Sear, D. A. (1994). River restoration and geomorphology. *Aquatic Conservation: Marine and Freshwater Ecosystems, 4*(2), 169−177. Available from https://doi.org/10.1002/aqc.3270040207.

Seliger, C., & Zeiringer, B. (2018). *River Connectivity, Habitat Fragmentation and Related Restoration Measures* (pp. 171−186). Springer Science and Business Media LLC. Available from https://doi.org/10.1007/978-3-319-73250-3_9.

Shahady, T. (2022). Sustainable water management with a focus on climate change. *Water and Climate Change: Sustainable Development, Environmental and Policy Issues* (pp. 293−316). United States: Elsevier. Available from https://www.sciencedirect.com/book/9780323998758, https://doi.org/10.1016/B978-0-323-99875-8.00020-3.

Shields, F. D., Copeland, R. R., Klingeman, P. C., Doyle, M. W., & Simon, A. (2003). Design for stream restoration. *Journal of Hydraulic Engineering, 129*(8), 575−584. Available from https://doi.org/10.1061/(ASCE)0733-9429(2003)129:8(575).

Simsek G. (2012). Urban River Rehabilitation as an Integrative Part of Sustainable Urban Water Systems, 2012 48th ISOCARP Congress 2012 Proceedings. 1-12.

Şimşek, G. (2014). River rehabilitation with cities in mind: The Eskişehir case. *METU JFA, 31*, 21−37. Available from https://doi.org/10.4305/metu.jfa.2014.1.2.

Souza, S. A. de, & Reis, D. S. (2022). Trend detection in annual streamflow extremes in Brazil. *Water, 14*(11). Available from https://doi.org/10.3390/w14111805.

Tang, T., Strokal, M., van Vliet, M. T. H., Seuntjens, P., Burek, P., Kroeze, C., Langan, S., & Wada, Y. (2019). Bridging global, basin and local-scale water quality modeling towards enhancing water quality management worldwide. *Current Opinion in Environmental Sustainability, 36*, 39−48. Available from https://doi.org/10.1016/j.cosust.2018.10.004.

Tang, X., Huang, S., Scholz, M., & Li, J. (2009). Nutrient removal in pilot-scale constructed wetlands treating eutrophic river water: Assessment of plants, intermittent artificial aeration and polyhedron hollow polypropylene balls. *Water, Air, and Soil Pollution, 197*(1-4), 61−73. Available from https://doi.org/10.1007/s11270-008-9791-z.

Tournebize, J., Chaumont, C., & Mander, Ü. (2017). Implications for constructed wetlands to mitigate nitrate and pesticide pollution in agricultural drained watersheds. *Ecol. Eng, 103*, 415−425. Available from https://doi.org/10.1016/j.ecoleng.2016.02.014.

Tzanakakis, V. A., Paranychianakis, N. V., & Angelakis, A. N. (2020). Water supply and water scarcity. *Water, 12*, 2347. Available from https://doi.org/10.3390/w12092347.

Uddin, M. J., & Jeong, Y. K. (2021). Urban river pollution in Bangladesh during last 40 years: Potential public health and ecological risk, present policy, and future prospects toward smart water management. *Heliyon, 7*(2). Available from https://doi.org/10.1016/j.heliyon.2021.e06107, http://www.journals.elsevier.com/heliyon/.

United Nation, 2023. International Decade for Action "Water for Life" 2005−2015. https://www.un.org/waterforlifedecade/; International Decade for Action on Water for Sustainable Development 2018−20128. https://www.un.org/en/events/waterdecade/

United Nations, 2020. The UN Decade on Ecosystem Restoration (2021−2030). https://wedocs.unep.org/bitstream/handle/20.500.11822/30919/UNDecade.pdf.

United States Environmental Protection Agency, 2023. Clean Water Act 1972 https://www.epa.gov/laws-regulations/summary-clean-water-act.

Vymazal, J., & Březinová, T. (2015). The use of constructed wetlands for removal of pesticides from agricultural runoff and drainage: A review. *Environment International, 75*, 11−20. Available from https://doi.org/10.1016/j.envint.2014.10.026.

Wada, (2010), Development of Asian river restoration network for knowledge sharing, 2010 13th International River Symposium.

Wade, P. M., Large, A. R. G., & de Wall, L. (1998). Rehabilitation of degraded river habitat: An introduction. In L. C. de Wall, A. R. G. Large, & P. Wade (Eds.), *Rehabilitation of Rivers: Principles and Implementation* (pp. 1−12). Chichester, UK: John Wiley & Sons.

Walker, D., Jakovljević, D., Savić, D., & Radovanović, M. (2015). Multi-criterion water quality analysis of the Danube River in Serbia: A visualisation approach. *Water Research, 79*, 158−172. Available from https://doi.org/10.1016/j.watres.2015.03.020.

Walton, C. R., Zak, D., Audet, J., Petersen, R. J., Lange, J., Oehmke, C., Wichtmann, W., Kreyling, J., Grygoruk, M., Jabłońska, E., Kotowski, W., Wiśniewska, M. M., Ziegler, R., & Hoffmann, C. C. (2020). Wetland buffer zones for nitrogen and phosphorus retention: Impacts of soil type, hydrology and vegetation. *Science of the Total Environment, 727*. Available from https://doi.org/10.1016/j.scitotenv.2020.138709, http://www.elsevier.com/locate/scitotenv.

Wang, H., Zhong, H., & Bo, G. (2018). Existing forms and changes of nitrogen inside of horizontal subsurface constructed wetlands. *Environmental Science and Pollution Research, 25*, 771−781. Available from https://doi.org/10.1007/s11356-017-0477-1.

Weigelhofer, G., Hein, T., & Bondar-Kunze, E. (2018). Phosphorus and nitrogen dynamics in riverine systems: Human impacts and management options. In S. Schmutz, & J. Sendzimir (Eds.), *Riverine Ecosystem Management: Science for Governing towards a Sustainable Future* (pp. 187−202). Cham: Springer Open.

Wohl, E., Angermeier, P. L., Bledsoe, B., Kondolf, G. M., MacDonnell, L., Merritt, D. M., Palmer, M. A., Poff, N. L. R., & Tarboton, D. (2005). River restoration. *Water Resources Research*, *41*(10), W10301. Available from https://doi.org/10.1029/2005WR003985.

Wright, I. A., McCarthy, B., Belmer, N., & Price, P. (2015). Subsidence from an underground coal mine and mine wastewater discharge causing water pollution and degradation of aquatic ecosystems. *Water, Air, and Soil Pollution*, *226*(10). Available from https://doi.org/10.1007/s11270-015-2598-9, http://www.kluweronline.com/issn/0049-6979/.

Xu, Z., Xu, J., Yin, H., Jin, W., Li, H., & He, Z. (2019). Urban river pollution control in developing countries. *Nature Sustainability*, *2*(3), 158−160. Available from https://doi.org/10.1038/s41893-019-0249-7.

Zahoor, I., & Mushtaq, A. (2023). Water pollution from agricultural activities: A critical global review. *International Journal of Chemical and Biochemical Sciences*, *23*(1), 164−176. Available from https://www.iscientific.org/wp-content/uploads/2023/05/19-IJCBS-23-23-24.pdf.

Zemanová, K., Picek, T., Dušek, J., Edwards, K., & Šantrůčková, H. (2010). Carbon, nitrogen and phosphorus tranformations are related to age of a constructed wetland. *Water, Air, and Soil Pollution*, *207*(1-4), 39−48. Available from https://doi.org/10.1007/s11270-009-0117-6.

Zerega, A., Simões, N. E., & Feio, M. J. (2021). How to improve the biological quality of urban streams? Reviewing the effect of hydromorphological alterations and rehabilitation measures on benthic invertebrates. *Water*, *13*, 2087. Available from https://doi.org/10.3390/w13152087.

Zhu, Z., Deng, Q., Zhou, H., Ouyang, T., Kuang, Y., Huang, N., & Qiao, Y. (2002). Water pollution and degradation in Pearl River Delta, South China. *Ambio*, *31*(3), 226−230. Available from https://doi.org/10.1579/0044-7447-31.3.226, http://www.bioone.org/loi/ambi.

Chapter 2

Assessment of site suitability for surface water retention and springshed rejuvenation using geospatial techniques: a case study of Mizoram State

Padam Jee Omar[1], Shashank Singh[2], Purushottam Kumar Mahato[2], Subash Prasad Rai[2], Harinarayan Tiwari[2] and Ravi Prakash Tripathi[3]

[1]Department of Civil Engineering, Babasaheb Bhimrao Ambedkar University, Lucknow, Uttar Pradesh, India, [2]Floodkon Consultants LLP., Sector 132, Noida, Uttar Pradesh, India, [3]Department of Civil Engineering, Rajkiya Engineering College, Sonbhadra, Uttar Pradesh, India

2.1 Introduction

Water scarcity presents a critical global challeng, necessitating the adoption of sustainable strategies to manage water resources effectively (Prasad et al., 2014). Among the key approaches to address this issue are surface water storage and the rejuvenation of springsheds. Surface water storage encompasses the capture and retention of water from various sources to satisfy water requirements during periods of reduced precipitation (Das et al., 2020). This practice offers numerous benefits, including heightened resilience to droughts, reduced flood risks, and the capacity to regulate water flow to accommodate diverse demands. Infrastructure such as dams and retention ponds play pivotal roles in storing excess rainfall or runoff (Biswas et al., 2023; Padmavathy et al., 1993).

The significance of surface water storage extends beyond mere water availability, impacting broader ecological and societal domains. Efficient storage mechanisms facilitate the conservation of water resources, thereby safeguarding ecosystems reliant on consistent water availability (Chand et al., 2023; Tripathi & Pandey, 2022). Moreover, by mitigating the impacts of water scarcity, surface water storage contributes to the sustenance of agricultural productivity and supports livelihoods dependent on irrigation and water-dependent industries (Omar & Kumar, 2022).

Similarly, the rejuvenation of springsheds represents a critical aspect of water resource management, particularly in regions reliant on groundwater sources. Springsheds serve as vital recharge areas for groundwater aquifers, influencing overall hydrological balance and water quality (Singh et al., 2023). Strategies aimed at rejuvenating springsheds encompass watershed management practices, reforestation initiatives, and groundwater replenishment schemes. By enhancing the resilience and sustainability of springsheds (Ranjan et al., 2024), these efforts bolster the reliability of groundwater sources and promote ecosystem health. The adoption of surface water storage and springshed rejuvenation strategies constitutes indispensable components of sustainable water resource management (Upreti et al., 2024). Addressing water scarcity requires holistic approaches that integrate technological innovations, policy interventions, and community engagement to ensure equitable access to water resources and safeguard environmental integrity (Gupta et al., 2022). Through concerted efforts and collaborative partnerships, societies can mitigate the impacts of water scarcity and build resilient water systems capable of meeting the diverse needs of present and future generations.

Springshed rejuvenation represents a concerted effort to restore and safeguard natural springs and their associated hydrological systems (Basha et al., 2024). These springs serve as vital sources of freshwater, crucial for drinking and irrigation purposes, while also fostering biodiversity and sustaining stream flow (Jamali et al., 2022). However, the unregulated extraction of groundwater from springsheds has precipitated diminished water discharge and contamination, imperiling their long-term viability (Kumar et al., 2008; Shivhare et al., 2014). Rejuvenation endeavors are geared

Hydrosystem Restoration Handbook. DOI: https://doi.org/10.1016/B978-0-443-29802-8.00002-9

toward replenishing groundwater reservoirs by employing diverse strategies such as reforestation, adoption of soil and water conservation practices, and implementation of groundwater recharge techniques (Bera & Mukhopadhyay, 2023). Through these measures, the aim is to restore the natural balance of springsheds and enhance their resilience to anthropogenic pressures and climatic variability.

Central to the success of rejuvenation initiatives is the active engagement of local communities and the application of scientific methodologies. Community involvement ensures the sustainability of efforts by fostering a sense of ownership and responsibility toward springshed conservation (Patra et al., 2023). Moreover, leveraging scientific insights enables the development of evidence-based interventions tailored to the specific needs and challenges of each springshed ecosystem. By rejuvenating springsheds, societies can safeguard critical water resources, mitigate the adverse effects of groundwater depletion, and preserve ecological integrity (Omar et al., 2023). Beyond the immediate benefits of enhanced water availability and quality, springshed rejuvenation contributes to broader objectives of environmental sustainability and climate resilience. Furthermore, the rejuvenation of springsheds holds promise for fostering socioeconomic development by facilitating access to reliable water sources for agriculture, livestock rearing, and domestic use. Additionally, healthy springsheds support ecotourism activities, promoting local economies and cultural heritage preservation.

Springshed rejuvenation represents a multifaceted approach to addressing the challenges of water scarcity and environmental degradation. Through collaborative efforts and interdisciplinary approaches, societies can nurture resilient springsheds that sustain ecosystems, support livelihoods, and enhance overall well-being. Embracing the principles of sustainability and stewardship, we can ensure the long-term vitality of springs and their invaluable contributions to human and ecological health.

Both surface water storage and springshed rejuvenation play integral roles in the comprehensive management of water resources. They contribute to water security, climate resilience, and sustainable development by ensuring a stable supply of water for multiple purposes while also safeguarding the environment and supporting local communities (Kumar et al., 2021; Omar & Kumar, 2021). Implementing these strategies yields numerous benefits. Surface water storage helps mitigate the impact of droughts, minimize the risks associated with flooding, and facilitate controlled water distribution. Springshed rejuvenation, on the other hand, replenishes groundwater resources, ensures a consistent flow of water from springs throughout the year, and promotes the preservation of ecosystems (Pradeepraju et al., 2022). Successful case studies from around the world demonstrate the potential effectiveness of these practices (Omar, 2015). By embracing surface water storage and springshed rejuvenation, we can actively work toward a sustainable and equitable future, guaranteeing access to clean water resources for everyone.

2.2   Study area

Mizoram State, nestled in the north-eastern part of India, is a picturesque region renowned for its lush greenery, undulating hills, and rich cultural heritage. It is a landlocked state bordered by Myanmar in the east and Bangladesh to the west, boasting a 722 km international frontier, while its northern periphery adjoins Manipur, Assam, and Tripura within the country. Spanning over 21,087 km^2, it ranks as the fifth smallest Indian state. Positioned between 21°56'N to 24°31'N latitude and 92°16'E to 93°26'E longitude, the state finds itself traversed by the Tropic of Cancer, nearly bisecting it. Mizoram's dimensions reveal a maximum north−south stretch of 285 km and an east−west extension of 115 km.

Mizoram is characterized by its rugged topography, with steep slopes and deep valleys defining its landscape. Defined by undulating hills, verdant valleys, meandering rivers, and serene lakes, Mizoram boasts a landscape punctuated by 21 major hill ranges or peaks of varying altitudes, interspersed with pockets of plains. Hills to the west typically stand around 1000 m, gradually ascending to 1300 m in the east, with certain areas reaching heights exceeding 2000 m. Among its majestic peaks, Phawngpui Tlang, colloquially known as the Blue Mountain, reigns supreme, towering at 2210 m in the southeastern reaches of the state. Enveloping approximately 76% of its expanse, forests cloak much of Mizoram's terrain, while around 8% comprises fallow land, and 3% lies barren and deemed uncultivable. The remainder constitutes cultivable and sown areas. Despite efforts to discourage traditional slash-and-burn or jhum cultivation, it persists, impacting the state's topography. Notably, a 2021 report by the Ministry of Environment, Forest and Climate Change underscores Mizoram's distinction as the Indian state with the highest forest cover, with an impressive 84.53% of its geographical area cloaked in forested splendor (Report (2021) *by the Ministry of Environment, Forest and Climate Change*, no date).

The state's climate is influenced by its proximity to the Bay of Bengal, resulting in a subtropical climate marked by moderate temperatures and abundant rainfall, particularly during the monsoon season. Mizoram is home to a diverse array of flora and fauna, with dense forests teeming with a variety of endemic species. The state's population primarily comprises various indigenous ethnic groups, each with its unique customs, languages, and traditions. A predominantly agrarian economy, Mizoram's agriculture sector sustains livelihoods through the cultivation of crops such as rice, maize, and fruits. Despite its natural beauty and cultural vibrancy, Mizoram has severe water difficulties because of its topography, rainfall patterns, and lack of surface water storage facilities. However, the state also has options for storing surface water and revitalizing springsheds, both of which can improve the sustainability and resilience of water resources.

The availability and distribution of water in Mizoram are significantly influenced by its topographical features, which include steep terrain, deep valleys, and dense forests. The terrain significantly influences all surface water runoff, river flow patterns, and the possibility for surface water storage. Mizoram experiences water scarcity problems during dry spells while getting an abundance of rainfall, with an average annual precipitation range between 1500 and 3000 mm. The difficulty of maintaining water resources all year long is made more difficult by the uneven distribution of rainfall.

Mizoram needs to focus on both groundwater resources and surface water storage in order to overcome the issue of water shortage. Maintaining a sustainable water supply necessitates evaluating the possible groundwater zones and putting recharge strategies into practice. Mizoram can gain from springshed development in addition to the potential for groundwater and storage of surface water. Natural springs are plentiful in the state and are important sources of freshwater. It is essential to develop and manage these springs in order to provide a reliable water supply for both residential and agricultural uses. Fig. 2.1 shows the Index Map of Mizoram state, and Fig. 2.2 shows administrative map of Mizoram state having district boundary.

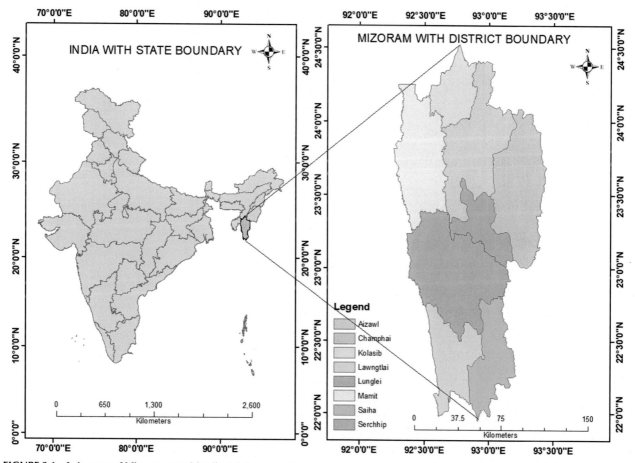

FIGURE 2.1   Index map of Mizoram state. Map lines delineate study areas and do not necessarily depict accepted national boundaries.

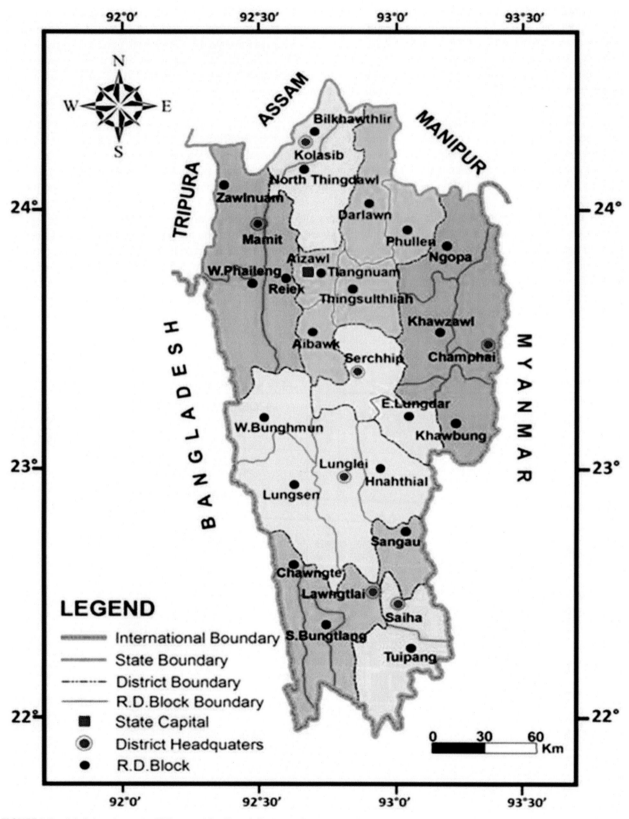

FIGURE 2.2   Administrative map of Mizoram. Map lines delineate study areas and do not necessarily depict accepted national boundaries.

2.3   Data used

Secondary data is utilized for the study and list of data used and their sources areas follows.

2.3.1   Rainfall data

Rainfall data has been taken from probable maximum precipitation (PMP) atlas of Brahmaputra Basin prepared by Central Water Commission. Return period rainfall for 50 and 100 years has been considered for this study. It has been observed that for Mizoram, the variation of 50-year return period rainfall is 180−270 mm and that for 100-year return period is 200−300 mm. Spatial variations of 50- and 100-year return period rainfall are shown in Fig. 2.3 and Fig. 2.4, respectively.

2.3.2   Digital elevation model

The Shuttle Radar Topographic Mission (SRTM) Digi Elevation Model (DEM) with a resolution of 30 m by 30 m has been acquired from the United States Geological Survey (USGS) Earth Explorer. This dataset, derived from radar data collected during the NASA SRTM mission in February 2000, provides comprehensive and high-resolution topographic information for various regions around the globe. The SRTM DEM offers valuable insights into the elevation profile of the Earth's surface, facilitating detailed analyses of terrain features, landforms, and hydrological characteristics. With

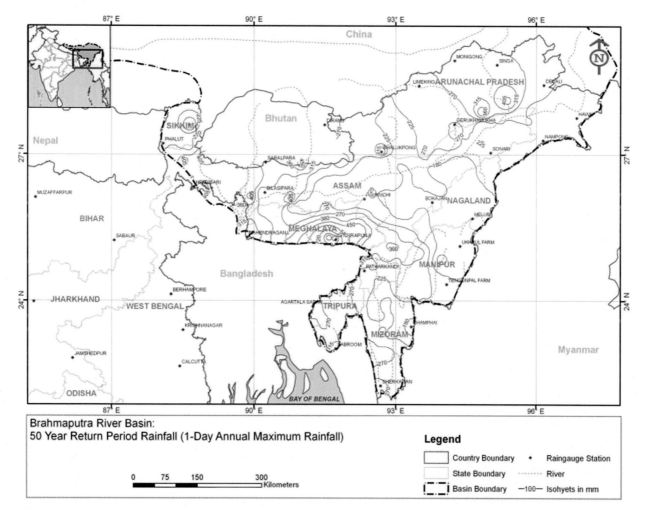

FIGURE 2.3   Fifty-year return period rainfall map. Map lines delineate study areas and do not necessarily depict accepted national boundaries. *From PMP Atlas.*

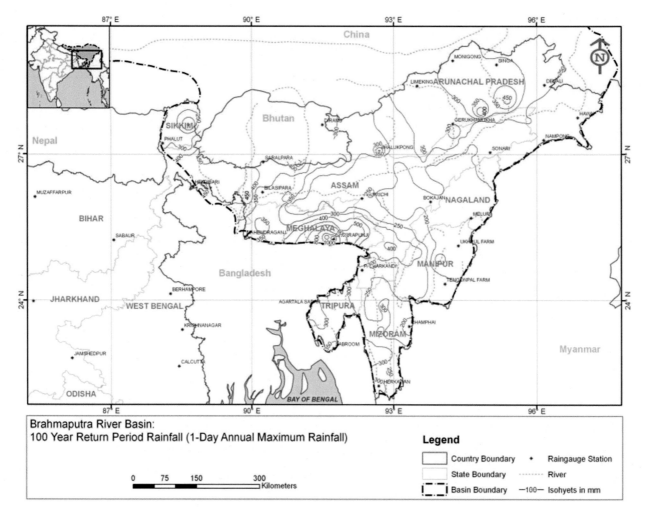

FIGURE 2.4   One hundred-year return period rainfall map. Map lines delineate study areas and do not necessarily depict accepted national boundaries. *From PMP Atlas*

its fine spatial resolution, the SRTM DEM serves as a critical resource for a wide range of applications, including geographic information system (GIS), environmental modeling, urban planning, and disaster management.

2.3.3   Drainage map

The drainage map of Mizoram has been obtained from the Mizoram Remote Sensing Application Center (MRSAC) and is integral to the present study. As a critical component of geospatial analysis, the drainage map provides comprehensive information on the natural drainage network of Mizoram, encompassing rivers, streams, and watercourses across the state. Acquiring the drainage map from MRSAC ensures access to reliable and up-to-date spatial data, essential for understanding the hydrological dynamics and water resource management in Mizoram state. Drainage map of Mizoram is shown in Fig. 2.5.

2.3.4   Land use land cover map

The land use and land cover (LULC) map of Mizoram state has been acquired from the MRSAC and is shown in Fig. 2.6. This dataset provides detailed information on the spatial distribution and composition of LULC categories across Mizoram state. Obtaining the LULC map from MRSAC ensures access to accurate and up-to-date geospatial data essential for analyzing patterns of land use change, monitoring environmental dynamics, and informing land management decisions. By integrating the LULC map into the study, different land use classes, including forests, agricultural lands, urban areas, and water bodies, within the state can be categorized. Furthermore, the LULC map serves as a

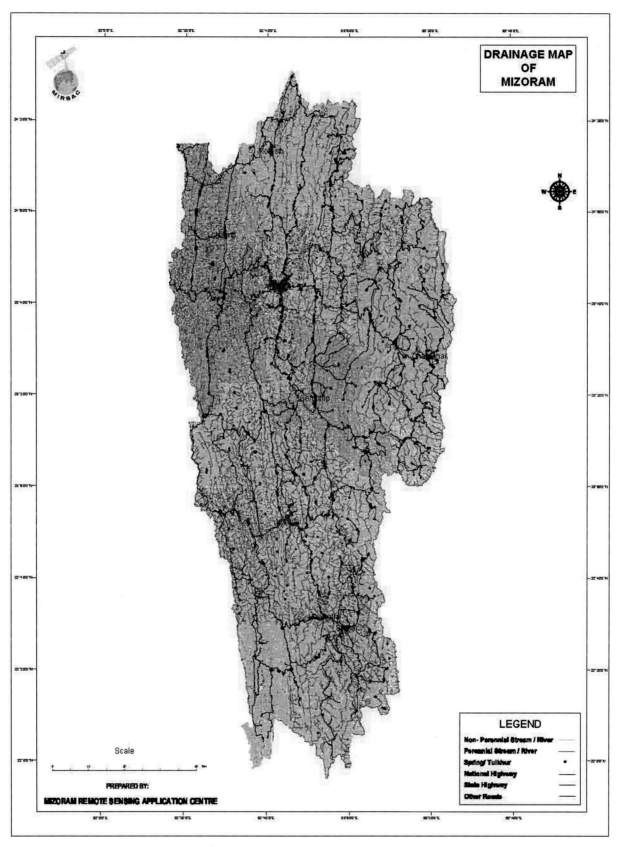

FIGURE 2.5 Drainage map of Mizoram state. Map lines delineate study areas and do not necessarily depict accepted national boundaries.

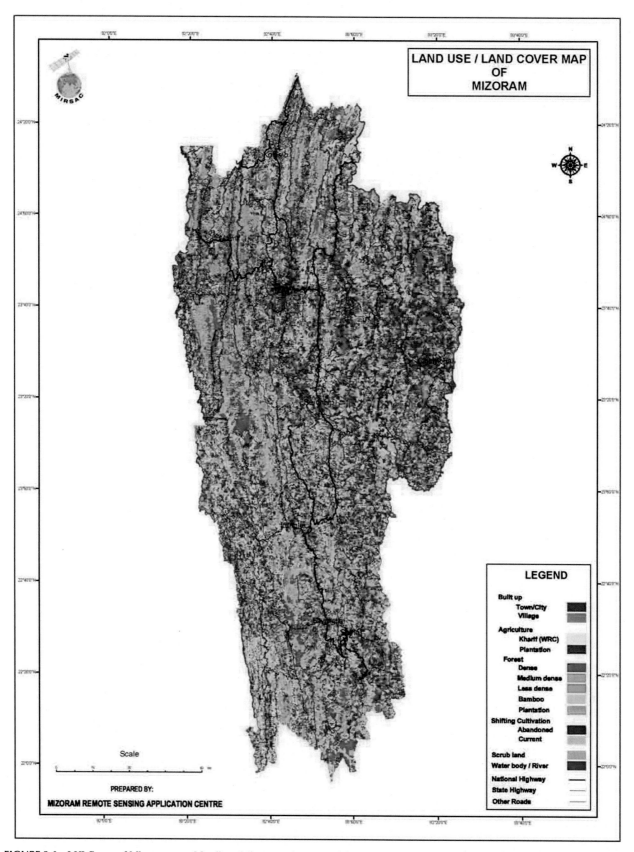

FIGURE 2.6   LULC map of Mizoram state. Map lines delineate study areas and do not necessarily depict accepted national boundaries.

valuable resource for studying the impact of human activities, climate change, and natural disturbances on the landscape of Mizoram state.

2.3.5 Soil map

The soil map, sourced from the MRSAC and shown in Fig. 2.7, provides crucial information on soil types within the study area. Understanding soil characteristics aids in assessing runoff generation. The map reveals soil texture, permeability, and moisture retention, guiding decisions on land use, agriculture, and water management. By utilizing the soil map data, the study enhances comprehension of hydrological processes and supports sustainable land and water management strategies in Mizoram state.

2.3.6 Groundwater potential map

The groundwater potential map, obtained from the MRSAC, is shown in Fig. 2.8. Fig. 2.8 illustrates the potential and yield of groundwater within the study area. This map serves to identify recharge zones and predict areas with high groundwater availability. By utilizing data from the groundwater potential map, the study gains insights into groundwater resources, facilitating informed decisions regarding water management and resource allocation. This comprehensive analysis aids in understanding groundwater dynamics and supports sustainable practices for groundwater utilization in Mizoram state.

2.4 Adopted methodology

The methodology used for the study involves a multidisciplinary approach tailored to the unique geographical and hydrological dynamics of the region. The study integrates geospatial techniques, including GIS and remote sensing, to analyze relevant datasets such as digital elevation maps, drainage patterns, land use/land cover maps, and groundwater potential maps. These datasets are taken from the MRSAC and other relevant agencies/organizations. Through meticulous analysis and interpretation of these datasets, the study aims to delineate potential sites for surface water retention structures and springshed rejuvation initiatives. The methodology involves the development of weighted maps, where factors such as slope, drainage density, land use/land cover, rainfall, and groundwater potential are assigned significance to identify suitable locations for intervention. Community participation and stakeholder engagement are also integral components of the methodology, ensuring the incorporation of local knowledge and perspectives in the decision-making process. By employing this comprehensive methodology, the study seeks to provide valuable insights into water resource management and conservation strategies tailored to the specific needs and challenges of Mizoram state, India. The adopted methodology for the study has been shown in the flowchart in Fig. 2.9. This visual representation illustrates the systematic approach undertaken to achieve the study's objectives.

2.4.1 Hydrological assessment

2.4.1.1 Digital Elevation Model preparation

The DEM downloaded from USGS Earth Explorer has been processed for the study area using the GIS tool. Mosaicking has been done to merge different raster files in order to combine them into one. DEM is then clipped according to the shapefile of the study area. The Digital Elevation Map of Mizoram is shown in Fig. 2.10.

2.4.2 Slope map preparation

A slope map using the GIS software from the DEM data has been prepared for the study area. This slope map categorizes the terrain based on the degree of steepness, providing valuable insights into the topographical characteristics of the region. The slope map allows us to identify areas with gentle slopes as well as those with steeper inclines, aiding in the assessment of terrain suitability for various purposes such as infrastructure development, land use planning, and natural resource management. The slope map of the study area is shown in Fig. 2.11, providing a clear depiction of the terrain's gradient and assisting in the analysis of spatial patterns and landscape dynamics.

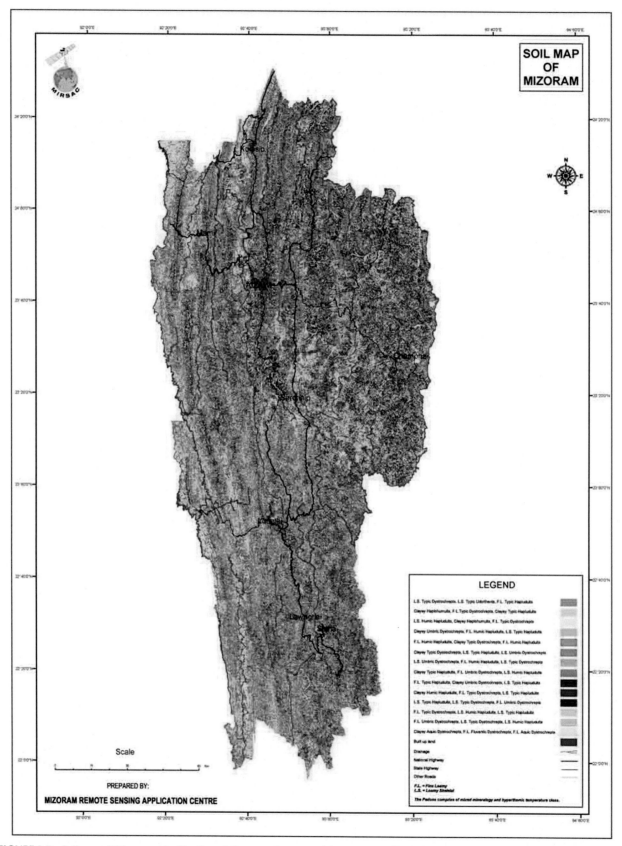

FIGURE 2.7 Soil map of Mizoram state. Map lines delineate study areas and do not necessarily depict accepted national boundaries.

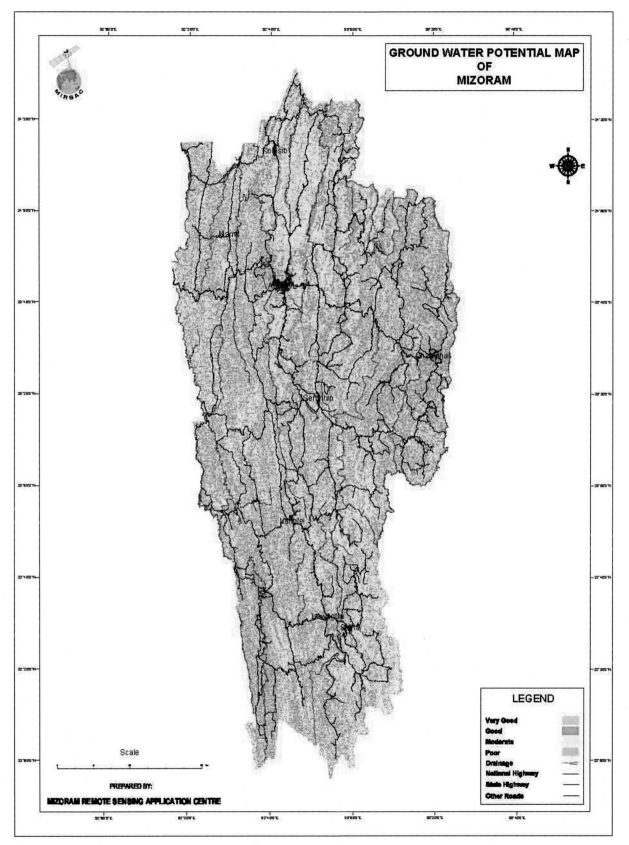

FIGURE 2.8 Groundwater potential map of Mizoram state. Map lines delineate study areas and do not necessarily depict accepted national boundaries.

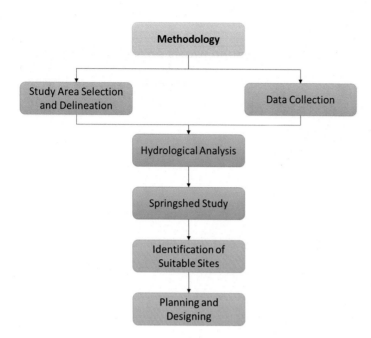

FIGURE 2.9   Flowchart adopted in the study.

2.4.3   Drainage density map generation

Based on the DEM, the drainage density map has been prepared in the GIS environment. This map has been classified as low-, medium-, and high-density zones. These classifications are done based on the length of drainage line per square kilometer of the area. Drainage density map is shown in Fig. 2.12. By classifying the drainage density, insights into the distribution and intensity of water flow across the study area has been gained. The drainage density map for the study area is shown in Fig. 2.12. The figure provides a clear representation of the varying degrees of drainage density throughout the region, aiding in the identification of areas prone to water accumulation or runoff and informing land management and water resource planning efforts accordingly.

2.5   Site suitability analysis

Based on above hydrological parameters and secondary data, suitable sites has been assessed in the GIS environment. Weightage has been given to slope, drainage density, LULC, and rainfall, and weighted map has been generated based on this. This map shows the potential location identified for the surface water storage structure. Site suitability map is shown in Fig. 2.13. Green color depicts the most suitable site location for storage purpose. The weightage has been given to least slope, high drainage density, LULC, and high rainfall.

2.6   Surface water storage

Based on site suitability analysis, the location of surface water storage structure has been identified and check dam has been proposed for the conservation of water. Ten meter height has been proposed for the check dam, and the total area that shall be submerged with water for this height has been assessed using GIS tools. The total submerged area is around 63 ha. Location of check dam and submerged area is shown in Fig. 2.14.

Similarly, for other suitable locations, the submergence area has been computed. The amount of water that can be stored with this proposed structure is of the order of 3.15 MCM. If five such structures are considered, the amount of water that can be stored is of the order of 15.75 MCM.

2.7   Groundwater rejuvenation: springshed rejuvenation

Springshed rejuvenation in Mizoram follows a comprehensive approach to restore and revitalize natural springs for long-term sustainability and prevent quality deterioration. The process begins with a thorough survey and mapping of the springs, identifying their locations, their recharge areas, sizes, water sources availability in time and space, and

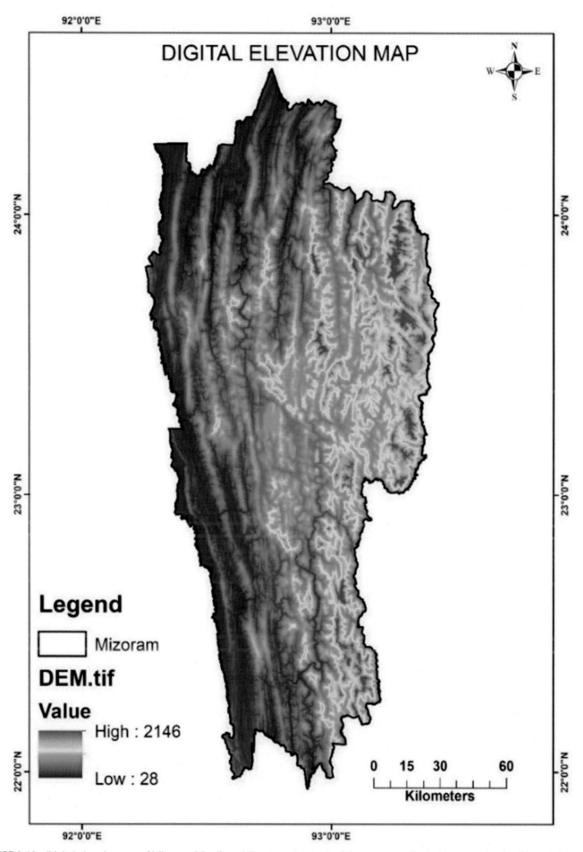

FIGURE 2.10   Digital elevation map of Mizoram. Map lines delineate study areas and do not necessarily depict accepted national boundaries.

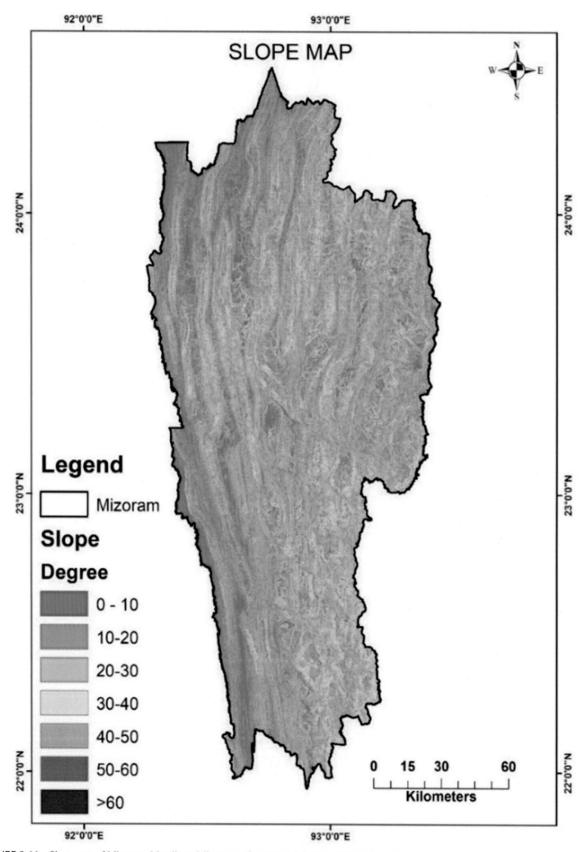

FIGURE 2.11 Slope map of Mizoram. Map lines delineate study areas and do not necessarily depict accepted national boundaries.

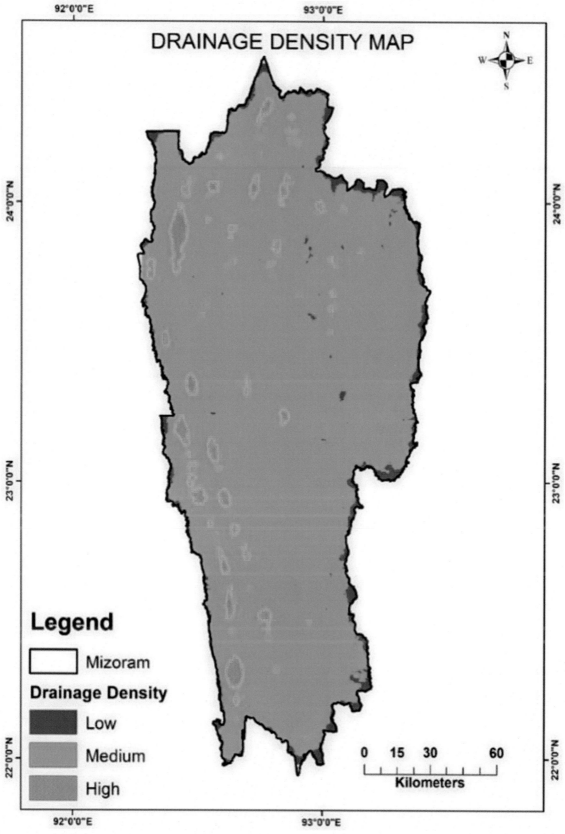

FIGURE 2.12   Drainage density map of Mizoram. Map lines delineate study areas and do not necessarily depict accepted national boundaries.

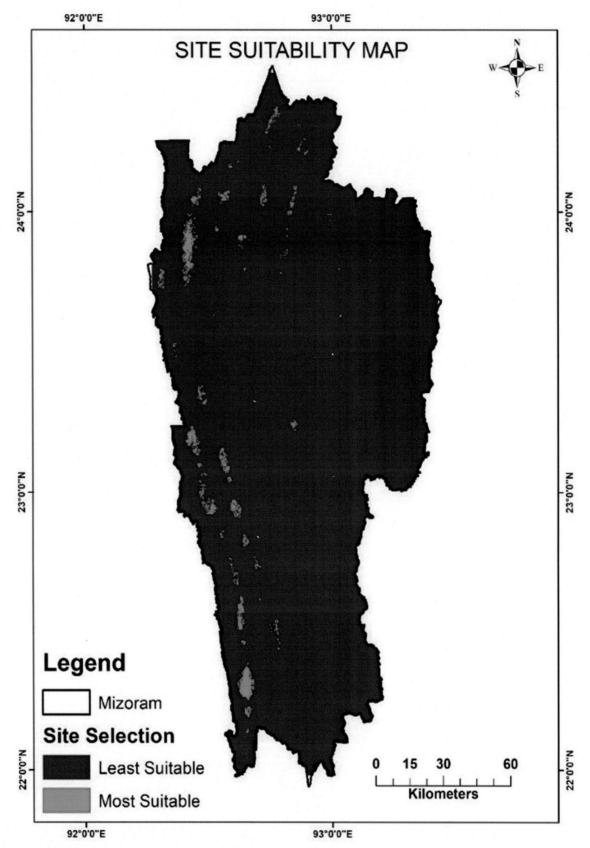

FIGURE 2.13 Site suitability map for surface water storage in Mizoram. Map lines delineate study areas and do not necessarily depict accepted national boundaries.

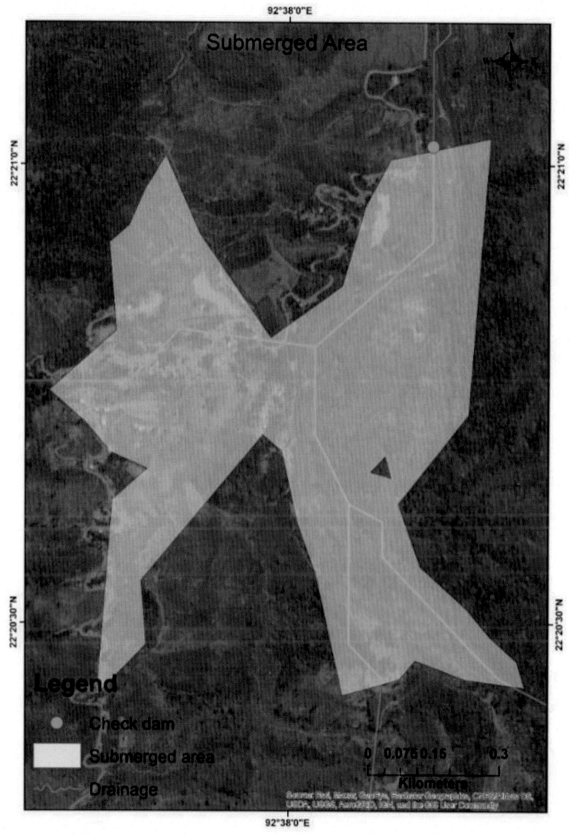

FIGURE 2.14 Location of proposed check dam.

water quality (Gashaw et al., 2022). Engaging local communities and stakeholders is crucial, as they play a vital role in the rejuvenation process maintaining the infrastructure created. Creating awareness about the significance of springs and garnering community support is essential. Protecting the catchment areas of the springs becomes a priority, with measures such as afforestation, soil and water conservation practices, and regulation of potentially harmful activities (Kadam et al., 2022).

Recharging the springshed is achieved through the construction of check dams, gabion structures, percolation tanks, contour trenches, among others, which facilitate the retention of rainwater and its further percolation that replenish the groundwater and revitalize the springs (Badhe et al., 2019). Science-based artificial recharge structures are also to be adopted. Efficient irrigation techniques and responsible water usage within households are to be adopted. Evaluating existing spring infrastructure, such as storage tanks and distribution systems, ensures their efficiency in water collection and distribution (Jamali et al., 2013; Ranjan et al., 2024).

Monitoring systems needs to be established to assess the quantity and quality of water in the springs, facilitating regular maintenance activities like desilting, disinfecting, and infrastructure repairs. Additionally, capacity building and training programs are conducted to empower local communities, village councils, and stakeholders, equipping them with the necessary knowledge and skills to actively participate in the rejuvenation process and ensure the long-term conservation of Mizoram's springs.

2.8 Results and discussion

The study conducted an assessment of site suitability for surface water retention and springshed rejuvenation in Mizoram state, India, using geospatial techniques. The digital elevation map generated using GIS revealed significant variations in elevation across the study area, ranging from 28 to 2146 m above the mean sea level. This variability reflects the diverse topography of Mizoram, encompassing both low-lying areas and rugged mountainous terrain. Furthermore, the analysis of slope gradients indicated a spectrum ranging from 0 to 60° across the study area, primarily attributed to the mountainous profile of the state. These slope variations play a crucial role in influencing the distribution and flow of surface water, presenting both challenges and opportunities for water management initiatives. The delineation of drainage lines through GIS techniques offered insights into the drainage density of the region. By assessing the density of drainage lines, the study aimed to understand the natural drainage network and identify potential sites for water storage structures. Fig. 2.14 shows the suggested location of check dam in the study area. By integrating factors such as slope, drainage density, LULC, and rainfall, a weighted map was created to identify suitable sites for surface water storage structures. This integrated approach facilitated the identification of suitable locations for surface water storage structures, with areas highlighted in green indicating the highest suitability, as shown in Fig. 2.15.

The study proposed the construction of check dams in these identified locations to capture surface water and facilitate its storage for subsequent use, particularly during periods of low rainfall and water scarcity. By strategically placing check dams in the identified locations, water resources can be effectively managed and utilized to support agricultural activities and mitigate water scarcity challenges. The findings of this study have significant implications for sustainable management and conservation of water resource in Mizoram state. By identifying suitable sites for surface water retention and advocating for the implementation of check dams, the study contributes to the sustainable utilization and preservation of water resources in Mizoram state.

The results of the study highlight the complex interplay between geographical factors and water resource dynamics in Mizoram state. The wide range of elevation and slope gradients observed across the region underscores the diverse topography, which poses both challenges and opportunities for water management initiatives. The presence of steep slopes, particularly in mountainous areas, can influence the flow and distribution of surface water, affecting drainage patterns and water retention capabilities. The analysis of drainage density provided valuable insights into the natural drainage network of the study area. By understanding the spatial distribution of drainage lines, stakeholders can identify areas prone to waterlogging or erosion, as well as potential sites for water storage infrastructure. Moreover, the integration of factors such as slope, drainage density, land use, and rainfall in the weighted map facilitated a comprehensive assessment of site suitability for surface water retention structures. The identification of suitable locations for check dams presents promising opportunities for enhancing water security and resilience in Mizoram state. Check dams can serve as effective means of capturing and storing surface water, thereby mitigating the impacts of seasonal fluctuations in rainfall and ensuring consistent water availability for agricultural and domestic purposes. Furthermore, the construction of check dams can contribute to groundwater recharge, supporting the sustainability of local aquifers and springsheds.

However, it is essential to recognize the potential socioeconomic and environmental implications associated with the implementation of water retention structures. The construction of check dams may require careful consideration of

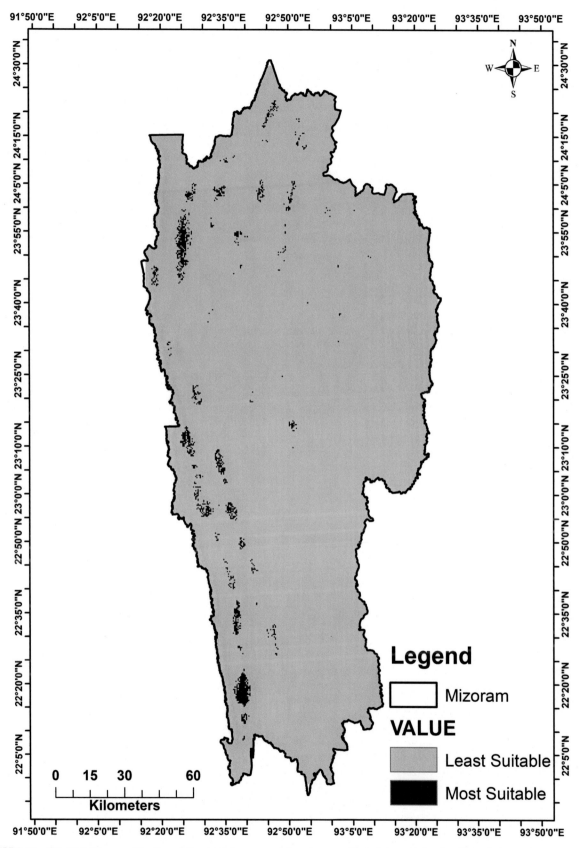

FIGURE 2.15   Site suitability map. Map lines delineate study areas and do not necessarily depict accepted national boundaries.

land, resource allocation, and community engagement to ensure equitable access to water resources and minimize negative impacts on ecosystems and livelihoods. Additionally, the long-term effectiveness of water retention initiatives may be influenced by factors such as climate change, land use dynamics, and sociopolitical factors, emphasizing the need for adaptive and participatory approaches to water resource management.

2.9 Conclusions

Site suitability of surface water storage structures can be predicted for any study area using GIS techniques. With the help of GIS, suitable site for storage structures in Mizoram has been assessed in this study, and it has been found that southern part of Mizoram is most suitable for surface water storage. Check dams has been proposed to collect the surface water, and it is found that 63 ha of land will be submerged if the height of check dam will be 10 m. The storage capacity corresponding to this check dam comes out to be 3.15 MCM. This amount of water can be reutilized in lean season when there is shortage of water.

These surface water storage structures will not only help in utilizing the water in lean season but also recharge the groundwater table in that area. Rise in groundwater level will decrease the demand of surface water and hence helps in reducing the water scarcity of Mizoram.

Acknowledgments

The authors would like to acknowledge the USGS Earth Explorer for providing the satellite imagery data for this study. The authors thank the Mizoram Remote Sensing Application Center for providing the other necessary data. They also thank the anonymous reviewer for a careful reading and providing constructive comments that helped us to improve the manuscript.

References

Badhe, Y., Medhe, R., & Shelar, T. (2019). Site suitability analysis for water conservation using AHP and GIS techniques: A case study of upper Sina River catchment, Ahmednagar (India). *Hydrospatial Analysis*, 3(2), 49–59. Available from https://doi.org/10.21523/gcj3.19030201.

Basha, U., Pandey, M., Nayak, D., Shukla, S., & Shukla, A. K. (2024). Spatial-temporal assessment of annual water yield and impact of land use changes on upper Ganga Basin, India, using InVEST model. *Journal of Hazardous, Toxic, and Radioactive Waste*, 28(2). Available from https://doi.org/10.1061/JHTRBP.HZENG-1245, http://ascelibrary.org/hzo/.

Bera, A., & Mukhopadhyay, B. P. (2023). Identification of suitable sites for surface rainwater harvesting in the drought prone Kumari River basin, India in the context of irrigation water management. *Journal of Hydrology*, 621. Available from https://doi.org/10.1016/j.jhydrol.2023.129655.

Biswas, B., Ghosh, A., & Sailo, B. L. (2023). Spring water suitable and vulnerable watershed demarcation using AHP-TOPSIS and AHP-VIKOR models: Study on Aizawl district of north-eastern hilly state of Mizoram, India. *Environmental Earth Sciences*, 82(3). Available from https://doi.org/10.1007/s12665-023-10766-w, https://www.springer.com/journal/12665.

Chand, D., Lata, R., Dhiman, R., & Kumar, K. (2023). *Groundwater Potential Assessment Using an Integrated AHP-Driven Geospatial Techniques in the High-Altitude Springs of Northwestern Himalaya, India Climate Change Adaptation, Risk Management and Sustainable Practices in the Himalaya* (pp. 337–360). India: Springer International Publishing. Available from https://link.springer.com/book/10.1007/978-3-031-24659-3, https://doi.org/10.1007/978-3-031-24659-3_15.

Das, N., Ohri, A., Agnihotri, A. K., Omar, P. J., & Mishra, S. (2020). Wetland dynamics using geo-spatial technology. *Lecture Notes in Civil Engineering*, 39. Available from https://doi.org/10.1007/978-981-13-8181-2_18, http://www.springer.com/series/15087.

Gashaw, D. Y., Suryabhagavan, K. V., Nedaw, D., & Gummadi, S. (2022). Rainwater harvesting in Modjo watershed, upper Awash River Basin, Ethiopia through remote sensing and fuzzy AHP. *Geocarto International*, 37(26), 14785–14810. Available from https://doi.org/10.1080/10106049.2022.2091158, http://www.tandfonline.com/toc/tgei20/current.

Gupta, N., Patel, J., Gond, S., Tripathi, R. P., Omar, P. J., & Dikshit, P. K. S. (2022). *Projecting Future Maximum Temperature Changes in River Ganges Basin Using Observations and Statistical Downscaling Model (SDSM)* (pp. 561–585). Springer Science and Business Media LLC. Available from https://doi.org/10.1007/978-981-19-7100-6_31.

Jamali, A. A., Arianpour, M., Pirasteh, S., & Eslamian, S. (2022). *Flood Handbook: Flood Analysis and ModelingGeospatial Techniques flood Hazard, Vulnerability and Risk Mapping in GIS Geodata Analytical Process in Boolean, AHP, and Fuzzy Models* (2). Boca Raton, FL: Taylor and Francis. Available from https://doi.org/10.1201/9780429463938-26.

Jamali, I. A., Olofsson, B., & Mörtberg, U. (2013). Locating suitable sites for the construction of subsurface dams using GIS. *Environmental Earth Sciences*, 70(6), 2511–2525. Available from https://doi.org/10.1007/s12665-013-2295-1.

Kadam, A. K., Kale, S. S., Umrikar, B. N., Sankhua, R. N., & Pawar, N. J. (2022). Assessing site suitability potential for soil and water conservation structures by using modified micro-watershed prioritization method: geomorphometric and geomatic approach. *Environment, Development and Sustainability*, 24(4), 4659–4683. Available from https://doi.org/10.1007/s10668-021-01627-2, http://www.wkap.nl/journalhome.htm/1387-585X.

Kumar, D., Dhaloiya, A., Singh Nain, A., Sharma, M. P., & Singh, A. (2021). Prioritization of watershed using remote sensing and geographic information system. *Sustainability*, 13(16). Available from https://doi.org/10.3390/su13169456.

Kumar, M. G., Agarwal, A. K., & Bali, R. (2008). Delineation of potential sites for water harvesting structures using remote sensing and GIS. *Journal of the Indian Society of Remote Sensing, 36*(4), 323−334. Available from https://doi.org/10.1007/s12524-008-0033-z.

Omar. (2015). Geomatics techniques-based significance of morphometric analysis in prioritization of watershed. *International Journal of Enhanced Research in Science Technology and Engineering, 4*(1), 2015.

Omar, P. J., Gupta, P., & Wang, Q. (2023). Exploring the rise of AI-based smart water management systems. *AQUA—Water Infrastructure, Ecosystems and Society, 72*(11), iii−iv. Available from https://doi.org/10.2166/aqua.2023.005, https://iwaponline.com/aqua/article/72/11/iii/98645/Exploring-the-rise-of-AI-based-smart-water.

Omar, P. J., & Kumar, V. (2021). Land surface temperature retrieval from TIRS data and its relationship with land surface indices. *Arabian Journal of Geosciences, 14*(18). Available from https://doi.org/10.1007/s12517-021-08255-0, http://www.springer.com/geosciences/journal/12517?cm_mmc = AD-_-enews-_-PSE1892-_-0.

Omar, P. J., & Kumar, V. (2022). Assessment of damage for dam break incident in Lao PDR using SAR data. *International Journal of Hydrology Science and Technology, 14*(4), 421−434. Available from https://doi.org/10.1504/ijhst.2022.126439, http://www.inderscience.com/ijhst.

Padmavathy, A. S., Ganesha Raj, K., Yogarajan, N., Thangavel, P., & Chandrasekhar, M. G. (1993). Checkdam site selection using GIS approach. *Advances in Space Research, 13*(11), 123−127. Available from https://doi.org/10.1016/0273-1177(93)90213-u.

Patra, S., Kumar, B., & Pandey, M. (2023). Experimental study on the turbulence characteristics in a vegetated channel. *Flow Measurement and Instrumentation, 94*. Available from https://doi.org/10.1016/j.flowmeasinst.2023.102464.

Pradeepraju, N., Pradeepraju, N., Nagaraju, D., Nagaraju, D., Sudeep, S. R., & Sudeep, S. R. (2022). Suitable site selections for artificial recharge structure in Bandalli watershed. Chamaraja Nagar District, Karnataka, India using remote sensing, and GIS techniques. *Current World Environment, 17*(3), 727−742. Available from https://doi.org/10.12944/cwe.17.3.20.

Prasad, H. C., Bhalla, P., & Palria, S. (2014). Site suitability analysis of water harvesting structures using remote sensing and GIS: A case study of Pisangan watershed, Ajmer district, Rajasthan. *The International Archives of the Photogrammetry, Remote Sensing and Spatial Information Sciences, XL-8*(8), 1471−1482. Available from https://doi.org/10.5194/isprsarchives-xl-8-1471-2014.

Ranjan, P., Pandey, P. K., & Pandey, V. (2024). Groundwater spring potential zonation using AHP and fuzzy-AHP in Eastern Himalayan region: Papum Pare district, Arunachal Pradesh, India. *Environmental Science and Pollution Research, 31*(7), 10317−10333. Available from https://doi.org/10.1007/s11356-023-26769-w, https://www.springer.com/journal/11356.

Report (2021) by the Ministry of Environment, Forest and Climate Change.

Shivhare, N., Rahul, A. K., Omar, P. J., Gaur, S., Dikshit, P. K. S., & Dwivedi, S. B. (2014). Utilizing SWAT for Surface water discharge Modeling: a case study of a watershed in Ganga basin.

Singh, A. N., Mudgal, A., Tripathi, R. P., & Omar, P. J. (2023). Assessment of wastewater treatment potential of sand beds of River Ganga at Varanasi, India. *Aqua Water Infrastructure, Ecosystems and Society, 72*(5), 690−700. Available from https://doi.org/10.2166/aqua.2023.200, https://iwaponline.com/aqua/article/72/5/690/95100/Assessment-of-wastewater-treatment-potential-of.

Tripathi, R. P., & Pandey, K. K. (2022). Numerical investigation of flow field around T-shaped spur dyke in a reverse-meandering channel. *Water Supply, 22*(1), 574−588. Available from https://doi.org/10.2166/ws.2021.253, https://watermark.silverchair.com/ws022010574.pdf?.

Upreti, M. R., Kayastha, S. P., & Bhuiyan, C. (2024). Water quality, criticality, and sustainability of mountain springs—A case study from the Nepal Himalaya. *Environmental Monitoring and Assessment, 196*(1). Available from https://doi.org/10.1007/s10661-023-12186-6, https://www.springer.com/journal/10661.

Chapter 3

Catchment areas in karst: advantages and disadvantages

Zoran Stevanović

Center for Karst Hydrogeology of the Department of Hydrogeology, Faculty of Mining & Geology, University of Belgrade, Belgrade, Serbia

3.1   Introduction

Karst covers more than 15% of the continental ice-free land (Goldscheider et al., 2020) while karst aquifers supply approximately 9.2% of the world's population, or more than 700 million people, with potable water (Stevanović, 2019).

Karst is a specific environment, different from many others (Cvijić, 1893; Ford & Williams, 2007; Gunn, 2004; LaMoreaux et al., 1984; Williams, 2008). It is full of contradictions and is characterized by the presence of hard but soluble rocks (carbonates and evaporites), compact impervious blocks and large cavities located not far from each other, shortage of water at high mountains and abundant springs at their foothills. Specific karst landscape and hydrographic networks, as well as the high permeability of karstified rocks, greatly influence both the distribution of flora and fauna and human life (Stevanović, 2015). Despite the dynamic water regimes in many places around the world, where there is karst—natural resources are limited. This is especially the case in arid areas, where karst and its landscapes and flora are very different from those in humid environments or zones that are permanently covered in ice. In marine coastal areas, karst is often very problematic as saltwater intrusion may disturb the fresh water supply (Bakalowicz, 2005).

Making karst and its nature sustainable for human lives is not an easy task, and various technical measures are sometimes needed to change such an ambience. These attempts have a very long history, starting with arranging the caves as human shelters. Later on, the focus moved to spring water use. Tapping structures are an ancient art, as old as the first civilizations: in what used to be ancient China, Babylon, and Persia, as well as in Israel and Egypt, there are many remnants of intake structures around large springs, and many cities had been founded in their vicinity (Kresic, 2013; Stevanović, 2010; Stevanović & Milanović, 2015). In the 20th century, especially its second half, many successful projects including the construction of dams and reservoirs were implemented in different karst regions around the globe (Milanović, 1981). However, numerous failures confirmed that karst is a risky environment for different aspects of karst water utilization and control (Stevanović & Milanović, 2015). One way or another, the karst environment is not always friendly and various engineering interventions are sometimes needed to adapt the ambience to human needs.

3.2   Characterization of karst systems

The two main characteristics of karstified rocks and karstic systems are *anisotropy* and *heterogeneity*. The anisotropy is very typical and has almost become the first association with the word "karst": compact blocks, small fissures and large cavities may be jointly present at short distances (Drogue, 1982; Mangin, 1984). While in a nonkarstified rock heterogeneity within a common area may be perhaps 1–50, in karstic rocks it can increase to perhaps 1–1 million (Ford & Williams, 2007).

Proper knowledge of an aquifer system is a prerequisite for its water utilization, protection from pollution and sustainable development. Recognizing and understanding a karstic system is a very long and difficult process, as the system is usually very complex. There are several properties that have to be evaluated (Stevanović, 2015, modified):

- Geometry (catchment and vertical distribution/thickness, while taking into consideration all the elements: epikarst, vadose zone, saturated zone, karstification base).
- Aquifer properties (porosity, permeability, conductivity, transmissivity, storativity).

Hydrosystem Restoration Handbook. DOI: https://doi.org/10.1016/B978-0-443-29802-8.00003-0

- Recharge mechanism (autogenic, allogenic).
- Groundwater flow directions (high and low water seasons).
- Drainage (springs, subsurface drainage, GW extraction).
- Hydrodynamic conditions (confined, unconfined, piezometric pressure, hydraulic head, water tightness).
- Relationship of groundwater−surface waters (regime and interrelationship throughout the year, kind and quality of surface waters).
- Relationship with adjacent aquifers (boundary conditions, type of barrier, impacts under seasonal variations).
- Groundwater quality (hydrochemistry—macro and micro constituents, microbiology, gases, isotopes, physical properties).
- Aquifer's vulnerability to pollution (attenuation capacity, contaminant transport, sanitary protection zones of the sources).
- Groundwater regime (seasonal variation of groundwater quantity/depth to water table and discharge, and groundwater quality).

Selecting and undertaking appropriate methods for the characterization and conceptualization of the karst environment and aquifer system is an essential step for the success of any project, including the engineering ones (Milanović, 1981). The two main engineering projects in karst have to do with: (1) Utilization of groundwater for potable water supply, (2) Multi-purpose utilization of surface water and groundwater in reservoirs (energy production, irrigation, flood prevention, water supply). Other common practical projects include dewatering of mines, protection from pollution and remediation of karst waters. The latter regularly requires the definition of sanitary protection zones (Goldscheider, 2010), for which delineation of the catchment is a prerequisite.

The above list clearly shows that the very first step in the characterization and conceptualization of karst systems is the definition, or—to put it more clearly—the assessment of the karst system geometry.

3.3 Catchments in karst

The size of a karst system is defined by its boundaries, which are lateral (horizontal) and vertical. According to permeability, boundaries can be *permeable*, *semipermeable*, and *impermeable*. In the latter case, if an impermeable barrier is extended along the lowest level of the aquifer (erosional base), we can expect discharge through springs. If boundaries are permeable or even semipermeable, there is only a lithological break, but water circulation through such a boundary is possible (relative barrier) and may again result in drainage via springs, which in this case act as an overflow, in addition to subsurface drainage (Fig. 3.1).

The term *groundwater body* has been widely used in recent water practice, along with the term *aquifer system*. It has been introduced in the year 2000 by the Water Framework Directive of the European Union and entails possible linkage of two or more vertically or laterally interconnected aquifers with a single common water level.

The structure in which a karst aquifer has been formed and exists can be: (1) *open*; (2) *confined*, and (3) *semiconfined*. While in an open system (Fig. 3.1) recharge and discharge areas are open to the surface, with a so-called free water table, in confined systems karst is covered by other permeable, semipermeable or totally impervious overlying layers. In the last case, in a closed structure the groundwater resources are "static" (nonreplenishable) and can only be utilized by drilling wells or galleries, albeit during a limited period of time due to a lack of recharge. A semiconfined

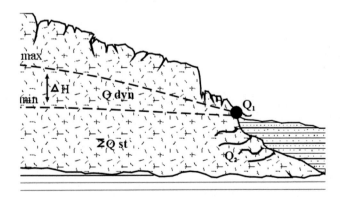

FIGURE 3.1 Drainage of a karst aquifer system. Q_1—drainage via spring, Q_2—subsurface drainage, ΔH—annual groundwater level amplitude between the minimal and maximal level, Q_{dyn}—dynamic groundwater reserves (replenishable), Q_{st}—static groundwater reserves (geological, below the drainage point).

aquifer has an open recharge and confined drainage area, and is characterized by a deep subsurface flow under the overlying rocks.

A single karst system can also be *unary* or *binary*. The former consists of an autogenic part, with the so-called autogenic recharge resulting from rainfalls or percolated sinking streams inside a delineated karstic catchment. If water for recharge arrives from contributing nonkarstic terrains, then we have an allogenic catchment and *allogenic or indirect recharge*. Therefore, a system that consists exclusively of karstic rocks with autogenic recharge is called a unary karst system, while a binary karst system includes an allogenic part with nonkarstic rocks and their contributing catchment (Bakalowicz, 2005; Marsaud, 1997). In mature karst it is common for active, concentric or diffuse ponors (swallow holes) to be located directly at, or near, the contact point between the impervious or low permeable rocks and karst, transforming perennial into sinking streams (Fig. 3.2).

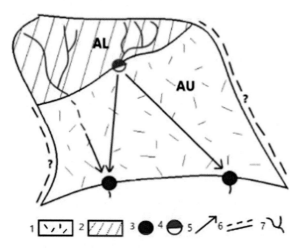

FIGURE 3.2 Conceptualization of a binary aquifer system. 1. Karst system, 2. Impervious rocks, 3. Karst spring, 4. Ponor (swallow hole), 5. Groundwater flow proven by a tracing test, 6. Uncertain lateral boundary, 7. Perennial stream. *AL*, Allogenic part; *AU*, autogenic part.

What makes karst different from other aquifer systems is the incompatibility of hydrogeological (hydrological) and topographic (orographic) boundaries. While in most aquifer systems (intergranular, fissure) these differences may also exist, in karst they can have a significant magnitude due to its large permeability. Herak et al. (1981) stated that the actual catchment area of the Cetina River in the Croatian part of the Dinaric karst is 2.7 times greater than its topographic frame. Bonacci (1987) also presented several examples of differences between subsurface and topographic boundaries (Fig. 3.3). He stated that "only a few terrains in karst have been studied well enough to make it possible to define the catchment precisely."

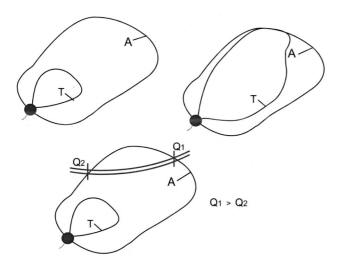

FIGURE 3.3 Possible relationship between a topographic (T) and real hydro(geo)logical catchment (A) in a karst spring. The bottom case shows an inflow that resulted from losing water between two sections of the sinking stream. *Modified from Bonacci, O. (1987). Karst hydrology with special reference to the Dinaric karst. Springer-Verlag.*

Incorrect calculation of the catchment surface area can result in erroneous technical solutions, for instance when estimating the volume of the water reservoir, flood probability, water reserves available for exploitation, or peak discharge.

The typical scheme of discrepancies between two boundaries is shown in Fig. 3.4. It is important to note that the position and orientation of the aquifer layer are factors that are important for the proper assessment of real boundaries, while the topographical dividing line is not.

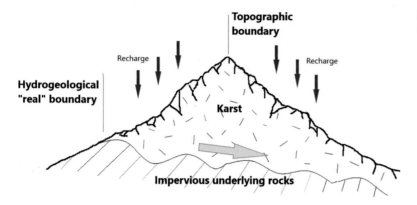

FIGURE 3.4 Conceptualization of a typical karst system—topographic and real hydrogeological boundaries.

Another common and specific property of a representative karst system is the unstable catchment size. It can vary depending on the volume of water accumulated in certain periods of the hydrological year. Therefore, the variability of lateral boundaries can be the result of variations in the water table during high and low water periods, or possible temporary reorientation of flow direction (Stevanović, 2015).

Fig. 3.5 shows changes in the groundwater dividing line in the two contrasting periods—low and high water seasons. During the low water season, the catchment (surface area) is of maximal size. In contrast, during the high water season the groundwater table (potentiometric level) rises, activating temporary springs on the other side of the catchment, which results in a reduced catchment area. Therefore, despite the increase in water resources, the catchment and recharge area will become smaller due to the overflow via the newly activated springs. Such fluctuations can be viewed as a sort of catchment "breathing."

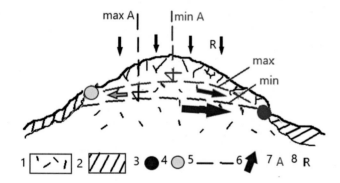

FIGURE 3.5 Variability of the dividing line in two different water periods. Maximal water table results in a minimal surface area (min A) because of the activation of a temporary spring on the other side of the aquifer, and vice versa, a minimal water table results in a maximal surface area (max A). 1. Karst aquifer, 2. Impervious rocks, 3. Karst spring, perennial 4. Karst spring, temporary, 5. Groundwater table (potentiometric surface), 6. Groundwater flow direction, 7. Surface area, 8. Recharge.

If the studied karst aquifer is laterally limited by adjacent catchments which consist of impervious or low permeable rocks, determination of the karst surface area should not be a problem. But if adjacent catchments also consist of karst with permeable lateral boundaries, their boundaries will be very difficult to assess. Herak et al. (1981) concluded that most catchment areas are asymmetrical and only approximately determinable, even in regions where intensive surveys have been carried out.

Internal boundaries of subbasins within a large catchment can also change in the course of the hydrological year. Fig. 3.2 shows divergent groundwater flows from one ponor to two springs, proven by a tracing test. In low water season, groundwater flow may orient toward just one of these two springs. This is quite common and shows the need for detailed field research with repeated tracing tests and systematic monitoring. Fig. 3.6 shows a typical drainage karst area, clearly defined by the position of the spring but with an uncertain position of the lateral boundaries.

Another example of the "active life" of a catchment in karst is its evolution. Kresic (2013) showed how a karstification process with the deepening of the water level influences the development of a karst system (Fig. 3.7). This process is followed by a change in the surface and underground water patterns, piracy and the drying of inner streams. The erosional base can sometimes be very deep at the bottom of the incised canyon (Fig. 3.8).

FIGURE 3.6 Drainage area of the Sosan spring in southern Iran. The catchment's lateral limits can only be supposed.

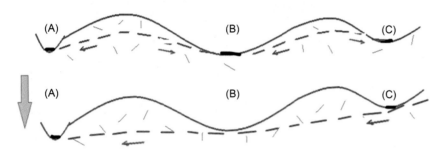

FIGURE 3.7 An example of evolution of a catchment area in karst. With the progress of karstification, groundwater table lowers, resulting in the expansion of catchment which is draining at the main erosional base (A), drying of the inner stream (B) and conversion of the stream (C) from perennial to sinking. *Modified from Kresic, N. (2013).* Water in karst. Management, vulnerability and restoration. *McGraw Hill.*

FIGURE 3.8 Deeply incised canyon Verdone acts as the main erosional base, while most of the inner subbasins are completely disintegrated or remain as hanging and blind valleys (Provence, France).

The problem of determining the surface area in karst is additionally complicated in the case of composite or complex aquifer systems (groundwater bodies), where their connection is fully masked. In the Balkans region, for example, there are several large karst springs that receive a lot of water from adjacent and upper-positioned alluviums and intergranular aquifers situated along the riverbeds. This invisible recharge can be determined by several methods, such as tracing tests or simultaneous hydrometry of the rivers, or indirectly, by isotopic or hydrochemical surveys. Such a conceptual model has been confirmed in the case of karst spring Rašče which supplies Skopje, the capital city of North Macedonia, the largest Albanian spring Blue Eye (in Albanian: *Syri I Kalter*, Fig. 3.9), which is the source of the Bistrica River, and Bolje Sestre, the main water source for the entire coastal area of Montenegro (Fig. 3.10) (Eftimi, 2020; Stevanović, 2010).

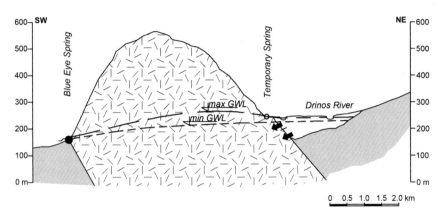

FIGURE 3.9 Composite karst—intergranular aquifer system and underground connection of the Blue Eye spring and the Drinos River alluvium. *Courtesy Eftimi R., published in Stevanović, Z. (2010). Utilization and regulation of springs. In N. Kresic & Stevanović, Z. (Eds.), Groundwater hydrology of springs: Engineering, theory, management and sustainability (pp. 339–388). Elsevier.*

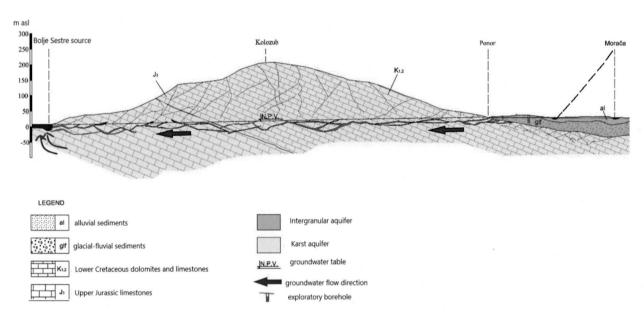

LEGEND

al	alluvial sediments	Intergranular aquifer
glf	glacial-fluvial sediments	Karst aquifer
K₁,₂	Lower Cretaceous dolomites and limestones	N.P.V. groundwater table
J₃	Upper Jurassic limestones	groundwater flow direction
		exploratory borehole

FIGURE 3.10 Composite karst—intergranular aquifer system and the underground connection of the Bolje Sestre spring and the Morača River alluvium. Groundwater connection between the ponor and the spring has been proven by a tracing experiment (Radulović et al., 2007). *Data from Radulović, M., Stevanović, Z., & Radulović, M. M. (2007). Report on the sanitary protection zones of the Bolje Sestre spring. Unpublished report Geoprojekt, Podgorica. Fund of IK Consulting and RWS Regional Waterworks for Montenegrin Coast, Budva, unpublished.*

3.4 Investigation of the catchment delineation methods

Hydrogeological and hydrological investigation methods must be chosen and adapted to local circumstances. Some of the methods can be applied in the hydrogeological research of any aquifer system, but some are highly specific and developed exclusively for karst (Bakalowicz, 2005; Goldscheider & Drew, 2007; White, 1969).

The most common methods that are used primarily to define catchment in karst are field hydrogeological surveys (mapping) and tracing tests.

Hydrogeological mapping in karst includes direct observation and analysis of karstified and impermeable rocks, their lithological composition, thickness and their mutual relationship, hydrogeological properties, recharge and drainage conditions. The number of observation points per unit area (km^2) depends on the degree of complexity of the terrain, as well as on the number of phenomena and objects in the investigated terrain. During mapping, one should strive to register all the phenomena and objects that can be of direct importance for understanding the hydrogeological characteristics of the terrain.

Hydrogeological mapping is a combination of:

1. Direct observation of visible geomorphological, geological, hydrological, hydrogeological symptoms and objects, and
2. Performing appropriate field and laboratory experiments and tests.

Considering the heterogeneity and anisotropy of the karst system, the study of the conditions of formation, the position of recharge and drainage zones and the underground networks of channels and caverns as "water conduits" is specific for almost every individual karst area and requires an approach that is adapted to local circumstances.

In general, mapping in karst terrain can be divided into two groups: *surface* and *underground*.

Surface mapping includes:

- Registering and studying karst water phenomena: karst springs (springs), ponors (swallow holes), thermal springs, estavelles.
- Investigating morphological forms: micro forms such as karrens, sinkholes, karst poljes, dry and blind valleys, soil collapses.

Underground research and mapping includes *direct* and *indirect methods*. Direct methods refer to the research of: potholes, caves, karst channels and caverns, to find out whether they are dry, occasionally active or permanently filled with water. It can be said that speleological and speleo-diving survey of the karst is the only method that provides direct insight into the knowledge of karst. Indirect methods include: geophysical research, drilling and testing of exploratory boreholes (logging, video-endoscopy), water sampling and chemical and microbiological analyses.

Groundwater tracing is a relatively simple, irreplaceable and, as a rule, harmless (in the sanitary sense) research method that can solve the complex issue of flow directions and circulation velocity of the "invisible resource," determine the connections of sinking surface waters and springs, limit the catchment area and obtain a number of other valuable information about the functioning of the hydrogeological system (Benischke, 2021; Käss, 1998).

According to physical-chemical properties, tracers can be insoluble (various substances) or soluble (fluorescent dyes, inorganic salts, radioactive isotopes).

Tracing is very important because a single inflow point (ponor) may drain to springs that are kilometers apart. Tracers are usually injected into ponors (Fig. 3.11), dolines or drilled holes, but can also be placed directly on or beneath the soil surface of karstified rocks to estimate connections with springs or to diffuse discharge zones.

FIGURE 3.11 Sodium fluoresceine injected into the ponor of Ključka River in Cerničko polje (Bosnia and Herzegovina). The system of cascade karstic poljes is well interconnected by springs on one, and a ponor on the other margin of each polje, with the main erosional base at the Adriatic Sea.

The appearance of tracers should be observed at all potential sites, and, together with the calculation of apparent velocity, the construction of a breakthrough curve enables quantitative analysis of the hydrodynamical, physicochemical and biological processes to which the tracer was subjected in the karst underground (Goldscheider et al., 2008).

Despite being the main and most exact method in karst hydrogeology, tracing results only acknowledge and prove the hydraulic connection between certain points, but not the system as a whole. Therefore, the problem of determining external and subbasins boundaries often remains unsolved and left for experts to approximate.

Many other methods can also be used to characterize a system. Their results can serve to understand the relationship between surface water and groundwater, between groundwaters from different aquifers, groundwater's origin and age, aquifer dynamics, and so on.

Hydrochemical and isotopic methods are the most common among them. These methods include a special group of natural tracers such as chemical elements, isotopes and other substances "from the environment." They are based on the process of following and measuring micro constituents or stable isotopes that are found in nature or are very specific for certain types of groundwater.

The dissolution of rocks and the duration of direct water-rock contact result in variable groundwater quality at discharge points. The mineral components of karst waters depend on the composition of the rocks through which water percolates. Calcium carbonate, which builds limestones, is highly soluble when carbon dioxide is present. Sodium chloride and calcium sulfate waters, caused by the dissolution of halite and gypsum, are undesirable in potable water because they change the organoleptic properties of water. Changes in water temperature or turbidity may indicate specific conditions in the studied catchment.

Most isotopic analyses concern stable environmental isotopes such as ^{2}H (deuterium—D), ^{16}O, ^{18}O (oxygen 16 and oxygen 18), and ^{13}C (carbon 13) and radioactive man-made isotopes such as ^{3}H (tritium). These analyses may significantly contribute to assessing the groundwater origin, age, contact time, recharge conditions and consequently even the catchment size (Clark & Fritz, 1999).

Groundwater budgeting (balancing) is one of the basic methods for assessing water availability and using it for various purposes (potable water supply, electricity production, irrigation, geothermal energy, hydrotherapy, mineral water bottling). To make these calculations possible or to establish a conceptual model (Kresic & Mikszewski, 2013; White, 1969), the first step is to determine the geometry of the concerned system. The obtained results of groundwater budgeting can also be indicators of the direction in which to orient the further survey or implement the engineering project (Bonacci, 1993). For instance, an imbalance between the high values of input parameters (e.g., high rainfall, sinking stream inflow) and the low values of output parameters (e.g., low runoff due to the absence of perennial streams, plus a small number of low discharging springs) may indicate the existence of certain invisible water balance parameters, such as e.g., active subsurface drainage. Such a situation can open a prospect of an additional survey to tap this underground flow. In contrast to the previous example, if output parameters are much larger than input parameters (e.g., drainage via springs that is much bigger than the volume of possibly infiltrated rainy water), it is a sign of "catchment deficit." Such a sign requires an additional survey to re-study the size of the catchment or identify other input sources, as in the case of two examples shown in Figs. 3.9 and 3.10 (Stevanović, 2010).

In hydrogeological practice, it is also common to apply the inverse method and estimate the recharge area based on the results obtained from water budget calculation (Kresic & Stevanović, 2010; Ristić Vakanjac, 2015). Although uncertain in many aspects, this method may nonetheless still provide a general view of the geometry of the studied aquifer, especially if similar karstic terrains and aquifers have been properly explored in terms of permeability and storativity.

Table 3.1 shows the commonly applied methods and their tasks in the process of karst catchment delineation.

TABLE 3.1 Methods that are commonly applied in hydrogeological practice to assess the size and boundaries of karst catchment areas.

Method	Task
Climatology, geology, and geomorphology surveying	Collect and evaluate data on geological structure and geomorphology, water budget elements, and assessment of water availability
Hydrography and hydrology surveying	Collect and evaluate data on hydrographic network, regime of surface waters
Hydrogeological mapping	Register all relevant water points, establish visual base for understanding hydrogeological settings, conduct in-field measurements and tests, and estimate catchment area and its boundaries
Tracing test	Study interrelationships: ponors—springs, composite aquifers (alluvial—karst groundwater), assess aquifer's properties (apparent velocity, residence time), estimate catchment area
Simultaneous hydrometry of the rivers	Provide data on interrelationship of alluvial aquifer and surface waters (rivers): subsurface drainage into riverbed (surplus flow) or percolation from riverbed into underlying and lateral aquifer (water losses)
Hydrochemistry analyses	Assess the quality of groundwater and surface waters, relationship between aquifer systems, and between surface waters (lake, river) and groundwater, their suitability for drinking purposes, content of pollutants, and their origin/fate
Isotopic analyses	Assess groundwater origin and age, relationship between aquifer systems, and between surface waters (lake, river) and groundwater
Groundwater regime	Collect and study information of groundwater regime, relationship with rainfalls and surface waters, anthropogenic impact on water regime
Water budgeting	Collect information on climate and hydrological elements (runoff, evapotranspiration, effective infiltration of rainfall), calculate water budget and assess relationship of groundwater and surface waters, groundwater regime, dynamic reserves, and availability, re-estimate size of catchment in accordance with obtained results
Groundwater stochastic modeling	Establish mathematical relation between various input and output parameters such as climate and hydrological elements, effective infiltration of rainfall and forecast groundwater regime, dynamic reserves and their availability, re-estimate size of catchment in accordance with obtained results
Geophysical parametrization	Conduct field surveying and logging to collect information on geological structure and properties along selected sections, estimate groundwater depth and flow intensity, position of karstification base and possibly drilling sites suitability
Geodetic surveying	Undertake a geodetic survey to precisely determine water levels along existing rivers, position and altitude of major springs and other water points
Water monitoring (quantity and quality)	Install automatic equipment (data loggers), rain and staff gauges, limnigraphs, pans, etc. at main water points and rivers. Process and evaluate collected data. Undertake field survey for additional measurements and readings (doppler radar surveying, current metering, sounding)

3.5 Case studies

A few case examples are used here to demonstrate different approaches and problems that experts may face in their attempt to assess catchment areas in a specific environment such as karst. All the study areas are in Southeast Europe, in mature, highly developed karst of Dinaric and Carpathian geological structures and mountain chains.

3.5.1 Dynamic karst aquifer regime with changeable and overlapped catchment areas—East Herzegovina (Bosnia and Herzegovina)

In an effort to understand the complexity of often-connected karst aquifers, the groundwater pattern, and to delineate their catchments, several hundred tests have been carried out in the Dinaric Karst, mostly using sodium-fluresceine (uranine).

Komatina (1983) examined the results from 380 tracing experiments in the Dinaric Karst and found that the frequency of groundwater apparent velocities was: 70% less than 0.005 m s^{-1}; 20% 0.005–0.01 m s^{-1}; 10% more than 0.01 m s^{-1}.

Based on the results of numerous experiments that were performed in the karst of eastern Herzegovina (Fig. 3.12), Milanović (1981) calculated the average flow velocity to be 0.05 m s^{-1}, with extremes in the wide range of 0.2 $\times$ 10^{-8} to 0.55 m s^{-1}. Whereas in the dry seasons water circulation in a karst system is slow, during the high-water season, water with tracer requires up to five times less time to travel the same distance. Milanović (1981) presents an example from eastern Herzegovina: when the groundwater elevation is low, it takes the underground flow 35 days to cover the 34 km from Gatačko Polje to the Trebišnjica Spring, while during high water levels water flow covers the same distance in just 5 days.

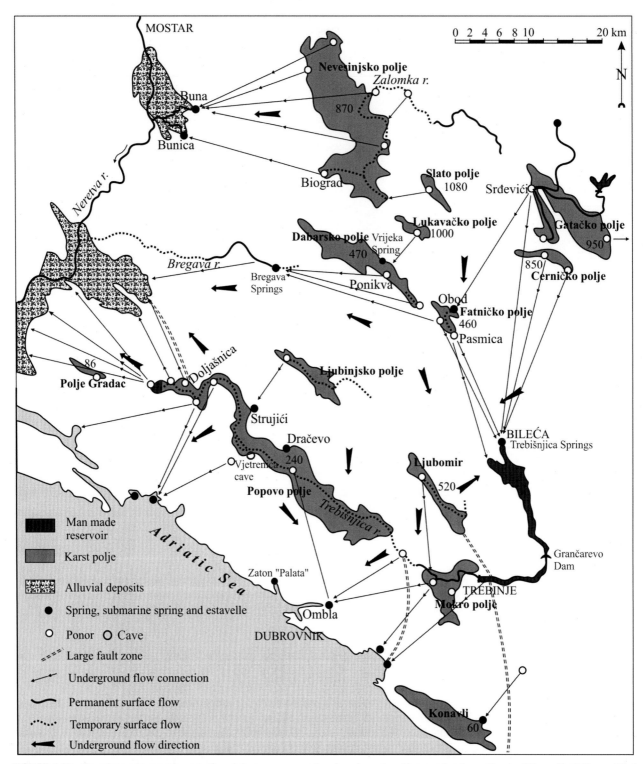

FIGURE 3.12 Complex underground connections between ponors and springs in eastern Herzegovina (according to Milanović, 1981, modified; printed with permission). One spring drains water from several ponors (convergent flow), and vice versa, water from one ponor flows toward several springs (divergent flow). The size of catchments is thus considerably changes throughout the hydrologic year.

3.5.2   An attempt to inversely assess surface area—Veliko Vrelo spring (Serbia)

Aiming to simulate the process of actual daily sums of evapotranspiration in water balancing, Ristić Vakanjac (in: Stevanović et al., 2010) introduced parameter Θ, which also helps to calibrate the previously roughly assessed surface of the studied catchment area. By taking the values of parameter Θ to be 0, 0.1, 0.2, ..., 0.8 and 0.9, the water balance equation for the catchment (F) of Veliko Vrelo (eastern Serbia) was established by calibrating the potential catchment size in such a way that it fulfilled the condition where the initial volume of stored water (V_0) is equal to the volume at the end of the predefined analytical period (V_k), that is, $V_0 = V_k$. The functional dependencies $\Theta = f(F)$ were formed using the obtained ranges of values of Θ and F (surface area), which fulfilled the set criterion. The graphic interpretation of the resulting function is shown in Fig. 3.13 (Stevanović et al., 2010). The analysis of the function $\Theta = f(F)$ shows that the correlation between these two parameters is nonlinear over the entire range of Θ. The low values of Θ do not have any substantial effect on the size of the catchment. In contrast, when the value of Θ is generally greater than 0.8, or when the variations in this parameter are small, there is a significant increase in the catchment size. The border point is also the point of inflection (apex) of the nonlinear function $\Theta = f(F)$. This approach was applied to define that the catchment size of Veliko Vrelo is 38.1 km^2 (Fig. 3.13), as well as to determine the actual values of evapotranspiration.

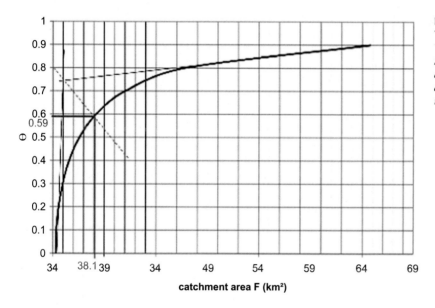

FIGURE 3.13   Function $\Theta = f(F)$ of the Veliko Vrelo catchment. *Conceptual solution after Ristić Vakanjac, published in Stevanović Z., Milanović S., & Ristić V. (2010). Supportive methods for assessing effective porosity and regulating karst aquifers.* Acta Carsologica, 39(2), 313–329, *reprinted with permission of Acta Carsologica.*

3.5.3   Problems in delineating sanitary protection zones and applying preventive protection measures in transboundary aquifer Ombla (Croatia and Bosnia and Herzegovina)

The Ombla Spring is located on the Adriatic coast, near the historic town of Dubrovnik. The spring drains a karstic aquifer that consists of karstified Mesozoic carbonates and appears at the contact with impervious flysch barrier, just two meters above the sea level (Fig. 3.14). The recorded minimum discharge rate was 2.3 m^3 s^{-1}, while its maximum rate is more than 130 m^3 s^{-1}. (Milanović, 2023).

FIGURE 3.14   The Ombla spring on the Adriatic Coast. Close behind the visible mountain peaks is the international boundary between Croatia and Bosnia and Herzegovina, the countries that share this important karst aquifer.

The Ombla spring is just one of the main drainage points of a large karst system whose total catchment area has been estimated at about 1630 km². This is the basin of the Trebišnjica River, the largest sinking river in Europe. From this total, Milanović (2023) distinguished the intermediate catchment of the Ombla spring, with the surface of around 600 km². The watershed between these two catchments does not "fluctuate," as is the case in many other karst aquifers, but is rather clearly marked by a lithological section built from dolomitic rocks (Milanović, 2023). After the construction of the Trebišnjica Hydrosystem and the channeling of the riverbed of Trebišnjica the average discharge of the Ombla spring has dropped from 33.8 to 24.4 m³ s⁻¹.

The main problem for water experts and managers is to delineate sanitary protection zones and impose water protection measures for this very important aquifer, which provides potable water to Dubrovnik and many other nearby settlements along this tourist area in Croatia. The reason for this is that the aquifer is shared between Croatia, where the Ombla spring is situated, and the neighboring Bosnia and Herzegovina. Less than 10% of the catchment is in Croatia, while most of the Trebišnjica River basin is in Bosnia and Herzegovina. This problem is studied within the framework of the international project DIKTAS (Dinaric Karst Transboundary Aquifer System (Stevanović et al., 2016).

3.5.4 Failure and correction in assessing the surface area—Mrljiš Spring (Serbia)

Mrljiš is one of the largest springs of the karst system of the Kučaj-Beljanica Mt. in eastern Serbia. This spring and its catchment, built mostly from karstified Lower Cretaceous limestones, have been studied for the purpose of constructing a regional water system "Bogovina" to ensure water for several towns of the Timok region. Full attention has been paid to the proper assessment of water resources, the creation of a viable water budget, the engineering regulation of the aquifer, and to the introduction of adequate monitoring to ensure sustainable water use and protective measures against pollution (Fig. 3.15). The project was completed successfully by drilling a battery of wells, which ensured tapping of three times more groundwater than the minimal recorded natural springflow of Mrljiš. So instead of 80 L s⁻¹, some 250 L s⁻¹ of water are now being pumped on a regular basis (Stevanović, 2010).

In the initial stage of the project, when the plan to establish meteorological and hydrological network was created, preliminary expertise indicated that Mrljiš spring covered some 60 km² (the area is approximately drawn in Fig. 3.15 and presented under no. 1). However, the water budget analysis and the field survey suggested that the catchment could be larger, which was confirmed soon after by repeated tracing tests performed in one of the caves with a system of concentric underground flow. Consequently, it was found that an additional surface of about 40 km² would have to be attached, including the allogenic part in the upstream sections of the sinking streams connected to the traced cave (the area presented in Fig. 3.15 under no. 2).

3.6 Conclusions

Estimating and especially precisely delineating the boundaries of a catchment in karst are not easy tasks, even for professionals. There are many reasons for this. The main reason is that karst and its aquifers are specific, and are characterized by high permeability and considerable influence of external factors—climatic, morphological, geological, and hydrological. As a result, dynamic karst water regime causes a situation where, unlike in other aquifers (fissured, intergranular) there is no good correspondence between topography and hydrogeology, that is, between orographic and underground dividing lines. It is common for hydrogeological boundaries to be larger than topographic ones. Hence, the volume of stored groundwater could, in reality, be larger than those that were estimated by use of the conservative method based on delineate orographic dividing lines. *Vice versa* situations are extremely rare.

Another common situation in karst is that an aquifer may be "breathing," so the surface of the catchment area may change depending on the water table level: it can become larger during low-water periods, when most of the stored water is oriented in one direction, toward major water drainage points (springs), or smaller during high-water periods, when the water table rises, causing the temporary drainage points to also activate as overflows.

Apart from field and cave mapping and hydrochemical and isotopic surveys, tracing tests are an essential method that is applied in karst hydrogeology to estimate the size of the catchment. Due to the complexity of karst, tracing results can only prove hydraulic connection between certain points (ponors-springs), but not of the system as a whole. This is why the problem of assessing external and internal (subbasins) boundaries often remains open, and is left to experts and their approximations. A nonprofessional's attempt to do this is, of course, bound to fail.

Finally, special attention must be paid to assessing catchments, as miscalculation could result in failures related to engineering or water sources protection projects.

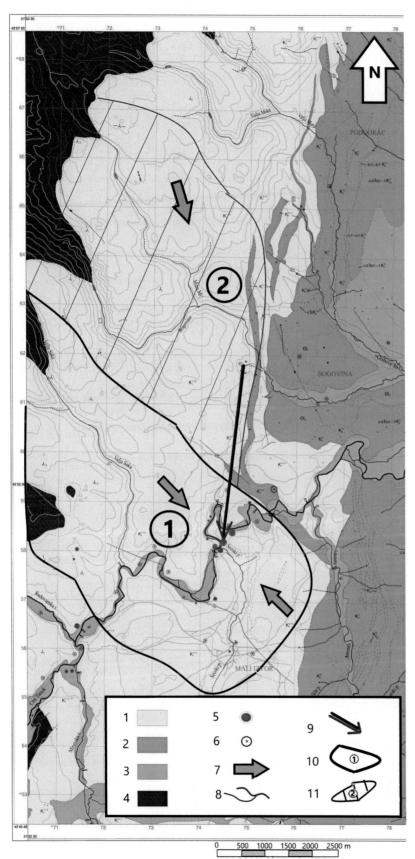

FIGURE 3.15 The Mrljiš spring catchment as a part of the regional water supply system "Bogovina" in eastern Serbia. The catchment consists of the initially assessed mostly autogenic karst aquifer labeled as no. 1 and the binary system labeled as no. 2, which was attached after a detailed survey and the tracing tests. 1. Karst aquifer, 2. Fissured aquifer, 3. Intergranular aquifer of the alluvium, 4. Impervious rocks, 5. Spring, 6. Borehole, 7. Groundwater flow direction, 8. River, 9. Proven connection between ponor and spring, 10. Subcatchment 1, 11. Subcatchment 2.

References

Bakalowicz, M. (2005). Karst groundwater: A challenge for new resources. *Hydrogeology Journal, 13*, 148−160. Available from https://doi.org/10.1007/s10040-004-0402-9, https://link.springer.com/article/.

Benischke, R. (2021). Review: Advances in the methodology and application of tracing in karst aquifers. *Hydrogeology Journal, 29*, 67−88. Available from https://doi.org/10.1007/s10040-020-02278-9, https://link.springer.com/article/.

Bonacci, O. (1987). *Karst hydrology with special reference to the Dinaric karst*. Springer-Verlag.

Bonacci, O. (1993). Karst spring hydrographs as indicators of karst aquifers. *Hydrological Sciences Journal, 38*(1), 51−62.

Clark, I., & Fritz, P. (1999). *Environmental isotopes in hydrogeology* (2nd ed.). Taylor & Francis Group.

Cvijić, J. (1893). Das Karstphaenomen. Versuch einer morphologischen monographie, Geograph. Abhandlungen Band, V, Heft 3, Wien.

Drogue, C. (1982). L'aquifère karstique: Un domain perméable original. *Le Courier du CNRS, 44*, 18−23.

Eftimi, R. (2020). Karst and karst water recourses of Albania and their management. *Carbonates and Evaporites, 35*, 1−14. Available from https://doi.org/10.1007/s13146-020-00599-0.

Ford, D., & Williams, P. (2007). *Karst hydrogeology and geomorphology*. Wiley.

Goldscheider, N. (2010). Delineation of spring protection zones. In N. Kresic, & Z. Stevanović (Eds.), *Groundwater hydrology of springs: Engineering, theory, management and sustainability* (pp. 302−338). Elsevier BH.

Goldscheider, N., & Drew, D. (Eds.), (2007). *Methods in karst hydrogeology. International Contribution to Hydrogeology, IAH* (26). Taylor & Francis/Balkema.

Goldscheider, N., Meiman, J., Pronk, M., & Smart, C. (2008). Tracer tests in karst hydrogeology and speleology. *International Journal of Speleology, 37*(1), 27−40. Available from https://doi.org/10.5038/1827-806X.37.1.3.

Goldscheider, N., Zhao, C., Auler, A., Bakalowicz, M., Broda, S., Drew, D., Hartmann, J., Jiang, G., Moosdorf, N., Stevanović, Z., & Veni, G. (2020). Global distribution of carbonate rocks and karst water resources. *Hydrogeology Journal, 28*(5), 1661−1677. Available from https://doi.org/10.1007/s10040-020-02139-5.

Gunn, J. (Ed.), (2004). *Encyclopedia of caves and karst science*. Fitzroy Dearborn.

Herak, M., Magdalenić, A., & Bahun, S. (1981). Karst hydrogeology. In Halasi & G. J. Kun (Eds.) *Pollution and water resources. Columbia University seminar series. Vol. XIV, part 1, Hydrogeology and other selected reports* (pp. 163−178). Pergamon Press.

Käss, W. (1998). *Tracing technique in geohydrology (Engl. transl. of Geohydrologische markierungstechnik)*. Brookfield. Available from https://doi.org/10.1016/S1462-0758(99)00011-4.

Komatina, M., (1983). Hydrogeologic features of Dinaric karst. In B. Mijatović (Ed.) *Hydrogeology of the Dinaric Karst, Field trip to the Dinaric karst*, Yugoslavia, May 15−28, 1983, "Geozavod" and SITRGMJ, Belgrade, 45−58.

Kresic, N., & Stevanović, Z. (Eds.), (2010). *Groundwater hydrology of springs: Engineering, theory, management and sustainability*. Elsevier Inc., BH.

Kresic, N. (2013). *Water in karst*. Management, vulnerability and restoration.. McGraw Hill.

Kresic, N., & Mikszewski, A. (2013). *Hydrogeological conceptual site model: Data analysis and visualization*. CRC Press.

LaMoreaux, P. E., Wilson, B. M., & Memon, B. A. (Eds.), (1984). *Guide to the hydrology of carbonate rocks. IHP studies and reports in hydrology* (41). UNESCO.

Mangin, A. (1984). Pour une meilleure connaissance des systèmes hydrologiques à partir des analyses corrélatoire et spectrale. *Journal of Hydrology, 67*, 25−43.

Marsaud., B. (1997). *Structure et fonctionnement de la zone noyée des karsts à partir des résultats expérimentaux* [Structure and functioning of the saturat-ed zone of karsts from experimental results]. Documents du BRGM 268, Editions de BRGM, Orleans.

Milanović, P. (1981). *Karst hydrogeology*. Water Resources Publications. Reprinted in 2018 by the Centre for Karst Hydrogeology, University of Belgrade.

Milanović, P. (2023). *Karst of eastern Herzegovina and Dubrovnik littoral. Edition: Karst and cave systems of the world*. Springer Nature Switzerland.

Radulović, M., Stevanović, Z., & Radulović M.M. (2007). Report on the sanitary protection zones of the Bolje Sestre spring. Unpublished report Geoprojekt, Podgorica. Fund of IK Consulting and RWS Regional Waterworks for Montenegrin Coast, Budva.

Ristić Vakanjac, V., Stevanović, Z., Maran Stevanović, A., Vakanjac, B., & Čokorilo Ilić, M. (2015). An example of karst catchment delineation for prioritizing the protection of an intact natural area. *Environmental Earth Science, 74*, 7643−7653. Available from https://link.springer.com/article/10.1007/s12665-015-4390-y.

Stevanović, Z. (2010). Utilization and regulation of springs. In N. Kresic, & Z. Stevanović (Eds.), *Groundwater hydrology of springs: Engineering, theory, management and sustainability* (pp. 339−388). Elsevier BH.

Stevanović, Z. (Ed.), (2015). *Karst aquifers−Characterization and engineering. Series: Professional practice in earth science*. Springer International Publisher Switzerland.

Stevanović, Z. (2019). Karst waters in potable water supply: A global scale overview. *Environmental Earth Science, 78*, 662. Available from https://doi.org/10.1007/s12665-019-8670-9.

Stevanović, Z., Kukurić, N., Pekaš, Ž., Jolović, B., Pambuku, A., & Radojević, D. (2016). Dinaric Karst Aquifer—One of the world's largest trans-boundary systems and an ideal location for applying innovative and integrated water management. In Z. Stevanović., N. Kresic, & N. Kukuric (Eds.), *Karst without boundaries* (pp. 3−25). CRC Press/Balkema, EH Leiden.

Stevanović, Z., & Milanović, P. (2015). Engineering challenges in karst. *Acta Carsologica, 44*(3), 381–399. Available from https://ojs.zrc-sazu.si/carsologica/article/view/2963.

Stevanović, Z., Milanović, S., & Ristić, V. (2010). Supportive methods for assessing effective porosity and regulating karst aquifers. *Acta Carsologica, 39*(2), 313–329.

White, W. B. (1969). Conceptual models for carbonate aquifers. *Ground Water, 7*(3), 15–21. Available from https://doi.org/10.1111/j.1745-6584.1969.tb01279.x.

Williams, P. (2008). *World heritage caves and karst*. IUCN.

Part II

Best management

Chapter 4

Managing streams through restored floodplains: a case of Ganga River in the middle Ganga plain

Ankit Modi[1], Saeid Eslamian[2] and Vishal Kapoor[3]

[1]Department of Civil Engineering, Indian Institute of Technology Kanpur, Kanpur, Uttar Pradesh, India, [2]Department of Water Sciences and Engineering, College of Agriculture, Isfahan University of Technology, Isfahan, Iran, [3]Indian Institute of Technology Kanpur, Kanpur, Uttar Pradesh, India

4.1 Introduction

Large rivers in their continuum perform various bio-geomorphological functions in the longitudinal and lateral dimensions along the river course, which involves the maintenance of river channels, riparian zones, floodplains, and connected wetlands (Dodge, 1989). The bio-geomorphological functions consist of four important functions, namely, hydrological, geomorphological, chemical, and biological functions, which provide sustainability to the flora—fauna of the river and the connected wetlands. The geomorphological functions (erosion, transport, and deposition), chemical functions (acid—base buffering), and biological functions (energy transformation, nutrient turnover, processing of organic matter, etc.) are byproducts of the hydrological functions (transporting water, viz., stormwater, rainwater, snowmelt water, or combination of these) of the river. These functions support a wide variety of rich life forms ranging from riparian vegetation to different aquatic and semiaquatic species apart from a plethora of goods and services for human needs including water, fisheries, agriculture, and sand.

In alluvial plains, the lateral dimension of a natural river shows a cyclic expansion and contraction due to variable flows every year and is highly unpredictable in the case of high unforeseen floods. In addition, due to the lower probability of high floods, the floodplains are highly susceptible to anthropogenic activities that cause significant damage not only to the riverine ecosystem but also to crops and human lives that encroach on the river's floodplain. The nonoptimal utilization of river floodplains adversely impacts its flora—fauna. The most common stress for river floodplains around the globe is the conversion of floodplains into agricultural land; for example, up to 90% of the floodplains were converted into agricultural land in the North American and European rivers. Along with floodplain conversion, dam structures significantly enhance the alteration of the floodplain inhabitants by reducing the number of flood peaks.

In India, the Ganga River, also known as the Ganges, is one of the most sacred and revered rivers. It holds immense cultural, religious, and historical significance for millions of people. Its vast floodplains are vital ecosystems that provide natural flood control, support rich biodiversity, and offer sustainable habitats for various flora and fauna. However, these floodplains face challenges such as encroachments and pollution, necessitating thoughtful management to protect the river's ecological balance and cultural heritage. Further, these floodplains are supported by various wetlands, for example, Upper Ganga Ramsar site (between towns Brijghat and Narora in Uttar Pradesh), which are essential components of this iconic river's ecosystem. The upper Ganga Ramsar site (hereafter UGRS) was designated as a Ramsar site in the year 2005 for the prudent use of river-floodplain resources and to support several critically endangered aquatic and semiaquatic flora—fauna, for example, the Ganges River dolphin, wolly-necked stork, gharial, white-rumped vulture, and red-crowned roofed turtle (Garg et al., 2015; Khan & Khan, 2013; MoEF&CC, 2020a). The UGRS meets five criteria out of nine of this convention for species and ecological communities such as waterbirds and fish (MoEF&CC, 2020a). Since 2005, all anthropogenic activities, such as fishing, intensive farming, and domestic—agriculture—industrial effluents dumping, have been managed in a "command and control mode" by the Ministry of Environment Forest and Climate Change (MoEF&CC), Government of India (MoEF&CC, 2020b). The "command and control mode" management was done to preserve the native species and ecological communities.

Hydrosystem Restoration Handbook. DOI: https://doi.org/10.1016/B978-0-443-29802-8.00004-2

Many studies on Ramsar wetlands have been published worldwide, but few studies on India's wetlands, notably the UGRS, have been undertaken. For example, Murthy et al. (2013) prepared a wetland atlas of India using the Resourcesat-1 LISS III satellite imagery data. They obtained various outputs, such as wetland extent, bio-geographic zones, and agroclimatic zones for the 12 km buffer area. Khan and Khan (2013) used Landsat data from 2000 and 2013 to examine the land use and land cover (LULC) of the UGRS for a 5 km buffer area on both sides of the river and detected the changes for different attributes of the LULC. They revealed that the river and the built-up regions had grown by 11% and 9%, respectively, by replacing wasteland (usually wetlands) and forest cover. Using phytoremediation techniques, Garg et al. (2015) discovered plant species in the UGRS that might remove heavy metals from soil and/or water. Following a review of the literature, it was revealed that research was conducted mostly for LULC changes in the UGRS, with no studies so far evaluating the positive impact of regulation on the UGRS's floodplain-wetlands.

Hence, it is critical to investigate the impact of UGRS conservation by comparing the hydrometeorological parameters of UGRS to the downstream floodplains. In addition, all floodplains were compared from pre-2005 to post-2005 to highlight UGRS's significance, as UGRS was designated a Ramsar site in 2005. Three research questions are explored, as follows:

1. How UGRS improved monsoonal flood response for the event 2010 in comparison to the 1978?
2. Did it influence water quality parameters and suspended sediment load for postflood months (August−September) during 1995−2010?
3. Did it mediate biodiversity using secondary field survey data?

Section 4.2 of this chapter provides an overview of the study area. In Section 4.3, the methodology for the present assessment is elaborated, comprising four subsections. These subsections furnish essential details regarding hydrology, water quality, sediment, and biodiversity. Section 4.4 presents the results and discussion, and is further divided into four subsections that explain the contributions of wetlands to hydrology, water quality, sediment, and biodiversity, which are crucial for maintaining a healthy river system.

4.2 Study area and data

The study region includes seven Central Water Commission (CWC) hydrometeorological sites in the middle Ganga plains between the cities of Haridwar and Prayagraj. The Ramsar site contains one CWC site (Garhmukteshwar CWC), while there are six CWC sites downstream of the Ramsar site. Throughout the stretch, three barrages for the domestic−agriculture−industrial water supply do not affect monsoon floods (Modi et al., 2020).

The Ganga is a snow-fed perennial river with substantial variability in discharge due to heavy rainfall during the monsoon season (July−September), which is primarily responsible for river flooding. CWC data on streamflow, water quality, and sediment were compiled for 1975−2010 (flow) and 1995−2010 (water quality/sediment). The biodiversity statistics were drawn from various sources, including a recent report published by India's Central Inland Fisheries Research Institute, based on a quarterly field survey conducted from January 2016 to December 2018. Fig. 4.1 and Table 4.1 depict the key features of the study region and some pertinent information.

The LULC classes for the floodplain's average valley width are assessed and tabulated in Table 4.2. The LULC data were downloaded from the Soil Water Assessment Tool website (https://swat.tamu.edu/data/), initially prepared by the International Water Management Institute using Moderate Resolution Imaging Spectroradiometer (MODIS) 500-m and Advanced Very High Resolution Radiometer (AVHRR) 10-km satellite sensor data merged with NRSC land use/land cover map 2007−2008. According to Fig. 4.1 and Table 4.2, agriculture is the most common land use in all stretches, and the first three stretches have more forest cover than the rest of the downstream sections (Fig. 4.2).

4.3 Materials and methods

Physical, chemical, biological, and socioeconomic components are the core for analyzing complex environmental stress (Pahl-Wostl, 2007; Ragulina et al., 2022; Whitfield & Elliott, 2002). Similarly, an assessment of water, sediment, and flora−fauna characteristics has been performed for this study to determine the environmental stress of the floodplains in the study area. In Fig. 4.3, a methodology is framed to illustrate the significant elements of the data source, data processing, objective, and assessment criteria. Furthermore, the critical criteria are defined exclusively in the subsections.

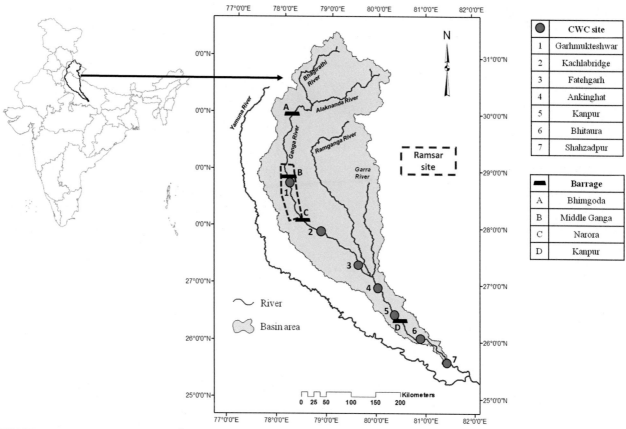

FIGURE 4.1 The study area showing the Ganga river, Ramsar site, and selected CWC sites. Map lines delineate study areas and do not necessarily depict accepted national boundaries. *Modified after MoEFCC, Ramsar Sites in 75th Year of Independence 75 Ramsar Sites 75th Year Independence. (2022), 2022.*

TABLE 4.1 Stretches in the study area based on CWC observation sites and their salient features.

ID	Stretch	CWC site taken into assessment	Length of the stretch, km	Average active floodplain width, km[a]	Average valley width, km[+]
S0	Garhmukteshwar upstream	Garhmukteshwar	–	–	–
S1	Garhmukteshwar–Kachlabridge	Kachlabridge	130	10	15
S2	Kachlabridge–Fatehgarh	Fatehgarh	110	08	29
S3	Fatehgarh–Ankinghat	Ankinghat	70	10	16
S4	Ankinghat–Kanpur	Kanpur	70	10	16
S5	Kanpur–Bhitaura	Bhitaura	70	02	05
S6	Bhitaura–Shahzadpur	Shahzadpur	90	02	05

[a]*MoEFCC (2022).*

TABLE 4.2 Land use and land cover for the selected stretches (in % of the total area).

Class	S0	S1	S2	S3	S4	S5	S6
Current fallow	0.48	3.02	5.05	1.57	1.50	1.88	2.47
Agriculture	5.30	14.45	18.29	12.00	9.25	4.61	4.51
Forest	0.46	0.30	0.27	0.08	0.00	0.02	0.07
Rainfed	0.01	0.01	0.15	0.06	0.05	0.02	0.01
Brush	0.23	0.61	2.29	0.75	0.47	0.41	0.21
Grass	0.08	0.29	1.28	0.59	0.51	0.56	0.79
Urban	0.01	0.04	0.10	0.01	0.19	0.10	0.01
Water	0.43	0.34	0.82	0.45	0.33	0.34	0.39
Sum	7.00	19.06	28.25	15.51	12.30	7.94	8.46

Total area: 14752 km².

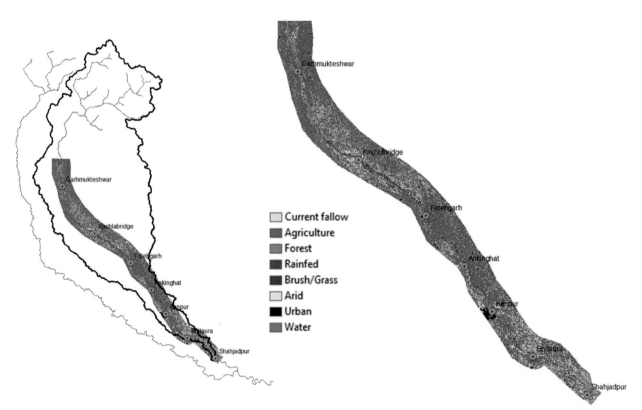

FIGURE 4.2 Land use and land cover profile of the study area. Map lines delineate study areas and do not necessarily depict accepted national boundaries.

4.4   Flood hydrograph

A flood hydrograph is a proxy indicator of the floodplain's state during high-water events. Keeping climate parameters the same, one can undoubtedly estimate the basin's features, such as size and slope, drainage density, and land use, by

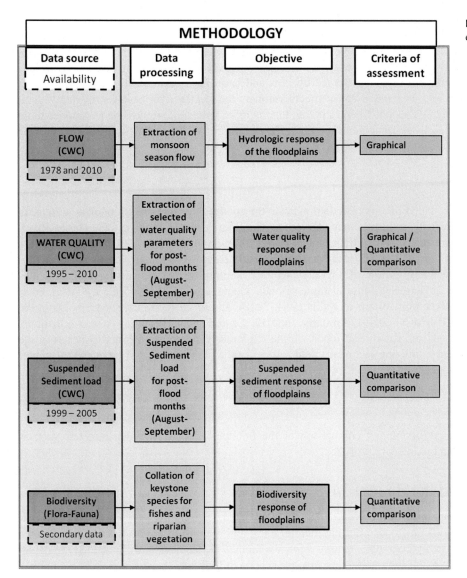

FIGURE 4.3 Methodology for the present study.

looking at the hydrograph shape (Subramanya, 1994). The hydrograph shape will differ between the natural floodplain and the floodplain roughened through land use. As a result, hydrographs for two flood events were analyzed for all CWC sites, and a spatio-temporal comparison for floodplain responses was made.

4.5   Water quality

Flood water quality has particular characteristics and has received much research for the world's major rivers for various reasons (Doll et al., 2020; Howitt et al., 2007). In general, natural floodplains add nutrients (nitrates, ammonia, phosphates, etc.), salts (ions of chloride, sulfate, potassium, bicarbonate, calcium, etc), and sediments to rivers during the monsoon and postmonsoon seasons, whereas urban floodplains contaminate rivers and increase the pollution load (Lyubimova et al., 2016). Nevertheless, nutrient removal from floodplains was also observed by Gordon et al. (2020), who revealed nutrients removing floodplains across North America and Europe. Floods are responsible for increased fertilizers, pesticides, and other pollutants in a river that runoff through agricultural floodplains (Mateo-Sagasta et al., 2017). Agriculture is essential in the middle Ganga floodplains, so water quality during the monsoon and postmonsoon seasons was researched to expose floodplain features.

4.6   Sediment load

The geomorphology of an alluvial river is primarily governed by floods and is closely related to sediment aggradation—degradation (Kale, 2003). The floodplain characteristics can be determined using sediment load during floods (Ahilan et al., 2018; Nones, 2019). Sediment load is governed by flow and is primarily affected by catchment factors such as bed slope and LULC. Furthermore, vegetation cover affects sediment load in the river because riparian vegetation binds the soils and works against bank erosion, whereas barren land is more vulnerable to erosion and erodes more due to agricultural activity (Khan et al., 2018). Hence, in this study, the assessment of the sediment load is used as the geomorphic characteristics of the floodplains.

4.7   Biodiversity profile

The presence and diversity of the flora—fauna in a river system indicate the condition of the river-floodplain system. In India, the unprecedented loss of biodiversity and ecosystem services of the river system, particularly the Ganga River, has become a national concern (Bhaskar & Karthick, 2015; Jameel et al., 2020; Singh & Singh, 2019). To identify the status quo of the biodiversity of the Ganga River, government agencies have recently completed many rapid surveys and ecological assessments as the biodiversity profile of a river system are determined by field surveys and samplings (Hughes, 1997; WII-NMCG, 2019). In 2018, the Wildlife Institute of India, Dehradun (WII) produced a technical report on the wildlife values of the Ganga River from Bijnor to Ballia (WII, 2018), and in 2019, it published a book on the Ganga River's biodiversity profile (WII-NMCG, 2019). Similarly, in 2019, Central Inland Fisheries Research Institute, Barrackpore, West Bengal (CIFRI) produced a mid-term report on the exploratory survey of fish and fisheries in the Ganga River (ICAR-CIFRI, 2019). Hence, utilizing secondary field survey data, an assessment was produced to support the role of floodplains in biodiversity mediation.

4.8   Results and discussion

The following subsections present a spatio-temporal comparison of the floodplains for different parameters such as flood, water quality, sediment load, and biodiversity profile.

4.9   Hydrologic response of the floodplains

Figs. 4.4 and 4.5 illustrate flood hydrographs for the years 1978 and 2010. These two floods were chosen for two reasons: first, to assess the impact of UGRS on floods after and before 2005, and second, to ensure that the floods were equivalent in volume and frequency. The hydrographs are drawn for two months (September 1 to October 31) to depict the flood event with its three segments: rising limb, crest segment, and recession limb.

The hydrograph for the Garhmukteshwar CWC site is positively skewed, whereas the hydrograph for the Shahzadpur CWC site is negatively skewed, indicating that a smaller upstream floodplain region provided flood occurrences for the Garhmukteshwar CWC than for the Shahzadpur CWC. Fleischmann et al. (2016) also demonstrated that the negative skewness of the hydrograph was related to the higher upstream floodplain inundation in their study. On comparing the hydrographs for all CWC sites from 1978 and 2010, it was observed that, except for Garhmukteshwar CWC, all hydrographs had changed or did not follow the same pattern for both years. It signifies that floodplain characteristics have altered throughout time. In addition, at all CWC sites where flow recessed from peak flood to 1-year return period flood, a piece of the hygrograph's recession curve is depicted in the inset (refer to Figs. 4.4 and 4.5). The 1978 hydrographs indicated two types of patterns: concave upward and concave downward. Garhmukteshwar, Kachlabridge, and Fatehgarh displayed concave upward, indicating that these floodplains had similar properties. On the other hand, downstream floodplains Kanpur, Bhitaura, and Shahzadpur displayed concave downward, indicating that these floodplains had equivalent traits dissimilar to upstream floodplains (Eslamian, 2008).

Nevertheless, this is not the case for the hydrographs from 2010, as each hydrograph exhibits a unique pattern. The recession curve in the Garhmukteshwar CWC site follows an exponential pattern, but it exhibits a quadratic expression at the Kanpur CWC site. Several studies have revealed that the quadratic equation is

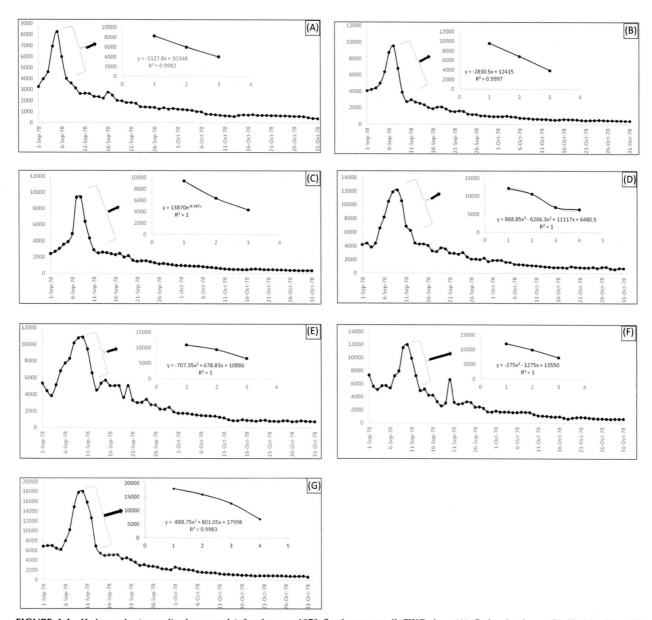

FIGURE 4.4 Hydrographs (normalized to a scale) for the year 1978 flood event at all CWC sites: (A) Garhmukteshwar, (B) Kachlabridge, (C) Fatehgarh, (D) Ankinghat, (E) Kanpur, (F) Bhitaura, and (G) Shahzadpur. The y- and x-axes represent discharge in cumecs and corresponding dates, respectively. Each hydrograph shows a separate graph for peak flood to 1-year return period flood discharge in the inset.

appropriate for spring discharge, whereas the exponential decay represents natural river discharge recession (Barnes, 1939; Hall, 1968; Subramanya, 1994). The recession phase after the flood peak serves as a channel or conduit flow because the contributions from interflow and baseflow are small compared to the surface flow (Adji et al., 2016).

Table 4.3 shows that flood length was nearly the same for all CWC sites for the 1978 flood event but increased for the 2010 flood from upstream to downstream. This could be attributed to the fact that the floodplain condition was consistent in the preceding scenario, but when floodplains were used more for agriculture by replacing heavy vegetation, flood duration increased due to a decrease in floodplain roughness.

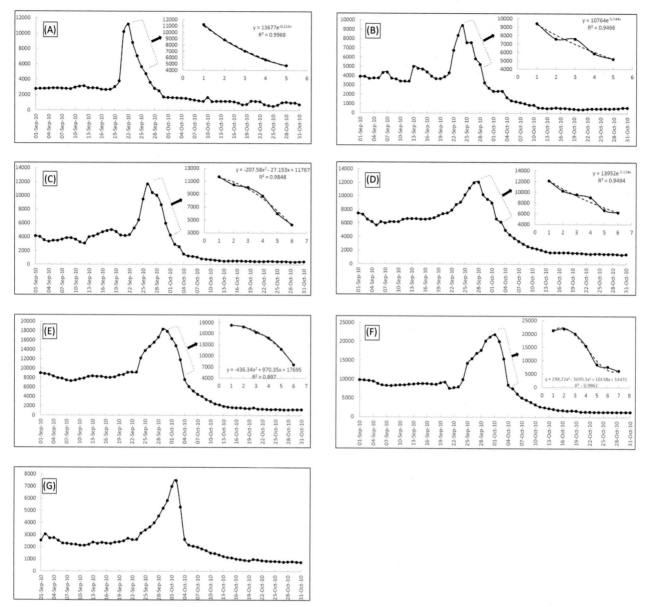

FIGURE 4.5 Hydrographs (normalized to a scale) for the year 2010 flood event at all CWC sites: (A) Garhmukteshwar, (B) Kachlabridge, (C) Fatehgarh, (D) Ankinghat, (E) Kanpur, (F) Bhitaura, and (G) Shahzadpur. The *y*- and *x*-axes represent discharge in cumecs and corresponding dates, respectively. Each hydrograph shows a separate graph for peak flood to 1-year return period flood discharge in the inset.

4.10   Water quality response of the floodplains

The average water quality parameters are assessed for postflood months (August−September) for 1995−2010 at all CWC sites, as reported in Table 4.4 According to the table, the average concentration of the parameters increased from upstream to downstream up to Bhitaura, indicating an increased agricultural runoff contribution to the river. The parameters' average concentration was lower in the S6 stretch than in the S1−S5 due to lower agriculture−urban and higher forest−grass areas. Overall, floods increase the concentration of nutrients (nitrates, ammonia, and phosphates), salts (ions of chloride, sulfate, potassium, bicarbonate, and calcium), pesticides, and sediments in river segments where floodplains are heavily impacted by agriculture. In contrast, lower concentrations were found on floodplains with a higher forest−grass cover.

TABLE 4.3 Salient features of 1978 and 2010 flood events.

CWC site	Date of peak flood	Normalized peak flood, cumecs	Flood duration from peak discharge to bankfull discharge, days
Garhmukteshwar	05-09-1978	8273	03
	22-09-2010	11,223	05
Kachlabridge	07-09-1978	9554	03
	24-09-2010	9426	05
Fatehgarh	08-09-1978	9418	03
	26-09-2010	11,609	06
Ankinghat	09-09-1978	12,200	04
	28-09-2010	12,100	06
Kanpur	10-09-1978	10,857	03
	29-09-2010	18,324	06
Bhitaura	10-09-1978	12,000	03
	30-09-2010	21,254	07
Shahzadpur	10-09-1978	17,987	04
	02-10-2010	7475	01

TABLE 4.4 Average values of water quality parameters for post-flood months (August-September) for the duration 1995−2010.

CWC site	Q (cumec)	pH	Total dissolved solid	Temperature (°C)	Total alkalinity, as $CaCO_3$ (mg/L)	Na (mg/L)	NO_3-N, as N (mg/L)[a]	SO_4 (mg/L)
Garhmukteshwar	1816	8	127	24	102	5	0.098	11
Kachlabridge	2457	8	135	28	100	5	0.185	13
Fatehgarh	2219	8	132	27	101	5	0.331	17
Ankinghat	3001	8	190	29	124	9	0.970	22
Kanpur	3219	8	183	31	125	8	0.517	21
Bhitaura	2333	8	200	28	135	10	1.293	17
Shahzadpur	2359	8	200	30	121	15	0.263	7

[a]Based on 2005−2010.

Temporal changes for some water quality parameters are given in Figs. 4.6 and 4.7 for the Garhmukteshwar CWC and Kanpur CWC sites, respectively, to examine the influence of UGRS management on downstream floodplains. Fig. 4.6 shows that post-2005 temperatures have been lowered to 19.0−24.0°C from 23.6−28.0°C in pre-2005. Similarly, from pre-2005 to post-2005, the pH range was changed from 8.0−8.2 to 7.3−8.1. Other parameters, such as salt and sulfate, exhibited less fluctuation in post-2005 duration compared to pre-2005 duration.

Fig. 4.3 depicts an increasing trend in temperature at Kanpur CWC while the pH remains constant. The remaining parameters, sodium and sulfate, show no trend other than significant variability, primarily connected with floodplain agriculture runoff because floodplains are heavily impacted by agricultural activities (refer to Table 4.2).

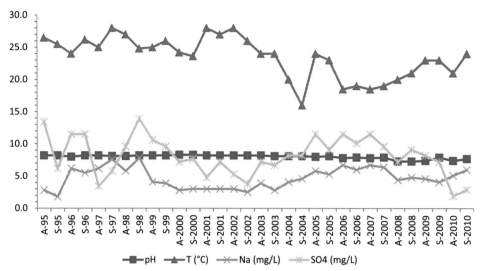

FIGURE 4.6   Water quality parameters at Garhmukteshwar CWC for postflood months (August−September) for the duration 1995−2010.

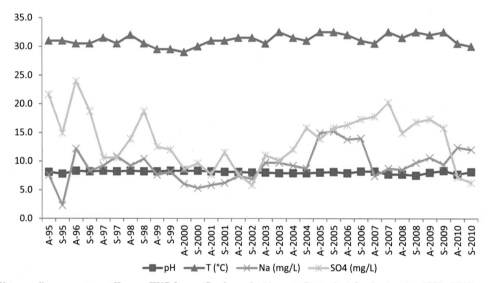

FIGURE 4.7   Water quality parameters at Kanpur CWC for postflood months (August−September) for the duration 1995−2010.

The alluvial soil in this stretch of the Ganga basin is endowed with rich soil nutrients. However, due to the overuse of pesticides and fertilizer in agriculture, the river water exhibits increased alkalinity, salinity, calcareousness, and acidity. As a result, it may be argued that floodplains are primarily used for agricultural purposes as rivers migrate from upstream to downstream, continuously losing their native bio-geomorphological functions.

4.11   Sediment response of floodplains

Table 4.4 shows the average suspended sediment concentrations in the Ganga River during the monsoon season (August−September) at five CWC sites. The average values were determined using CWC's 10-day data from 1995 to 2010. The sediment data for the Kachlabridge CWC site was unavailable and hence not included in the analysis. During the monsoon season, the S3 stretch had the highest sediment transport, while the S0 stretch had the lowest. Except for S0, the order of magnitude for sediment transport was nearly identical for all sections. Table 4.4 shows a significant difference in sediment concentrations between S0 and downstream stretches. The S0 stretch has the lowest value, indicating that the riparian forest cover is quite significant (refer to Fig. 4.1 and Table 4.2), and floodplain

TABLE 4.5 Average sediment load for the selected stretches.

CWC site (Stretch)	Average annual (grams per liter) after (Khan et al., 2018)	Average monsoon sediment concentrations calculated in the present study (grams per liter)
Garhmukteshwar (S0)	0.11	0.05
Kachlabridge (S1)	Not available	Not available
Fatehgarh (S2)	0.09	0.46
Ankinghat (S3)	0.09	0.71
Kanpur (S4)	0.11	0.67
Bhitaura (S5)	0.08	0.65
Shahzadpur (S6)	Not available	0.52

TABLE 4.6 A list of secondary sources used for an impact assessment on flora−fauna in the middle Ganga stretch.

Source	Reference
Spatial distribution and characterization of floral and faunal species	Kapoor and Mathur (2019)
Status of fish and fisheries	Mathur et al. (2019)
Assessment of fish and fisheries of the Ganga river system for developing suitable conservation and restoration plan	ICAR-CIFRI (2019)
Assessment of the wildlife values of the Ganga river from Bijnor to Ballia, including Turtle Wildlife Sanctuary, Uttar Pradesh	WII (2018)
Faunal resources of Ganga. Part I. Zoological Survey of India	"Faunal resources of Ganga Part 1 (General Introduction, and Vertebrate Fauna," 1991)
Status of Ganges River Dolphin (*Platanista gangetica gangetica*) in the Ganga River Basin, India: A review	Behera et al. (2013)
Current Status of Ganges River Dolphin (*Platanista gangetica gangetica*) in the Rivers of Uttar Pradesh, India	Behera et al. (2014)
Biodiversity profile of the Ganga River: Planning aquatic species restoration for Ganga River	WII-NMCG (2019)

management has improved after the Ramsar convention. Regarding slope profile and other geomorphic properties, the S0 stretch is nearly identical to the other stretches (Singh et al., 2007; Sinha et al., 2017).

The average yearly sediment concentrations for all CWC locations were also compared to the average monsoon sediment concentrations. In comparison, it was noticed that the monsoon sediment concentrations differ for all CWC sites, whereas the average annual sediment concentrations were nearly identical. The difference in monsoon sediment concentrations between S2 and S3 reaches may be attributed to riparian forest cover (refer to Fig. 4.1 and Table 4.2). As a result, afforestation is the solution for better floodplain management since riparian buffer/vegetation cover plays an important role in sediment retention (Hughes, 1997; Vigiak et al., 2016). Also, Khan et al. (2018) and Savita et al. (2019) recommended afforestation in the barren lands/agricultural lands in the middle Ganga plains to curb soil erosion (Table 4.5).

4.12 The cumulative impact on biodiversity

Based on the information provided in the most recent studies (refer to Table 4.6), the following subsections summarize a comparative assessment of flora and fauna such as algae, zooplankton, fishes, reptiles, and higher vertebrates within and downstream of the UGRS.

4.13 A comparative study of floral and faunal diversity in different habitat zones (S0−S1) and (S1−S6)

The habitat zone S0−S1 supports diverse aquatic and riparian ecosystems but has been changed drastically in S1−S6. According to Kapoor and Mathur (2019), the middle Ganga stretch's algal diversity has been altered as the ratio of diatoms, green algae, and blue-green algae has been changed. The ratio of diatoms, green algae, and blue-green algae has been reported as 100:36:15 and 100:67:36 for the stretches S0−S1 and S1−S6, respectively (Kapoor and Mathur, 2019). The algal composition reported upstream and downstream of the UGRS indicated that the stretches between S1−S6 have been degraded by anthropogenic environmental factors such as domestic sewage, pollutant discharge from the industries, agricultural runoff, more groundwater extraction, and agricultural encroachment of floodplains. Changes in the composition, distribution, and proportion of algal communities (phytoplankton and periphyton) have been reported, indicating a change in the water quality of the lotic and lentic systems for S1−S6. The oligotrophic/ultraoligotrophic to hypereutrophic conditions in S1−S6 has been reported by the presence of *Anabaena*, *Asterionella*, *Aulacoseira*, *Stephanodiscus*, *Ankistrodesmus*, and *Sphaerocystis* spp.

Zooplankton is a free-floating assemblage of microscopic animal forms, including protozoa, rotifers, small crustaceans, copepods, cladocerans, insect larvae, and pupae, and a significant food supply for planktivorous fish. Succession in the zooplankton population and a shift in their composition have increased the tiny forms such as rotifers and cladocera (*Ceriodaphnia* and *Bosmina*), indicating anthropogenic transformations and changes in the trophic structure (Vehmaa et al., 2018). The dominance of different Rotifera and Cladocera genera in S1−S6 revealed considerable spatial shifts in the ecosystem. In the S1−S6, 13 Rotifer and 12 Cladocera species were reported, compared to two Rotifer and Cladocera species in the S0−S1 (Kapoor & Mathur, 2019). Previous research on lowland rivers also supports the current results, where considerable community changes have been documented in response to changes in resource usage within the catchment region (Czerniawski & Kowalska-Góralska, 2018; Sługocki et al., 2021).

Anthropogenic changes in the S1−S6 stretch have become a severe problem, frequently accompanied by changes in phytoplankton biomass and composition, riparian corridor alterations, changes in water quality due to industrial and domestic discharges from urban centers, and nutrient enrichment from floodplain agriculture runoff, all of which have implications for fish abundance and community structure. The reported carps and catfish ratio in the S0−S1 is 3.36, which decreased to 1.29 in the S1−S6. The dominating catfish families and their genera reported in the S0−S1 and S1−S6 were Sisoridae (*Bagarius*, *Garra*, and *Glyptothorax*), Bagridae (*Mystus*, *Rita*, and *Sperata*), Schilbeidae (*Clupisoma*), Heteropneustidae (*Heteropneustidae*), Ambyceptidae (*Amblyceps*) and Sisoridae (*Bagarius*, *Glyptothorax*, *Sisor*, *Glyptothorax*, *Garra*, and *Nangara*), Siluridae (*Ompok* and *Wallago*), Pangasiidae (*Pangasius*), Bagridae (*Leiocassis*, *Mystus*, *Rita*, and *Sperata*), Clariidae (*Clarias*), Heteropneustidae (*Heteropneustes*), Chacidae (*Chaca*), and Schilbeidae (*Ailia*, *Clupisoma*, *Eutropiichthys*, *Pseudotropius*, and *Silonia*), respectively (Mathur et al., 2019).

Several other studies in the same study area yielded comparable results. According to a recent mid-term report from the Central Inland Fisheries Research Institute, the highest number of fish species was surveyed at Bijnor (nearby Garhmukteshwar CWC site), while Kanpur had the lowest (ICAR-CIFRI, 2019). *Tor putitora* (golden mahseer), a keystone species of the lower Himalayas, is found between Harsil and Bijnor in the monsoon months (ICAR-CIFRI, 2019). The study also revealed the dominance of exotic carp in the stretch along with Kanpur, indicating habitat alterations due to massive anthropogenic activities.

The increased levels of chlorophyll content in the stretch also indicate poor water quality (ICAR-CIFRI, 2019). The WII Dehradun undertook a rapid ecological assessment from Bijnor to Ballia in 2018 and found deteriorating habitat in several stretches between Kannauj to Kanpur (WII, 2018).

The zoological survey of India documented 27 species of reptiles in the Ganga River beside an endangered mammal Ganges river dolphin (*Platanista gangetica gangetica*) ("Faunal resources of Ganga Part 1 (General Introduction & Vertebrate Fauna)," 1991). Jhingran and Ghosh (1978) and Jhingran (1991) described the presence of fishes, amphibians, and reptiles in the Ganga River system. Among the critical higher vertebrates reported in the middle Ganga are Ganga river dolphins, gharyals, soft-shelled turtles, and hard-shelled turtles. Protecting the "flagship" species, the Ganga river dolphin ultimately leads to protecting other wildlife sharing a common niche (MoEFCC, 2022). Ganga river dolphin has been included as a Schedule-I animal of Wildlife (Protection) Act of India (1972) and listed under the "Endangered" category by International Union for Conservation of Nature (IUCN) (Braulik & Smith, 2017). The Government of India declared the Ganga river dolphin a "National Aquatic Animal" in 2009. The species had been sighted in the S0−S1, where 56 dolphins were documented in 2008 (Behera et al., 2013). The presence of regular deep pools (6.9 m) and shallow river depths (1.3 m), together with an adequate food supply (small fishes), favored the dolphins. Dolphins were occasionally sighted downstream of Narora up to Kanpur (S2−S4), but was believed to be in a good number in the S5—S6 (Behera et al., 2013). The S5—S6 region is a good breeding habitat for the river Dolphins, as around nine calves were observed in an IUCN survey done in October 2012 (Behera et al., 2014).

The riparian corridor in the middle Ganga stretch has been subjected to various anthropogenic pressures, such as sand mining, agriculture, overgrazing, land reclamation, and human habitation. One of the highly disturbed sites in the middle Ganga stretch (Kanpur and Allahabad) reported by (WII-NMCG, 2019), having a high percentage (>50%) of exotic species, indicates the positive correlation between disturbance and growth of invasive and tolerant species. Savita et al. (2019) also emphasized the high requirement of forestry treatment in the middle Ganga plains.

Overall, the loss of riverine connectivity in terms of lateral and longitudinal in the upper and middle stretches of river Ganga adversely affected its natural biota and production functions. Some of the other critical direct or indirect drivers for changing the ecosystem dynamics in the middle Ganga stretch and subsequently reflecting in its ranking under most polluted global rivers are pollution pressures, unplanned urban and industrial sectors, construction of dams and barrages across the river reaches, land-use changes, discharge of untreated municipal and industrial wastes, floral and religious offerings, and cremation of the dead bodies on the river banks (Khwaja et al., 2001). River water temperature also has affected the life of aquatic flora—fauna, which is directly linked to the riparian vegetation cover, water depth, turbidity, and channel morphology (Dallas, 2009). The average river temperature is highest for Kanpur among all CWC sites (refer to Table 4.4), which is also negatively associated with the aquatic biodiversity.

4.14 Further discussion

In plains, the lateral dimension of a natural river shows a cyclic expansion and contraction due to variable flows every year and is highly unpredictable in the case of unforeseen high floods. In addition, due to the lower probability of high floods, the floodplains are most susceptible to anthropogenic activities that cause significant damage to the riverine ecosystem and the crops and human lives due to the encroachment of the river's floodplain. Floodplains, as ecological hotspots, are the source of multiple ecosystem services. The alteration in the fluvial dynamics directly or indirectly impacts the region's ecological balance. The current study area around UGRS is home to a wide range of aquatic and terrestrial life forms with a rich assemblage of phytoplanktons, zooplanktons, fishes, reptiles, and higher vertebrates. The change in the river dynamics afterward imposed significant biodiversity alteration in terms of richness and assemblages. The manifold increase in anthropogenic activities is responsible for the loss or migration of several aquatic and riparian species in the subsequent lower stretches. It is noted that UGRS had also been exposed to various anthropogenic waste, such as domestic—industrial—agricultural wastewater/effluents from alongside towns, industries, and agricultural land (Goyal et al., 2022). However, as discussed in the previous sections, these activities do not significantly impact the river water's physical, chemical, or biological properties compared to the downstream stretches. The main reason behind this is the diverse forest cover and riparian vegetation in the floodplains of the UGRS. The forest cover and vegetation capture pollutants through phytoremediation, bind sands through roots, and reduce floods' peak intensity (Garg et al., 2015).

Hence, riparian and floodplain vegetation is a crucial part of a river system that gets an inundation through lateral migration of channels during floods. Previously, for the Upper Ganga basin, Tare et al. (2017) stated that the E-Flows in the river should be maintained for at least 18 days (i.e., around 20% dependable flow in a flow duration curve) in a year to inundate the riparian vegetation in river floodplain in case of a regulated river system. Similarly, Wolman and Leopold (1957) concluded that the lateral migration of channels is essential in controlling the elevation of the floodplain.

4.15 Recommendations for a floodplain management

Based on the present study, the following recommendations may be proposed to develop future river basin management protocols.

- A natural river-floodplain system safeguards bio-geomorphic functions in the ecosystem (Dodge, 1989), so the river should have minimum space to fulfill its numerous functions (Modi et al., 2022). The width can be determined by the return period floods and their relationship to the river-floodplain ecosystem, for example, baseflow as fish maintenance flows, bankfull flow or 1-year return period flow as channel maintenance flows, 1.5-year to 10-year return period flood as riparian maintenance flows, and more than 10-year return period flood as valley-forming flows (Hill et al., 1991).

- Floodplain management in the river's minimal space can be accomplished by categorizing it into two zones: (1) no-encroachment zone and (2) encroachment zone. The UGRS "command and control mode" management can be used as a prototype for the Himalayan rivers. The no-encroachment should equal the width of the 1-year return period flood. Anything should be prohibited in this zone (even seasonal agriculture) because active floodplains feature marginal vegetation, such as shrubs and trees, that control the morphology of active channels and provide a safeguard for the aquatic fauna (Nepf & Vivoni, 2000). At the banks of the river, there should be a riparian buffer strip. Prior research indicated that the riparian zone or forested wetlands contributed to removing nutrients/toxic elements generated from agricultural−domestic−industrial activity, preventing excess silt, and supporting vital ecosystem functions (Lu et al., 2009; Nóbrega et al., 2020; Rabeni & Smale, 1995). Agriculture should only be practiced on barren areas in the encroachment zone. Wetlands, oxbow lakes, and point bars should all be protected since they are critical biodiversity hotspots.
- Biodiversity parks should be developed in the river floodplain to rejuvenate the river and its ecosystems. Recently, the Central Pollution Control Board (CPCB) issued a compliance report and submitted it to the National Green Tribunal (NGT, an environmental tribunal in India), advocating the establishment of a biodiversity park in the river floodplain to revitalize the degraded rivers and ecosystems (CPCB (2020) Compliance Report on behalf of CPCB in compliance to Hon'ble NGT Order dated 12th December, 2019 in the matter of M.C. Mehta Vs. Union of India, & Ors, 2014). The scale of biodiversity parks should be determined by the floodplain extent, riparian zone breadth, existence of wetlands, and highland area.

4.16 Summary and conclusions

The present study focuses on providing valuable insights into the response of the middle Ganga floodplain and aims to compare these responses with other downstream stretches. The investigation encompasses crucial aspects such as hydrology, water quality, sediment, and biodiversity, which collectively underscore the significant importance of the floodplain system. By analyzing these aspects, the study sheds light on the overall ecological significance of floodplains within the middle Ganga River basin.

One of the key findings of the study is that the management approach known as the "command and control technique" has played a pivotal role in restoring and conserving the floodplains of the upper Ganga River (UGRS) after the area was designated as a Ramsar site in 2005. The Ramsar site designation highlights the international recognition of the site's ecological importance, making it a crucial area for conservation efforts. Based on the information gathered in the study, there is a strong recommendation for implementing the Ramsar framework to manage the river-floodplain system, specifically for the floodplains of the middle Ganga River. The framework is deemed beneficial for a variety of reasons, primarily due to its potential to enhance floodplain-wetland functionality. Floodplains are known for their vital role in flood buffering, mitigating the impacts of heavy rainfall and floods by acting as natural sponges that absorb and store excess water. By doing so, they help reduce the intensity of floodwaters that reach downstream areas, minimizing potential damage to human settlements and infrastructure.

Additionally, floodplains serve as essential sites for groundwater recharge. During periods of flooding, the excess water percolates into the ground, recharging groundwater aquifers and contributing to the overall water balance of the region. This process helps sustain water availability during dry periods and supports the needs of both human populations and ecosystems. Moreover, floodplains play a critical role in nutrient recycling. As floodwaters recede, they leave behind nutrient-rich sediments on the floodplain surface. This replenishes the soil with essential nutrients, making it fertile and conducive to supporting diverse flora and fauna. Consequently, floodplains become suitable habitats for native plants and animals, contributing significantly to regional biodiversity.

However, the study also highlights certain limitations in the Ramsar framework. Notably, it does not explicitly define a minimum floodplain width requirement. As a result, it may not fully address the needs of specific floodplain-wetland types, potentially leading to inconsistencies in management practices. To address this, future research could explore and determine an optimal floodplain width to better manage floodplain-wetland-type Ramsar sites effectively.

In conclusion, the present study provides valuable first-hand information about the middle Ganga floodplain response and its comparison with downstream stretches. The study emphasizes the importance of floodplains in maintaining a healthy river system, considering their contributions to hydrology, water quality, sediment, and biodiversity. The successful implementation of the "command and control technique" in restoring and conserving the UGRS floodplains post-Ramsar designation is noteworthy. The recommendation to adopt the Ramsar framework for managing the river-floodplain system in the middle Ganga River region holds promise for enhancing floodplain-wetland functionality, benefitting both ecological and human communities. Nonetheless, addressing the undefined minimum floodplain width within the framework presents an area for future research and improvement. Overall, this study offers valuable insights to guide sustainable management practices for floodplain-wetlands and the preservation of riverine ecosystems.

References

Adji, T. N., Haryono, E., Fatchurohman, H., & Oktama, R. (2016). Diffuse flow characteristics and their relation to hydrochemistry conditions in the Petoyan Spring, Gunungsewu Karst, Java, Indonesia. *Geosciences Journal, 20*(3), 381−390. Available from https://doi.org/10.1007/s12303-015-0048-8, http://www.springerlink.com/content/1226-4806.

Ahilan, S., Guan, M., Sleigh, A., Wright, N., & Chang, H. (2018). The influence of floodplain restoration on flow and sediment dynamics in an urban river. *Journal of Flood Risk Management, 11*, S986−S1001. Available from https://doi.org/10.1111/jfr3.12251, http://www.interscience.wiley.com/jpages/1753-318X.

Barnes, B. S. (1939). The structure of discharge-recession curves. *Eos, Transactions American Geophysical Union, 20*(4), 721−725. Available from https://doi.org/10.1029/TR020i004p00721.

Behera, S. K., Singh, H., & Sagar, V. (2013). Status of Ganges river dolphin (*Platanista gangetica gangetica*) in the Ganga River basin, India: A review. *Aquatic Ecosystem Health Management., 16*, 425−432. Available from https://doi.org/10.1080/14634988.2013.845069.

Behera,S. K., Singh, H., Sagar, V. (2014). Current Status of Ganges River Dolphin (*Platanista gangetica gangetica*) in the Rivers of Uttar Pradesh, India Rivers for life: Proceedings of the International Symposium on River Biodiversity : Ganges-Brahmaputra-Meghna River System. International Union for Conservation of Nature. IUCN, IUCN, 2014.

Bhaskar, A., & Karthick, N. M. (2015). Riparian forests for healthy rivers. *Current Science, 108*, 1788−1789.

Braulik, G., Smith, B. D, (2017). Available from https://doi.org/10.2305/IUCN.UK.2017-3.RLTS.T41758A50383612.en2017.

CPCB (2020) Compliance Report on behalf of CPCB in compliance to Hon'ble NGT Order dated 12th December, 2019 in the matter of M.C. Mehta Vs. Union of India & Ors. (2014), 2014.

Czerniawski, R., & Kowalska-Góralska, M. (2018). Spatial changes in zooplankton communities in a strong human-mediated river ecosystem. *PeerJ, 6*(7). Available from https://doi.org/10.7717/peerj.5087.

Dallas, (2009). The effect of water temperature on aquatic organisms: a review of knowledge and methods for assessing biotic responses to temperature. The Freshwater Consulting Group, 2009.

Dodge, (1989). The Morphology of Large Rivers: Characterization and Management. In: 1989 Proceedings of the International Large River Symposium (LARS).

Doll, B. A., Kurki-Fox, J. J., Page, J. L., Nelson, N. G., & Johnson, J. P. (2020). Flood flow frequency analysis to estimate potential floodplain nitrogen treatment during overbank flow events in urban stream restoration projects. *Water (Switzerland), 12*(6). Available from https://doi.org/10.3390/W12061568, https://www.mdpi.com/2073-4441/12/6/1568.

Eslamian, S. S., (2008). Stream Ecology and Low Flows (SELF), *International Journal of Ecological Economic & Statistics*, Ed., Special Issue Volume, CESER, Vol. 12, No. F08, pp. 1−97.

Faunal Resources of Ganga Part 1 (General Introduction and Vertebrate Fauna). ZSI. (1991), 1991.

Fleischmann, A. S., Paiva, R. C. D., Collischonn, W., Sorribas, M. V., & Pontes, P. R. M. (2016). On river-floodplain interaction and hydrograph skewness. *Water Resources Research, 52*(10), 7615−7630. Available from https://doi.org/10.1002/2016WR019233, http://onlinelibrary.wiley.com/journal/10.1002/(ISSN)1944-7973.

Garg, A., Joshi, B., & Singh, R. (2015). Floral-faunal mutualism and its role in sustenance of the Upper Ganga Ramsar Site in Uttar Pradesh, India. *Geophytology, 45*, 87−94.

Gordon, B. A., Dorothy, O., & Lenhart, C. F. (2020). Nutrient retention in ecologically functional floodplains: A review. *Water, 12*. Available from https://doi.org/10.3390/w12102762.

Goyal, A., Upreti, M., Chowdary, V. M., & Jha, C. S. (2022). *Delineation and Monitoring of Wetlands Using Time Series Earth Observation Data and Machine Learning Algorithm: A Case Study in Upper Ganga River Stretch* (pp. 123−139). Springer Science and Business Media LLC. Available from https://doi.org/10.1007/978-3-030-98981-1_5.

Hall, F. R. (1968). Base-flow recessions—A review. *Water Resources Research, 4*(5), 973−983. Available from https://doi.org/10.1029/WR004i005p00973.

Hill, M. T., Platts, W. S., & Beschta, R. L. (1991). Ecological and geomorphological concepts for instream and out-of-channel flow requirements. *Rivers, 2*, 198−210.

Howitt, J. A., Baldwin, D. S., Rees, G. N., & Williams, J. L. (2007). Modelling blackwater: Predicting water quality during flooding of lowland river forests. *Ecological Modelling, 203*(3-4), 229−242. Available from https://doi.org/10.1016/j.ecolmodel.2006.11.017.

Hughes, F. M. R. (1997). Floodplain biogeomorphology. *Progress in Physical Geography: Earth and Environment, 21*(4), 501−529. Available from https://doi.org/10.1177/030913339702100402.

ICAR-CIFRI, (2019). Assessment of fish and fisheries of the Ganga river system for developing suitable conservation and restoration plan., 2019.

Jameel, Y., Stahl, M., Ahmad, S., et al. (2020). India needs an effective flood policy. *Science (New York, N.Y.), 369*, 1575. Available from https://doi.org/10.1126/science.abe2962.

Jhingran, A. G., & Ghosh, K. K. (1978). The fisheries of the Ganga River System in the context of Indian aquaculture. *Aquaculture (Amsterdam, Netherlands), 14*(2), 141−162. Available from https://doi.org/10.1016/0044-8486(78)90026-1.

Jhingran, V. G. (1991). *Fish and Fisheries of India* (p. 1991) Hindustan Publishing Corporation.

Kale, V. S. (2003). Geomorphic effects of monsoon floods on Indian rivers. *Natural Hazards, 28*(1), 65−84. Available from https://doi.org/10.1023/A:1021121815395.

Kapoor, V., & Mathur, R. P. (2019). Spatial distribution and characterization of floral and faunal species. In V. Tare, & R. P. Mathur (Eds.), *Compendium of biodiversity in Ganga river system* (pp. 51−112). Mauritius: Lambert Academic Publishing.

Khan, M. S., & Khan, M. M. A. (2013). Quantifying land use land cover change along upper Ganga River (Brijghat to Narora Stretch). *Using Landsat TM. The Geographer, 60*, 2013.

Khan, S., Sinha, R., Whitehead, P., Sarkar, S., Jin, L., & Futter, M. N. (2018). Flows and sediment dynamics in the Ganga River under present and future climate scenarios. *Hydrological Sciences Journal, 63*(5), 763−782. Available from https://doi.org/10.1080/02626667.2018.1447113.

Khwaja, A., Singh, R., & Tandon, S. (2001). Monitoring of Ganga water and sediments vis-a-vis tannery pollution at Kanpur (India): A case study. *Environmental Monitoring and Assessment, 68*, 19−35.

Lu, S., Zhang, P., Jin, X., Xiang, C., Gui, M., Zhang, J., & Li, F. (2009). Nitrogen removal from agricultural runoff by full-scale constructed wetland in China. *Hydrobiologia, 621*(1), 115−126. Available from https://doi.org/10.1007/s10750-008-9636-1.

Lyubimova, T., Lepikhin, A., Parshakova, Y., & Tiunov, A. (2016). The risk of river pollution due to washout from contaminated floodplain water bodies during periods of high magnitude floods. *Journal of Hydrology, 534*, 579−589. Available from https://doi.org/10.1016/j.jhydrol.2016.01.030.

Mateo-Sagasta, J., Zadeh., Turral., & Burke, J. (2017). *Water Pollution from Agriculture: A Global Review. Executive Summary*. WLE, 2017.

Mathur, R. P., Sharma, A. P., Joshi, K. D., & Kapoor, V. (2019). Status of fish and fisheries. In V. Tare, & R. P. Mathur (Eds.), *Compendium of Biodiversity in Ganga River System* (pp. 51−112). Mauritius: Lambert Academic Publishing.

Modi, A., Kapoor, V., & Tare, V. (2022). River space: A hydro-bio-geomorphic framework for sustainable river-floodplain management. *The Science of the Total Environment, 812*151470. Available from https://doi.org/10.1016/j.scitotenv.2021.151470.

Modi, A., Tare, V., Medhi, H., & Rai, P. K. (2020). A framework for the hydrological assessment of at-site bankfull discharge-width for (semi-) incised Ganga river in Middle Ganga plains. *Journal of Hydrology, 586*. Available from https://doi.org/10.1016/j.jhydrol.2020.124912.

MoEF&CC (2020a) Ramsar Sites of India—Factsheets.

MoEF&CC (2020b) Guidelines for Implementing Wetlands (Conservation and Management) Rules, 2017.

MoEFCC, (2022). Ramsar Sites in 75th Year of Independence 75 Ramsar Sites 75th Year Independence, 2022.

Murthy, T. V. R., Patel, J. G., Panigrahy, S., & Parihar, J. S. (2013). National wetland atlas: Wetlands of international importance under Ramsar convention. *Space Applications Centre (ISRO), 2013*.

Nepf, H. M., & Vivoni, E. R. (2000). Flow structure in depth-limited, vegetated flow. *Journal of Geophysical Research: Oceans, 105*(12), 28547−28557. Available from https://doi.org/10.1002/2000JC900145, http://onlinelibrary.wiley.com/journal/10.1002/(ISSN)2169-9291.

Nones, M. (2019). Dealing with sediment transport in flood risk management. *Acta Geophysica, 67*(2), 677−685. Available from https://doi.org/10.1007/s11600-019-00273-7.

Nóbrega, R. L. B., Ziembowicz, T., Torres, G. N., Guzha, A. C., Amorim, R. S. S., Cardoso, D., Johnson, M. S., Santos, T. G., Couto, E., & Gerold, G. (2020). Ecosystem services of a functionally diverse riparian zone in the Amazon−Cerrado agricultural frontier. *Global Ecology and Conservation, 21*. Available from https://doi.org/10.1016/j.gecco.2019.e00819.

Pahl-Wostl, C. (2007). Transitions towards adaptive management of water facing climate and global change. *Water Resources Management, 21*(1), 49−62. Available from https://doi.org/10.1007/s11269-006-9040-4.

Rabeni, C. F., & Smale, M. A. (1995). Effects of siltation on stream fishes and the potential mitigating role of the buffering riparian zone. *Hydrobiologia, 303*(1-3), 211−219. Available from https://doi.org/10.1007/BF00034058.

Ragulina, Y. V., Merdesheva, E. V., & Titova, O. V. (2022). Economic foundations of integrated environmental monitoring. *Geo-Economy of the Future* (pp. 127−135). Springer.

Savita, P., Sharma, L. K., & Kumar, M. (2019). Forestry interventions for Ganga rejuvenation: A geospatial analysis for prioritizing sites. *Indian Forester, 144*, 1127−1135.

Singh, M., Singh, I. B., & Müller, G. (2007). Sediment characteristics and transportation dynamics of the Ganga River. *Geomorphology, 86*, 144−175. Available from https://doi.org/10.1016/j.geomorph.2006.08.011.

Singh, R., & Singh, G. S. (2019). Integrated management of the Ganga River: An ecohydrological approach. *Ecohydrology & Hydrobiology*. Available from https://doi.org/10.1016/j.ecohyd.2019.10.007.

Sinha, R., Mohanta, H., Jain, V., & Tandon, S. K. (2017). Geomorphic diversity as a river management tool and its application to the Ganga River, India. *River Research and Applications, 33*(7), 1156−1176. Available from https://doi.org/10.1002/rra.3154, http://onlinelibrary.wiley.com/journal/10.1002/(ISSN)1535-1467.

Subramanya, K. (1994), 1994.

Sługocki, Ł., Czerniawski, R., Kowalska-Góralska, M., & Teixeira, C. A. (2021). Hydro-modifications matter: Influence of vale transformation on microinvertebrate communities (Rotifera, Cladocera, and Copepoda) of upland rivers. *Ecological Indicators, 122*. Available from https://doi.org/10.1016/j.ecolind.2020.107259.

Tare, V., Gurjar, S. K., Mohanta, H., Kapoor, V., Modi, A., Mathur, R. P., & Sinha, R. (2017). Eco-geomorphological approach for environmental flows assessment in monsoon-driven highland rivers: A case study of Upper Ganga, India. *Journal of Hydrology: Regional Studies, 13*, 110−121. Available from https://doi.org/10.1016/j.ejrh.2017.07.005.

Vehmaa, A., Katajisto, T., & Candolin, U. (2018). Long-term changes in a zooplankton community revealed by the sediment archive. *Limnology and Oceanography, 63*, 2126−2139.

Vigiak, O., Malagó, A., Bouraoui, F., Grizzetti, B., Weissteiner, C. J., & Pastori, M. (2016). Impact of current riparian land on sediment retention in the Danube River Basin. *Sustainability of Water Quality and Ecology, 8*, 30−49. Available from https://doi.org/10.1016/j.swaqe.2016.08.001.

Whitfield, A. K., & Elliott, M. (2002). Fishes as indicators of environmental and ecological changes within estuaries: A review of progress and some suggestions for the future. *Journal of Fish Biology, 61*(sA), 229−250. Available from https://doi.org/10.1111/j.1095-8649.2002.tb01773.x.

WII, (2018). Assessment of the Wildlife Values of the Ganga River from Bijnor to Ballia Including Turtle Wildlife Sanctuary, 2018.

WII-NMCG, Biodiversity Profile of the Ganga River: Planning Aquatic Species Restoration for Ganga River. Wildlife Institute of India. (2019), 2019.

Wolman, M. G., Leopold, L. B. (1957). River Flood Plains: Some Observations on Their Formation, 1957.

Chapter 5

Best management practices in stream: debris and runoff reduction, riparian buffers and plantings, and stabilizing stream banks

Sagarika Patowary[1], Mridusmita Debnath[2] and Arup K. Sarma[3]

[1]North Eastern Regional Institute of Water and Land Management, Tezpur, Assam, India, [2]Scientist, Indian Council of Agricultural Research-Research Complex for Eastern Region, Patna, Bihar, India, [3]Indian Institute of Technology Guwahati, Guwahati, Assam, India

5.1 Introduction

Globally, anthropogenic pressure caused by various activities related to deforestation, flow regulation, cleaning of plants in riparian zones, and channelization has affected the fluvial systems. Geomorphically, the impacts of such perturbations vary from regional-level scouring to large-scale changes in channel patterns. Streams are defined as a unidirectional water body and are further classified into two groups depending on the hydrological behavior: perennial and ephemeral stream. Perennial streams flow continuously by baseflow and rainfall, whereas nonperennial streams are streams flowing periodically. They are further divided into ephemeral, intermittent, episodic, and seasonal based on flow quantity. However, they are commonly termed as intermittent streams due to the complexity of distinction. The upper reaches of streams, flow in phases, like flowing and non flowing phase. During the flowing phase , upper reaches are commonly classified as headwater and tailwater streams. The upper reaches are thus mostly found as intermittent streams (Zhou & Cartwright, 2021). The headwater stream comprises 70% and above of the whole channel length (Datry et al., 2014). Also, intermittent streams' length consists of more than half of the total water networks worldwide (Shanafield et al., 2021). Climate change and demographic growth pose risks of gradually transforming perennial streams to nonperennial streams (Datry et al., 2014), thus leading to more nonperennial streams globally.

Water is one of the crucial elements among the various environmental resources and is a part of studies concerning environmental justice. In countries like United States, the assessment of water resources is traditionally based on the chemical and physical characteristics of water. The nationwide water quality assessment of riverine ecosystems performed by the US Environmental Protection Agency (USEPA) showed that even with the implementation of regulations for water quality, 40% of the country's streams depicted ill biological conditions. Intermittent streams especially function for various processes such as sediment transport, nutrient removal and cycling, connecting catchments, detention of particles, and organic matter transportation (Castelar et al., 2022; Di Pillo et al., 2023; Fritz et al., 2019) (Di). They are also used to recharge groundwater (Beetle-Moorcroft et al., 2021), along with accelerated nutrient conversion due to high surface area-to-volume ratios (Addy et al., 2019; Arce et al., 2021). Therefore it helps in accumulating and removing suspended sediments from water and detaining contaminants before they travel to downstream reaches. Further, it was found that several ephemeral and intermittent streams are finally directed toward large reservoirs (tanks) and are also included for rainwater harvesting, especially in dry arid zones. For example, cascade-type irrigation systems in India that function to date (Srivastava & Chinnasamy, 2021) are one good example. Similarly, cases can be observed in many regions (e.g., Sri Lanka (Sakalasooriya, 2021), Yemen (Aklan et al., 2023), etc.). Therefore the estimation of variations in water quality parameters and factors inducing the changes in streams is of utmost priority in the case of reservoirs.

Hydrosystem Restoration Handbook. DOI: https://doi.org/10.1016/B978-0-443-29802-8.00005-4

5.1.1 Overview of best management practices

The underlying physical and chemical characteristics of streams alter according to the prevailing weather and flow, which can be investigated with regular monitoring of their quality. At the time of flow cessation, substrates gather and physicochemical properties change, eventually releasing organic matter and dissolved nutrients at the time of resumption of flow and rewetting (Shumilova et al., 2019). The water quality of a stream is dependent on the quantity of water flowing, which varies with seasons. Datry et al. (2017) reported that microbial population and diversity are related to the frequency and intensity of the hydrological flow regime and that alternate wetting and drying affect the functions of microbes and thus the biogeochemical processes. However, continuous evaluation of stream health indices especially in a larger area requires extensive capital and time. Thus several modeling techniques have been utilized to measure stream health indices (Tung & Yaseen, 2020). The inputs generally used for these models are landscape characteristics, instream water quantity, and quality parameters. Most commonly, for larger catchment areas, methods such as linear regression and multivariate analysis (Azhar et al., 2015; Ewaid et al., 2018; Fathi et al., 2018; Kadam et al., 2019) were used for stream health model development. Later, nonlinear techniques such as fuzzy logic (Gharibi et al., 2012; Patel & Chitnis, 2022), artificial neural network (Gupta et al., 2019; Tung & Yaseen, 2020), and adaptive neuro-fuzzy inference system (ANFIS) (Azad et al., 2019) were further used to develop a robust model with less uncertainty. In spite of all these efforts, the predictive power of stream health models is moderate. This is mainly due to the complexity of natural systems.

5.1.2 Debris and runoff reduction strategies

Mountainous and hilly areas are faced with natural disasters such as landslides and debris flow, which have negative impacts on structures, residential areas, and other elements. Transport of debris is mainly dependent on two factors: high intensity of rainfall and earthquakes (Al-Jubouri et al., 2022; Tsunetaka et al., 2021). Deterioration of stream water quality caused by the conversion of forest land into urbanized areas is a global issue nowadays (Sarma et al., 2013). Urbanization-driven deforestation has significant effects on the hydrological cycle and consequently water chemistry. Primarily, the removal of forest cover causes higher sediment and water yield from the catchment. Deforestation-led sediment loss is a serious issue with obvious consequences on water quality and peak flow. Runoff and sediment yields increase multiple times when a forested watershed is altered to an urban watershed (Corbett et al., 1997; Nelson & Booth, 2002). Urbanization even in a small area may double the annual average discharge (Barron et al., 2013). Also, forest and removal of forest cover influence transportation and deposition of sediments through erosion. These sediments considerably affect the composition of runoff coming in contact with them. Again, nonpoint source pollution through runoff contributes a major share to the deterioration of water quality in rivers and streams. Nonpoint source pollutants from urban areas originate primarily from vehicular traffic and atmospheric deposition (Kim et al., 2016; Maniquiz et al., 2010). In addition to this, urban, road debris in the form of soil, plant matter, and waste materials that get collected on the side of roads may be cleared mechanically by sweeping activities to maintain in the sense of visual aesthetics (Kim et al., 2019). This debris sometimes contains environmental pollutants, which may be eventually transported to nearby streams and canals. Agriculture contributes to nonpoint source pollution with the excess application of chemical fertilizers, herbicides, and pesticides (Debnath et al., 2020). Stormwater control measures (SCMs) are planned to treat, minimize, and slow down runoff transportation to sink waterways. SCMs, in addition to treatment and holding of runoff, also infiltrate significant volume of surface runoff for more natural flow patterns (Blecken et al., 2017). SCMs are mainly classified according to their hydrological purpose. (i) Infiltration-based systems that are used both for infiltration and holding of runoff. These include infiltration ditches, swales, infiltration basins, unlined biofiltration systems, and pervious pavements. (ii) Retention-based systems (e.g., green roofs, tanks, and lined biofiltration systems, ponds, and wetlands) reduce the overall runoff volume primarily through retention. Distribution of SCMs throughout the urban basin provides an appropriate and proportionate share of infiltration and retention for runoff reduction and restoration of baseflows. Trees planted in urban environment plays a vital role in interception and eventually reducing runoff resulting from rainfall (Bartens et al., 2008; Berland et al., 2017). Also, trees planted, use the infiltrated water to transpire back into the atmosphere (Ponte et al., 2021). SCMs combined with trees may overall achieve stormwater retention through both evapotranspiration and exfiltration. Rivers are the lifeline of mankind supporting the livelihood of millions of people residing in the basin. Rivers also play a crucial role in the water−food−energy nexus, as it is used to meet water demand in sectors such as agriculture, industries, and flood control. Even though rivers are beneficial to mankind, anthropogenic pressure on the river endangers river health; for instance, deforestation activity along the river bank and river regulation through dam construction. In addition to such major anthropogenic disturbances, regional-level disturbances also occur, for instance channelization, removal of instream vegetation, and sand mining along the fluvial systems. This results in variations of flow—sediment pattern, salt—water intrusion, alteration in channel reach, coastal erosion, and loss of wetland in riparian zones. However, studies about the process—form relationship in these poorly gauged—regulated rivers are very scarce and urge a multidisciplinary approach for a robust understanding of the phenomenon.

5.1.2.1 Riparian buffers

Riparian zones are endangered by various developmental, natural, and anthropogenic activities such as urbanization, agriculture, alteration in the flow direction, climate change, overexploitation, and pollution (Alam et al., 2017; Arif et al., 2021; Miller et al., 2003). A best-managed riparian area is vital for the sustainability of ecosystem services of streams and rivers, which is finally for the benefit of mankind. Ecosystem services of riparian buffers include bank stabilization, groundwater recharge, temperature control, pollutant and sediment trapping, wildlife habitat, and food web for various species.

5.1.2.1.1 Nature's filtration system sediment basins and traps: controlling runoff contaminants

Mountainous regions and the streams present in those regions are characterized by high geomorphic activities involving increased rates of hillslope erosion and thereby subsequently transporting bedload materials. The large quantity of sediment transport that occurs in mountain streams at the time of heavy rainfall and subsequent floods often leads to economic impacts for instream structures. Sediment traps in those areas are utilized for the quantification of sediment types and sedimentation cycles, identification of sources of sediment for the prevention of sedimentary processes (Mathers et al., 2021), and estimation of sediment accumulation rates. Agricultural fields in mountainous areas contribute to erosion, which can be controlled by sediment traps (Haribowo et al., 2019). Similarly, nowadays, evaluation of microplastic flux rates is required for investigating the accumulation rate of microplastics in various environments and for knowing the microplastic concentrations change over time. Open check dams are a type of sediment trap and are mainly comprised of a slot or slit (beam) dam (Piton & Recking, 2016; Xie et al., 2023). Sediment traps perform in a conceptual manner by holding back sediment during floods when the open check dam is bounded by coarse-sized grains, which is known as mechanical control or in cases where the flood water surpluses the discharge capacity of check dam outlets known as hydraulic control. The material retained by the check dam is stored in sediment traps located upstream of the dam, thus causing minimal effect on downstream of the stream.

5.1.2.1.2 Erosion control blankets: safeguarding soil integrity harnessing the power of vegetation

Soil loss is encountered from both agricultural and noncultivable lands. On average in India, it is reported that soil loss occurs annually at a rate of 2.5 to 12.5 Mg ha^{-1}, which is dependent on the soil quality. About 57% of soil has loss occurring at a rate of 10.0 Mg ha^{-1} per year, which can be treated with various management practices, although soil loss occurring at a rate of 2.5 Mg ha^{-1} per year needs to be prioritized first (Mandal & Sharda, 2011). Soil loss poses a threat to the productivity of crops. Further, during soil loss, the soil particles along with fertilizers, contaminants, and chemicals are gradually transferred to local water bodies, thereby affecting the stream health. According to USEPA, soil erosion is the major contributor to nonpoint source pollution (USEPA, 1997).

5.1.2.1.3 Runoff and erosion control with organic materials

Mulching of soil and soil amendment using organic materials are found to prevent soil erosion to a considerable degree. Organic amendments like shredded bark and mulches provide a layer to the soil that intercepts the raindrops and, as a result, energy dissipation occurs, which in turn prevents soil erosion (Table 5.1). The act of better contact with soil and reduced movement of soil particles because of water and wind shows that mulches are better compared to straw mats and hay (Lyle, 1987). They also obstruct the overland flow of runoff and allow water to infiltrate the soil surface. Korkanç and Şahin (2021) reported that total nitrogen transport was the least in plots where straw mulching was done at 4 t ha^{-1}, whereas total organic carbon was the least in plots with mulching at the rate of 2 t ha^{-1}. Mulching is usually found to increase the pH and EC values.

   Sadeghi et al. (2015) estimated the scale effects of mulching or amendments in the soil for plot sizes of 0.25 and 6 m^2, and it was found that mulching was more effective in soil loss reduction for smaller plot sizes of 0.25 m^2. Lucas-Borja et al. (2019) assessed the effect of mulching on runoff reduction after a forest fire and has been found to reduce

TABLE 5.1 Details of various organic mulch used by researchers.

S. no.	Type of organic mulch	Function	References	Location	Plot size
1.	Wood fiber mulch and using subsurface irrigation system.	For retention of water, soil temperature, and growth of plants	Gruda (2008)	Pot experiment	
2.	Poultry litter	Acts as a barrier to soil loss in runoff	(Faucette et al., 2004)		
3.	Compost wood mulch, wood-based hydromulch	98% and 75% reduction of sediment discharge on the compost/mulch blend plots hydromulch treatment as compared to bare soil	Eck et al. (2010)	Parker County, Texas	12.2 × 2.4 m^2
4.	Rice straw mulch	Rice straw mulch acts as soil conditioner and prevents soil degradation through runoff and erosion	(Parhizkar et al., 2021)	Laboratory	0.5 × 1 m^2
5.	Mulch from cover crops	A legume cover crop (Hairy vetch) gives a yield of 4900 to 6000 kg ha^{-1}, providing 140 to 225 kg N ha^{-1}, which is sufficient to meet the N requirement of maize. Soil moisture retention was more for mulch from legume species relative to winter annual grain species (Silva, 2014)	Mirsky et al. (2017)	Eastern United States Wisconsin, Upper Midwest	1.5 × 6.1 m^2 9.14 × 15.24 m^2

No permission required.

soil erosion rate for logging operations done immediately after a wildfire, as mulching reduced significantly the total suspended solid from a 200 m^2 plot as mulching provides a protective cover to the first rain occurring after the fire.

5.1.2.1.4 Ecological management practices for runoff and erosion control

Sustainable development of urban land surfaces with ecological management practices (EMPs) is effective for controlling the sediment yield and runoff caused by the conversion of natural land surfaces to urban covers (Sarma, 2011). EMPs are the combination of different structural and vegetative measures like tree/grass plantation, detention and retention ponds, contour terracing, buffer zone with vegetation strips, sediment traps, rainwater harvesting systems, vegetated waterways, perforated covers with pebble and stone, etc. that can be used for reducing the ecological disturbances by controlling the sediment and water yield from a degraded watershed (Sarma et al., 2015). Sarma et al. (2015) introduced the optimal EMP model with Linear Programming for Single Ownership (OPTEMP-LS) to determine the optimum allocation of EMPs in a hilly urban watershed to control the sediment and runoff yield from the watershed within a permissible limit but with a minimum possible cost. The model was constrained by several conditions like slope, soil character, land availability, existing land cover type, and ease of maintenance. The Revised Universal Soil Loss Equation was used for defining the sediment yield constraint and the Rational Method was used to address the discharge constraint. Again, Patoway and Sarma (2018) proved that in case of Geographic Information System (GIS) based soil loss estimation from urban hilly watersheds, the estimated value can be less by 28% than the observed value for nonconsideration of the sediment loss from the steep hill cut area. Accordingly, Patowary et al. (2019) revised the OPTEMP-LS by incorporating the steep hill cut area not visible in satellite images. The revised OPTEMP-LS model can be computationally complex, but it gives a more accurate picture of ecological management through the control of the sediment and peak runoff within permissible limits.

5.2 Riparian plantings

The zone along the side of the river, which is covered with vegetative strips, is generally identified as a riparian zone. However, regulation of river has affected riparian plant species diversity and abundance (Lozanovska et al., 2020), thus resulting in the shrinking of the total riparian area. Similarly, the channelization of streams has also led to a narrower and less diversified riparian plant population (Blake & Rhanor, 2020). The main purpose of riparian zones is to protect and demarcate the aquatic habitat from land uses in the boundary of rivers. The smaller riparian area, links the aquatic and terrestrial conditions, thus having a great influence on stream health as compared to the catchment area. Other benefits of riparian zones are primarily regulation of stream temperature, reduction of flood flow, inclusion of organic matter as feed for the aquatic life, and sedimentation. The relationship that exists between the riparian zone and in-stream systems provides restoration of streams, thereby mitigating the harmful effects of land use change (Majumdar & Avishek, 2023).

5.2.1 Aquatic plants for water quality and habitat enhancement

Phytoremediation of water bodies are cost-effective approach to solve the contaminated sites along with providing aesthetic scene. In tropical and subtropical climate, invasive instream vegetation for example the water hyacinth (*Eichhornia crassipes*) and water lettuce (*Pistia stratiotes* L.) are utilized as phytoremediation plants. These plants have higher nutrient removal efficiency, as they can uptake nutrient at a faster rate and have greater biomass production. Antimony and arsenic are contaminants that have hazardous effects and need to be studied exhaustively for their environmental risks. The study reported by Qin et al. (2021) showed that bioaccumulation and uptake of arsenic-contaminated soil by acidic mine water was high in leaves of water spinach. Submerged plant species such as epiphytic biofilms are a potential indicator of toxic elements present in the water (Geng et al., 2019).

5.2.2 Floating wetlands: innovative solutions for pollutant removal

Improving stream health using sustainable management options is faced with two challenges: to prohibit the addition of pollutants from various sources from entering the water bodies and to clear the already existing pollutants. The pollutants are removed from water bodies using innumerable techniques, and the use of constructed floating wetlands is one of them. Constructed floating wetlands (CFWs) are a cost-effective and promising ecological engineering technique for restoration of stream health. It consists of buoyant carriers that include the plantation of macrophytes. The roots extend into the water bodies in order to take up dissolved pollutants. However, timely harvest of the plant biomass and sowing it again helps to remove the build-up pollutants in stream in an efficient manner. In addition to this, the microbial film in the root zone converts the pollutants through root exudation into compounds that can be easily uptaken by plants. The symbiosis that exists between the plant roots and microbes is one of the major pathways for clearing some pollutants, for example, pesticide and organic pollutants (Srivastava & Chinnasamy, 2021). CFWs also supply biodiversity and provide shade to the water, which regulate the water temperature and prevents algal growth by preventing the penetration of sunlight.

5.2.3 Stabilizing stream banks: engineering and ecology

5.2.3.1 Bioengineering techniques

The plantation of vegetation is a vital measure to prevent soil erosion along the river bank and thus provide bed stabilization. Bioengineering is a cost-effective and eco-friendly approach as compared to other conventional methods of soil slope stabilization and erosion control. This method includes bush layering, such as fascines, vegetated gabions, etc., that uses both plants and inert materials. An effective bioengineering technique design includes an appropriate evaluation of the plant root and the interaction of the root with the soil. For this, several methods have been devised to investigate the complex phenomenon for the analysis of the interaction of the soil—root system. Apparently, long-term river bank protection is formed by this technique, provided there is self-generation potentiality of the plants. In addition to these, plants require adequate amount of sunlight and water for their sturdy growth that can resist shear stress occurring during flooding events.

5.2.3.2 Rock and vegetative revetments

Revetments are structures designed along the road or trails nearing a watercourse to prevent bank erosion and provide bank stability. They are made from a variety of materials. Revetment made with rock is referred to as riprap. The presence of revetment confines the natural flow movement of streams, thus obstructing the natural fluvial processes like the formation of flood plains and spawning bed for fishes.

5.2.3.3 Gabions and riprap

Gabions are mainly wire-bounded structures that are filled with rock of various sizes also called gabion baskets. These structures also consist of stacks of logs named as bank-armor or stacks of cement bags.

5.2.3.4 Regrading and terracing

Alteration of landscapes in river banks for its stabilization reduces landslides in steep slopes of mountainous regions. Terraces are known to reduce slope steepness by cutting them into several sections. Thus terracing is known to significantly reduce soil erosion rate when constructed, keeping in mind the soil fertility for cultivation purposes and proper installation (Rutebuka et al., 2021).

5.2.4 Maintaining and monitoring stream management practices

Best management practices (BMPs) are designing on-farm management techniques to minimize nutrient losses in drainage water bodies to a permissible limit. Management practices that incur huge cost for farm implementation and put financial pressure on cultivators cannot be regarded as BMPs. Only in situations when the benefit returned is at acceptable value, the management practice can be regarded as BMP. Higher-cost projects come with strong arrangements and legal regulations for successful implementation.

It is noteworthy that any effective BMP program calls for timely monitoring of both the BMP implementation along with its effect on water quality. The authors evaluate that for standard-based discharge regulatory program and to prevent non point source pollution (NPS) into the stream, the water quality monitoring program would cost around $5-50$ US\$ ha^{-1} per year per t, which is a function of the size and type of operation.

5.2.5 Adapting strategies to changing conditions

Testing of the effectiveness of BMPs is usually performed on a watershed scale. This allows for the optimal site selection. In addition to this, further exploration of the effectiveness of BMPs for longer period urges for considering change in rainfall patterns under climate change. High flows due to intense rainfall events may perform differently when compared with existing BMP performance, which is devised based on current rainfall events (Eslamian & Eslamian, 2022). Climate change impacts the quantity as well as quality of water; however, proper BMPs considering the climate effect would help in mitigation. Therefore evaluation of climate change impacts on BMP effectiveness is necessary for decision-makers and engineers to build a cost-effective BMP scenario.

Sociohydrologic modeling is a novel modeling technique developed at the watershed level, which is formed by an economic model integrated into the water quality model. The results of such integrated modeling techniques provide the trade-off between sustainable environment and farmers' profit (Debnath et al., 2024; Dziubanski et al., 2020; Roobavannan et al., 2020). Pouladi et al. (2019) investigated multifaceted decision-making in terms of crop choice in farmers' field and water conservation strategies in the river basin by examining the effects of socioeconomic as well as environmental factors in the Zarrineh River Basin, Iran, which contributes 49% of water in Urmia Lake, where agent-based model and theory of planned behavior were integrated to devise a sociohydrological framework.

5.2.6 Community engagement and collaborative conservation

Farm management operations in agricultural fields contribute to NPS, which in turn is regulated by farmers and stakeholders. BMP implementation in the real field is done voluntarily with the active participation of stakeholders. A range of BMPs devised through modeling techniques can be tested in the ground, thereby accelerating the implementation process. Engaging stakeholders during decision-making gives more transparency to the scientific process, thus allowing for more efficient policy formulation (Kalcic et al., 2016). Positive feedback is generated due to the participatory approach by deciphering the best solution to complex regional-level issues using the modeling approach due to the representation of the near-natural condition of the practical field. Participation of stakeholders or farmers resolves conflict among them, thereby facilitating recommended policy implementation (Debnath et al., 2023). Interaction between people residing in the basin and hydrologist is crucial as the behavior of the society of the people affects the hydrological model output. The development of sociohydrological model in line with the river basin modeling provides two-way feedback between natural systems and human, which, scientists are trying to learn and gain knowledge not only about the physical processes occurring in the basin but also about issues such as social, political, and economic in the integrated modeling technique. Hence, active engagement of farmers and stakeholders in practical implementation of watershed models for improvement of river health status is essential. However, a rigorous and proper primary data collection on the society is required for such type of watershed modeling, which needs both time and effort.

5.3 Conclusions

Stream water management defines managing flood, controlling erosion, and improving water quality. This can be achieved by implementing practices that are mostly known as BMPs. BMPs are structural, vegetative, or managerial practices used for management of water resources. In this study, an exclusive review of different BMPs has been carried out to explore their effectiveness. However, analysis of the efficiency of BMPs for a longer time period urges for considering change in rainfall patterns under climate change. At the same time, management practices or the measures should be such that they can maintain or restore the water resources as much as possible in the preurbanized or natural state. In such a case, implementations of optimal combinations of EMPs are extremely inevitable, as they can control the sediment and peak runoff from watersheds within permissible limits with minimum possible cost.

References

Addy, K., Gold, A. J., Welsh, M. K., August, P. V., Stolt, M. H., Arango, C. P., & Groffman, P. M. (2019). Connectivity and nitrate uptake potential of intermittent streams in the Northeast USA. *Frontiers in Ecology and Evolution, 7.* Available from https://doi.org/10.3389/fevo.2019.00225, https://doi.org/10.3389/fevo.2019.00225.

Aklan, M., Al-Komaim, M., & de Fraiture, C. (2023). Site suitability analysis of indigenous rainwater harvesting systems in arid and data-poor environments: A case study of Sana'a Basin, Yemen. *Environment, Development and Sustainability, 25*(8), 8319−8342. Available from https://doi.org/10.1007/s10668-022-02402-7.

Al-Jubouri, S. M., Waisi, B. I., & Eslamian, S. (2022). *Debris and Solid Wastes in Flood Plain Management* (pp. 421−434). Informa UK Limited. Available from https://doi.org/10.1201/9780429463327-26.

Alam, G. M., Alam, K., Mushtaq, S., & Clarke, M. L. (2017). Vulnerability to climatic change in riparian char and river-bank households in Bangladesh: Implication for policy, livelihoods and social development. *Ecological Indicators., 72,* 23−32.

Arce, M. I., Bengtsson, M. M., von Schiller, D., Zak, D., Täumer, J., Urich, T., & Singer, G. (2021). Desiccation time and rainfall control gaseous carbon fluxes in an intermittent stream. *Biogeochemistry, 155*(3), 381−400. Available from https://doi.org/10.1007/s10533-021-00831-6, http://www.wkap.nl/journalhome.htm/0168-2563.

Arif, M., Jie, Z., Wokadala, C., Songlin, Z., Zhongxun, Y., Zhangting, C., Zhi, D., Xinrui, H., & Changxiao, L. (2021). Assessing riparian zone changes under the influence of stress factors in higher-order streams and tributaries: Implications for the management of massive dams and reservoirs. *Science of The Total Environment, 776.* Available from https://doi.org/10.1016/j.scitotenv.2021.146011.

Azad, A., Karami, H., Farzin, S., Mousavi, S. F., & Kisi, O. (2019). Modeling river water quality parameters using modified adaptive neuro fuzzy inference system. *Editorial Office of Water Science and Engineering, Iran Water Science and Engineering, 12*(1), 45−54. Available from https://doi.org/10.1016/j.wse.2018.11.001. Available from: http://kkb.hhu.edu.cn/.

Azhar, S. C., Aris, A. Z., Yusoff, M. K., Ramli, M. F., & Juahir, H. (2015). Classification of river water quality using multivariate analysis. *Procedia Environmental Sciences, 30,* 79−84. Available from https://doi.org/10.1016/j.proenv.2015.10.014.

Barron, O. V., Barr, A. D., & Donn, M. J. (2013). Effect of urbanisation on the water balance of a catchment with shallow groundwater. *Journal of Hydrology, 485,* 162−176. Available from https://doi.org/10.1016/j.jhydrol.2012.04.027.

Bartens, J., Day, S. D., Harris, J. R., Dove, J. E., & Wynn, T. M. (2008). Can urban tree roots improve infiltration through compacted subsoils for stormwater management? *Journal of Environmental Quality., 37*(6), 2048−2057. Available from https://doi.org/10.2134/jeq2008.0117UnitedStates, http://jeq.scijournals.org/cgi/reprint/37/6/2048.

Beetle-Moorcroft, F., Shanafield, M., & Singha, K. (2021). Exploring conceptual models of infiltration and groundwater recharge on an intermittent river: The role of geologic controls. *Journal of Hydrology: Regional Studies, 35,* 100814.

Berland, A., Shiflett, S. A., Shuster, W. D., Garmestani, A. S., Goddard, H. C., Herrmann, D. L., & Hopton, M. E. (2017). The role of trees in urban stormwater management. *Landscape and Urban Planning, 162,* 167−177. Available from https://doi.org/10.1016/j.landurbplan.2017.02.017, http://www.elsevier.com/inca/publications/store/5/0/3/3/4/7.

Blake, C, & Rhanor, A. K. (2020). The impact of channelization on macroinvertebrate bioindicators in small order Illinois streams: Insights from long-term citizen science research. *Aquatic Sciences, 82*(2), 35, In this issue.

Blecken, G. T., Hunt, W. F., Al-Rubaei, A. M., Viklander, M., & Lord, W. G. (2017). Stormwater control measure (SCM) maintenance considerations to ensure designed functionality. *Urban Water Journal, 14*(3), 278−290. Available from https://doi.org/10.1080/1573062X.2015.1111913, http://www.tandf.co.uk/journals/titles/1573062X.asp.

Castelar, S., Bernal, S., Ribot, M., Merbt, S. N., Tobella, M., Sabater, F., Ledesma, J. L. J., Guasch, H., Lupon, A., Gacia, E., Drummond, J. D., & Martí, E. (2022). Wastewater treatment plant effluent inputs influence the temporal variability of nutrient uptake in an intermittent stream. *Urban Ecosystems, 25*(4), 1313−1326. Available from https://doi.org/10.1007/s11252-022-01228-5, http://www.kluweronline.com/issn/1083-8155.

Corbett, C. W., Wahl, M., Porter, D. E., Edwards, D., & Moise, C. (1997). Nonpoint source runoff modeling. A comparison of a forested watershed and an urban watershed on the South Carolina coast. *Journal of Experimental Marine Biology and Ecology*, *213*(1), 133−149. Available from https://doi.org/10.1016/S0022-0981(97)00013-0.

Datry, T., Larned, S. T., & Tockner, K. (2014). Intermittent rivers: A challenge for freshwater ecology. *Bioscience.*, *64*, 229−235.

Datry, T., Singer, G., Sauquet, E., Capdevilla, D. J., Von Schiller, D., Subbington, R., et al. (2017). Science and management of intermittent rivers and ephemeral streams (SMIRES). *Research Ideas and Outcomes.*, *3*, 23 p.

Debnath, M., Mahanta, C., & Sarma, A. K. (2020). Nutrient fluxes from agriculture: Reducing environmental impact through optimum application. *Environmental Processes and Management: Tools and Practices*, 37−51.

Debnath, M., Mahanta, C., Sarma, A. K., Upadhyaya, A., & Das, A. (2023). Participatory design to investigate the effects of farmers' fertilization practices under unsubmerged conditions toward efficient nutrient uptake in rainfed rice. *South African Journal of Botany.*, *163*, 338−347.

Debnath, M., Sarma, A. K., & Mahanta, C. (2024). Optimizing crop planning in the winter fallow season using residual soil nutrients and irrigation water allocation in India. *Heliyon*, *10.*.

Di Pillo., De Girolamo, A. M., Porto, A. L., & Todisco, M. T. (2023). Detecting the drivers of suspended sediment transport in an intermittent river: An event-based analysis. *Catena*, *222*, 2023.

Dziubanski, D., Franz, K. J., & Gutowski, W. (2020). Linking economic and social factors to peak flows in an agricultural watershed using socio-hydrologic modeling. Copernicus GmbH, United States. *Hydrology and Earth System Sciences*, *24*(6), 2873−2894. Available from https://doi.org/10.5194/hess-24-2873-2020, http://www.hydrol-earth-syst-sci.net/volumes_and_issues.html.

Eck, B., Barrett, M., McFarland, A., & Hauck, L. (2010). Hydrologic and water quality aspects of using a compost/mulch blend for erosion control. *Journal of Irrigation and Drainage Engineering*, *136*(9), 646−655. Available from https://doi.org/10.1061/(asce)ir.1943-4774.0000223.

Eslamian, S., & Eslamian, F. (2022). *Flood Handbook: Impacts and Management*. CRC Press. Available from https://doi.org/10.1201/9780429463327.

Ewaid, S. H., Abed, S. A., & Kadhum, S. A. (2018). Predicting the Tigris River water quality within Baghdad, Iraq by using water quality index and regression analysis. *Environmental Technology and Innovation*, *11*, 390−398. Available from https://doi.org/10.1016/j.eti.2018.06.013, http://www.journals.elsevier.com/environmental-technology-and-innovation/.

Fathi, E., Zamani-Ahmadmahmoodi, R., & Zare-Bidaki, R. (2018). Water quality evaluation using water quality index and multivariate methods, Beheshtabad River, Iran. *Applied Water Science*, *8*(7). Available from https://doi.org/10.1007/s13201-018-0859-7.

Faucette, L., Risse, L., Nearing, M., Gaskin, J., & West, L. (2004). Runoff, erosion, and nutrient losses from compost and mulch blankets under simulated rainfall. *Journal of soil and water conservation*, *59*, 154−160.

Fritz, K. M., Pond, G. J., Johnson, B. R., & Barton, C. D. (2019). Coarse particulate organic matter dynamics in ephemeral tributaries of a Central Appalachian stream network. *Ecosphere*, *10*(3). Available from https://doi.org/10.1002/ecs2.2654, http://esajournals.onlinelibrary.wiley.com/hub/journal/10.1002/(ISSN)2150-8925/.

Geng, N., Wu, Y., Zhang, M., Tsang, D. C. W., Rinklebe, J., Xia, Y., Lu, D., Zhu, L., Palansooriya, K. N., Kim, K.-H., & Ok, Y. S. (2019). Bioaccumulation of potentially toxic elements by submerged plants and biofilms: A critical review. *Environment International*, *131*. Available from https://doi.org/10.1016/j.envint.2019.105015.

Gharibi, H., Mahvi, A. H., Nabizadeh, R., Arabalibeik, H., Yunesian, M., & Sowlat, M. H. (2012). A novel approach in water quality assessment based on fuzzy logic. *Iran Journal of Environmental Management*, *112*, 87−95. Available from https://doi.org/10.1016/j.jenvman.2012.07.007, https://www.sciencedirect.com/journal/journal-of-environmental-management.

Gruda, N. (2008). The effect of wood fiber mulch on water retention, soil temperature and growth of vegetable plants. *Journal of Sustainable Agriculture*, *32*, 629−643.

Gupta, R., Singh, A. N., & Singhal, A. (2019). Application of ANN for water quality index. *International Journal of Machine Learning and Computing*, *9*(5), 688−693. Available from https://doi.org/10.18178/ijmlc.2019.9.5.859.

Haribowo, R., Andawayanti, U., & Lufira, R. D. (2019). Effectivity test of an eco-friendly sediment trap model as a strategy to control erosion on agricultural land. *Journal of Water and Land Development*, *42*(1), 76−82. Available from https://doi.org/10.2478/jwld-2019-0047, http://versita.com/science/environment/jwld/.

Kadam, A. K., Wagh, V. M., Muley, A. A., Umrikar, B. N., & Sankhua, R. N. (2019). Prediction of water quality index using artificial neural network and multiple linear regression modelling approach in Shivganga River basin, India. *Modeling Earth Systems and Environment*, *5*(3), 951−962. Available from https://doi.org/10.1007/s40808-019-00581-3, springer.com/journal/40808.

Kalcic, M. M., Kirchhoff, C., Bosch, N., Muenich, R. L., Murray, M., Griffith Gardner, J., & Scavia, D. (2016). Engaging stakeholders to define feasible and desirable agricultural conservation in western Lake Erie watersheds. *Environmental Science & Technology*, *50*(15), 8135−8145.

Kim, D.-G., Kang, H.-M., & Ko, S.-O. (2019). Reduction of non-point source contaminants associated with road-deposited sediments by sweeping. *Environmental Science and Pollution Research.*, *26*, 1192−1207.

Kim, H.-Y., Seok, H.-W., Kwon, H.-O., Choi, S.-D., Seok, K.-S., & Oh, J. E. (2016). A national discharge load of perfluoroalkyl acids derived from industrial wastewater treatment plants in Korea. *Science of The Total Environment.*, *563*, 530−537.

Korkanç, S. Y., & Şahin, H. (2021). The effects of mulching with organic materials on the soil nutrient and carbon transport by runoff under simulated rainfall conditions. *Journal of African Earth Sciences.*, *176*, 104152.

Lozanovska, I, Rivaes, R, Vieira, C, Ferreira, M. T., & Aguiar, F. C. (2020). Streamflow regulation effects in the Mediterranean rivers: How far and to what extent are aquatic and riparian communities affected? *Science of the Total Environment*, *20*, 749.

Lucas-Borja, M., González-Romero, J., Plaza-Álvarez, P., Sagra, J., Gómez, M., Moya, D., et al. (2019). The impact of straw mulching and salvage logging on post-fire runoff and soil erosion generation under Mediterranean climate conditions. *Science of the Total Environment.*, *654*, 441−451.

Lyle, E. S. (1987). *Surface mine reclamation manual*. New York, NY: Elsevier.

Majumdar, A, & Avishek, K (2023). Riparian zone assessment and management: an integrated review using geospatial technology. *Water, Air, & Soil Pollution, 234*(5), 319.

Mandal, D., & Sharda, V. N. (2011). Assessment of permissible soil loss in India employing a quantitative bio-physical model. *Current Science, 100* (3), 383−390. Available from http://www.ias.ac.in/currsci/10feb2011/383.pdf.

Maniquiz, M. C., Lee, S.-Y., & Kim, L.-H. (2010). Long-term monitoring of infiltration trench for nonpoint source pollution control. *Water, Air, & Soil Pollution., 212*, 13−26.

Mathers, K. L., Kowarik, C., Rachelly, C., Robinson, C. T., & Weber, C. (2021). The effects of sediment traps on instream habitat and macroinvertebrates of mountain streams. *Journal of Environmental Management, 295*. Available from https://doi.org/10.1016/j.jenvman.2021.113066.

Miller, J. R., Wiens, J. A., Hobbs, N. T., & Theobald, D. M. (2003). Effects of human settlement on bird communities in lowland riparian areas of Colorado (USA). *Ecological Society of America, United States Ecological Applications, 13*(4), 1041−1059. Available from http://www.esajournals.org, https://doi.org/10.1890/1051-0761(2003)13[1041:EOHSOB]2.0.CO;2.

Mirsky, S. B., Ackroyd, V. J., Cordeau, S., Curran, W. S., Hashemi, M., Reberg-Horton, S. C., Ryan, M. R., & Spargo, J. T. (2017). Hairy vetch biomass across the Eastern United States: effects of latitude, seeding rate and date, and termination timing. *Agronomy Journal, 109*(4), 1510−1519.

Nelson, E. J., & Booth, D. B. (2002). Sediment sources in an urbanizing, mixed land-use watershed. *Journal of Hydrology., 264*(1), 51−68.

Parhizkar, M., Shabanpour, M., Lucas-Borja, M. E., Zema, D. A., Li, S., Tanaka, N., & Cerda, A. (2021). Effects of length and application rate of rice straw mulch on surface runoff and soil loss under laboratory simulated rainfall. *International Journal of Sediment Research, 36*(4), 468−478.

Patel, A., & Chitnis, K. (2022). Application of fuzzy logic in river water quality modelling for analysis of industrialization and climate change impact on Sabarmati river. *Water Supply, 22*, 238−250.

Patowary, S., & Sarma, A. K. (2018). GIS-based estimation of soil loss from hilly urban area incorporating hill cut factor into RUSLE. *Water Resources Management, 32*(10), 3535−3547. Available from https://doi.org/10.1007/s11269-018-2006-5, http://www.wkap.nl/journalhome.htm/0920-4741.

Patowary, S., Sarma, B., & Sarma, A. K. (2019). A revision of OPTEMP-LS model for selecting optimal EMP combination for minimizing sediment and water yield from hilly urban watersheds. *Water Resources Management, 33*, 1249−1264.

Piton, G., & Recking, A. (2016). Design of sediment traps with open check dams. I: Hydraulic and deposition processes. *Journal of Hydraulic Engineering., 142*, 04015045.

Ponte, S., Sonti, N. F., Phillips, T. H., & Pavao-Zuckerman, M. A. (2021). Transpiration rates of red maple (*Acer rubrum* L.) differ between management contexts in urban forests of Maryland, USA. *Scientific Reports, 11*(1). Available from https://doi.org/10.1038/s41598-021-01804-3, http://www.nature.com/srep/index.html.

Pouladi, P., Afshar, A., Afshar, M. H., Molajou, A., & Farahmand, H. (2019). Agent-based socio-hydrological modeling for restoration of Urmia Lake: Application of theory of planned behavior. *Journal of Hydrology, 576*, 736−748. Available from https://doi.org/10.1016/j.jhydrol.2019.06.080, http://www.elsevier.com/inca/publications/store/5/0/3/3/4/3.

Qin, J., Niu, A., Liu, Y., & Lin, C. (2021). Arsenic in leafy vegetable plants grown on mine water-contaminated soils: Uptake, human health risk and remedial effects of biochar. *Journal of Hazardous Materials, 402*. Available from https://doi.org/10.1016/j.jhazmat.2020.123488.

Roobavannan, M., Kandasamy, J., Pande, S., Vigneswaran, S., & Sivapalan, M. (2020). Sustainability of agricultural basin development under uncertain future climate and economic conditions: A socio-hydrological analysis. *Ecological Economics, 174*. Available from https://doi.org/10.1016/j.ecolecon.2020.106665.

Rutebuka, J., Munyeshuli Uwimanzi, A., Nkundwakazi, O., Mbarushimana Kagabo, D., Mbonigaba, J. J. M., Vermeir, P., & Verdoodt, A. (2021). Effectiveness of terracing techniques for controlling soil erosion by water in Rwanda. *Journal of Environmental Management, 277*. Available from https://doi.org/10.1016/j.jenvman.2020.111369, https://www.sciencedirect.com/journal/journal-of-environmental-management.

Sadeghi, S., Gholami, L., Sharifi, E., Khaledi Darvishan, A., & Homaee, M. (2015). Scale effect on runoff and soil loss control using rice straw mulch under laboratory conditions. *Solid Earth., 6*, 1−8.

Sakalasooriya, N. (2021). Climate-smart agriculture in cascade minor irrigation system: Status, scope and challenges in Sri Lanka: A case from Puttlam district. *Sri Lanka Journal of Social Sciences and Humanities, 1*, 109−122.

Sarma, B., Sarma, A. K., & Singh, V. P. (2013). Optimal ecological management practices (EMPs) for minimizing the impact of climate change and watershed degradation due to urbanization. *Water Resources Management, 27*(11), 4069−4082. Available from https://doi.org/10.1007/s11269-013-0396-y.

Sarma, B. (2011). *Optimal Ecological Management Practices for Controlling Sediment and Water Yield from a Hilly Urban System within Sustainable Limit (Doctoral dissertation)*. India: IIT Guwahati.

Sarma, B., Sarma, A. K., Mahanta, C., & Singh, V. P. (2015). Optimal ecological management practices for controlling sediment yield and peak discharge from hilly urban areas. American Society of Civil Engineers (ASCE), India. *Journal of Hydrologic Engineering, 20*(10). Available from https://doi.org/10.1061/(ASCE)HE.1943-5584.0001154, https://ascelibrary.org/journal/jhyeff.

Shanafield, M., Bourke, S. A., Zimmer, M. A., & Costigan, K. H. (2021). An overview of the hydrology of non-perennial rivers and streams. *Wiley Interdisciplinary Reviews: Water., 8*, e1504.

Shumilova, O., Zak, D., Datry, T., von Schiller, D., Corti, R., Foulquier, A., Obrador, B., Tockner, K., Allan, D. C., Altermatt, F., Arce, M. I., Arnon, S., Banas, D., Banegas-Medina, A., Beller, E., Blanchette, M. L., Blanco-Libreros, J. F., Blessing, J., Boëchat, I. G., . . . Zarfl, C. (2019). Simulating rewetting events in intermittent rivers and ephemeral streams: A global analysis of leached nutrients and organic matter. *Global Change Biology, 25*(5), 1591−1611. Available from https://doi.org/10.1111/gcb.14537.

Silva, E. M. (2014). Screening five fall-sown cover crops for use in organic no-till crop production in the Upper Midwest. *Agroecology and Sustainable Food Systems*, 38(7), 748.

Srivastava, A., & Chinnasamy, P. (2021). Water management using traditional tank cascade systems: A case study of semi-arid region of Southern India. *SN Applied Sciences*, 3, 1−23.

Tsunetaka, H., Hotta, N., Imaizumi, F., Hayakawa, Y. S., & Masui, T. (2021). Variation in rainfall patterns triggering debris flow in the initiation zone of the Ichino-sawa torrent, Ohya landslide, Japan. *Geomorphology*, 375. Available from https://doi.org/10.1016/j.geomorph.2020.107529.

Tung, T. M., & Yaseen, Z. M. (2020). A survey on river water quality modelling using artificial intelligence models: 2000−2020. *Journal of Hydrology*, 585, 124670.

Xie, X., Wang, X., Liu, Z., Liu, Z., & Zhao, S. (2023). Regulation effect of slit-check dam against woody debris flow: Laboratory test. *Frontiers in Earth Science*, 10. Available from https://doi.org/10.3389/feart.2022.1023652.

Zhou, Z., & Cartwright, I. (2021). Using geochemistry to identify and quantify the sources, distribution, and fluxes of baseflow to an intermittent river impacted by climate change: The upper Wimmera River, southeast Australia. *Science of Total Environment.*, 801, 149725.

USEPA. 1997. Innovative Uses of Compost: Erosion Control, Turf Remediation and Landscaping. EPA 530-F-97-043. Washington, DC: USEPA

Chapter 6

River restoration involves using riverside vegetation to enhance ecological health while implementing effective management strategies

Meenakshi Pawar[1] and Pushkar Pawar[2]

[1]Department of Architecture and Planning, Indira Gandhi Delhi Technical University for Women, Government of Delhi, New Delhi, India,
[2]Department of Environmental Planning, School of Planning & Architecture, New Delhi (SPAD), Ministry of Education, Government of India, New Delhi, Delhi, India

6.1 Introduction

Rivers and their associated riparian zones are vital ecosystems that provide numerous ecological services and support a diverse array of flora and fauna. Yuan et al. (2020) and Tanimoto et al. (1999) both emphasize the significance of taking into account the ecological setting and the stage of succession on the site. Tanimoto advises against the prevalence of non-native species. However, human activities, such as urbanization, agriculture, and industrialization, have often resulted in the degradation and loss of these critical habitats. To counteract these detrimental effects, riverside vegetation planting has emerged as a powerful river restoration and ecological revitalization tool. Strobl et al. (2015) underscores the importance of meticulous selection of restoration methods, given the potential for varied outcomes. Rowiński et al. (2018) highlights the capacity of vegetation as a nature-based solution in river management, particularly in addressing water quality and biodiversity concerns. These investigations collectively underscore the necessity for a comprehensive and adaptable strategy in riverside vegetation planting, considering ecological context, succession stage, and the precise objectives of restoration initiatives. This chapter aims to assess the ecological benefits of riverside vegetation planting and gain insights into the positive impact and develop informed approaches for sustainable river ecosystem management (Salehi Gahrizsangi et al., 2021). Riverside vegetation planting, also known as riparian or riverbank restoration, involves the deliberate establishment of native plant species along riverbanks to enhance ecosystem health and resilience. This practice offers a range of ecological benefits, including erosion control, water quality improvement, habitat creation, biodiversity promotion, and bank stabilization. By stabilizing the soil, vegetation plays a pivotal role in preventing erosion and reducing sediment runoff into the water, safeguarding both the riverbanks and the quality of water bodies. Furthermore, the presence of riverside vegetation contributes to preserving and restoring essential habitats for diverse wildlife species. Native trees, shrubs, and grasses provide nesting sites, food sources, and shelter for birds, mammals, insects, and aquatic organisms. This restoration approach helps conserve biodiversity and contributes to the overall ecological balance of the riparian zone.

6.2 Ecological benefits of riverside vegetation planting

6.2.1 Vegetation based Erosion control along river bank

Erosion control is the practice of preventing or controlling wind or water erosion in riverbanks, land development, coastal areas, and construction. An overview of how vegetation influences slope stability and its significance in controlling erosion. They explain the processes through which vegetation helps reduce erosion, highlighting the critical role of

Hydrosystem Restoration Handbook. DOI: https://doi.org/10.1016/B978-0-443-29802-8.00006-6

plant cover in preventing soil loss.The influence of vegetation on bank stability, bank erosion rates, and sediment dynamics, providing insights into the role of vegetation in mitigating erosion along riverbanks (Crouzy & Lane, 2017). The typologies of different vegetation have their effectiveness in erosion control and the underlying processes involved (Wang & Inoue, 2018). Vegetation influences bank stability, sediment transport, and channel morphology, highlighting the importance of vegetation for maintaining river system integrity (Gurnell et al., 2012).

As plants grow and mature, they contribute to bank reinforcement through the development of a dense root network and above-ground biomass. This growth enhances bank stability and resistance to erosion. For the risk associated with the severity of bank erosion along the streams and rivers, a tool (Bank Erosion Hazard Index [BEHI], 2001) was developed by the Natural Resources Conservation Service, which is a part of the US Department of Agriculture. The BEHI was created to assist landowners, land managers, and conservation professionals in identifying areas where bank erosion is likely to occur and to prioritize conservation and management actions accordingly.

6.2.1.1   Role of vegetation in soil stabilization

The role of vegetation in soil stabilization is well documented and recognized in various scientific studies. Vegetation, especially plants with extensive root systems, plays a crucial role in soil stabilization. The roots bind the soil particles together, creating a network of interlocking roots that help to resist soil erosion caused by water or wind. Liu et al. (2012) investigate the effect of plant roots on soil stabilization, examine the alteration of root architecture in *Zea mays* (maize) grown in reclaimed coal-mine soil and discuss how root systems contribute to soil stabilization. The role of root systems in facilitating the colonization of benthic invertebrates on artificial substrates is crucial and can be attributed to their positive influence on stabilizing sediment and creating favorable habitats for these organisms (Stokes et al., 2008). The roots of riparian vegetation penetrate and bind the sediment, reducing its mobility and susceptibility to erosion. This stabilization effect is particularly pronounced in the braided river environment, where high sediment transport rates and frequent channel changes are prevalent. The presence of roots helps to anchor sediment particles, preventing their downstream movement and promoting the formation of stable substrate conditions (Darby et al., 2016).

6.2.1.2   Prevention of bank erosion and collapse

Preventing bank erosion and collapse is a critical aspect of managing river systems and protecting adjacent areas from the detrimental effects of erosion. The major techniques and strategies employed to prevent bank erosion and collapse are planting and maintaining vegetation, particularly riparian plants that are native to that area. Bioengineering techniques using natural materials, such as logs, branches, and live stakes, in combination with vegetation to create structures that enhance bank stability. Riprap and gabions refer to the use of large rocks or stones placed along the bank of the river to protect against erosion. Gabions are wire mesh containers filled with stones or other materials. The extensive root systems of plants, such as trees, grasses, and shrubs, penetrate the soil, creating a network of roots that bind the soil particles together. This reinforcement helps to increase soil cohesion and shear strength, reducing the risk of bank failure. Gurnell et al. (2007) present a series of case studies from European river restoration projects, showcasing the role of vegetation in these initiatives. The case studies highlight different restoration techniques, such as the introduction of riparian vegetation, the creation of vegetated buffer zones, and the implementation of instream planting.

6.2.1.3   Reduction of sediment runoff

Reducing sediment runoff is an important aspect of managing land and water resources to minimize soil erosion and protect water quality. Flanagan et al. (2017) highlight the importance of rain-fed agriculture and its susceptibility to soil erosion and sediment runoff due to the lack of irrigation and reliance on natural precipitation. Several conservation practices like contour farming, terracing, grassed waterways, strip cropping, cover cropping, and conservation tillage enhance the comprehensive planning. The document titled "National Menu of Best Management Practices for Stormwater Phase II" by the US Environmental Protection Agency elaborates on the best practices for managing stormwater runoff from highways, roads, and bridges. It includes strategies for implementing green infrastructure practices, such as vegetated swales, permeable pavements, and bioretention areas, to enhance stormwater infiltration and pollutant removal.

6.2.2   Water quality improvement

Water quality improvement is a crucial aspect of river restoration, and the planting of vegetation along riversides can have several positive effects.

6.2.2.1   Natural filtration and nutrient absorption

Riverside vegetation, including plants such as reeds, grasses, and wetland species, plays a vital role in natural filtration. As water flows through the vegetation, it encounters roots, stems, and leaves, which act as physical barriers, filtering out suspended particles and sediment. Additionally, the roots of these plants absorb excess nutrients, such as nitrogen and phosphorus, from the water, helping to reduce nutrient pollution and promote water quality. "Wetlands" by Mitsch and Gosselink talks about the ecosystem services provided by wetlands, such as water purification, flood regulation, carbon sequestration, and provision of habitat for wildlife. It discusses the economic and societal value of wetlands and highlights the importance of their conservation and restoration efforts.

6.2.2.2   Reduction of pollutant and sediment load

The presence of vegetation along riverbanks can effectively reduce the load of pollutants and sediment reaching the water. Vegetation intercepts and traps sediment, preventing it from being transported downstream. This reduces turbidity and the accumulation of sediment in water bodies, which can adversely affect aquatic ecosystems. Moreover, the dense vegetation cover helps to intercept and retain pollutants such as pesticides, fertilizers, and other contaminants, preventing them from entering the water and thereby reducing pollution levels. vegetation-based ecological engineering techniques for wastewater treatment, as outlined by Qiu and Mitsch (2015), include vegetation-based treatment systems, such as constructed wetlands, floating treatment wetlands, and vertical flow wetlands. these systems utilize natural processes for the removal of pollutants. It discusses the mechanisms involved in wastewater treatment by vegetation, including physical processes like sedimentation and filtration, chemical processes such as nutrient absorption, and biological processes like microbial degradation.

6.2.2.3   Enhancement of water clarity and oxygen levels

By reducing sediment load and nutrient pollution, riverside vegetation planting can improve water clarity. Clearer water allows sunlight to penetrate deeper, promoting photosynthesis and the growth of aquatic plants. These plants contribute to oxygen production through photosynthesis, thereby increasing dissolved oxygen levels in the water. Higher oxygen levels are crucial for supporting aquatic life and maintaining a healthy ecosystem. "Nonpoint pollution of surface waters with phosphorus and nitrogen" explores the issue of nonpoint source pollution and its impact on surface water quality, specifically focusing on phosphorus and nitrogen pollution (Carpenter & Caraco, 1998). Nonpoint source pollution refers to the diffuse release of pollutants from various land-based activities such as agriculture, urban runoff, and atmospheric deposition. Nonpoint pollution, particularly phosphorus and nitrogen, has become a leading cause of water quality degradation of many surface waters.

6.2.3   Habitat creation

Riverside vegetation planting refers to the intentional establishment of vegetation, including trees, shrubs, grasses, and other plants, along the banks of rivers and streams. This practice is commonly employed in river restoration and habitat enhancement projects. Riverside vegetation planting offers numerous benefits and serves various ecological functions.

6.2.3.1   Importance of riparian zones as wildlife habitat

Riparian zones, the areas along the banks of rivers and streams, are crucial habitats for wildlife. Riverside vegetation planting enhances the quality and diversity of these habitats. The vegetation provides food sources, shelter, and breeding grounds for a wide range of species. Riparian zones act as corridors connecting different ecosystems, allowing wildlife to move between habitats, find mates, and access resources. Riparian corridors enhance biodiversity and provide critical ecosystem services in landscapes (Naiman et al., 2005). High species richness and abundance are often found in riparian zones, and they emphasize the ecological importance of maintaining and protecting these areas to preserve biodiversity. Vegetation and soils within riparian zones act as buffers, filtering pollutants, stabilizing stream banks, and reducing erosion.

6.2.3.2   Provision of nesting sites and shelter

Riverside vegetation offers nesting sites and shelter for various species. Trees and shrubs along the riverbanks provide nesting habitats for birds, such as herons, kingfishers, and songbirds. Hollowed-out tree trunks, branches, and dense vegetation offer shelter for small mammals, amphibians, and reptiles. The dense cover of vegetation also protects wildlife from predation, extreme weather events, and human disturbance.

6.2.3.3 Contribution of aquatic and terrestrial biodiversity

Riverside vegetation planting contributes to the overall biodiversity by supporting both aquatic and terrestrial species. Aquatic plants growing along the water's edge provide critical habitat for fish, amphibians, and invertebrates. These plants offer spawning sites, refuge, and feeding areas. The shade provided by the vegetation helps regulate water temperature, which is crucial for many aquatic organisms. Additionally, the diverse plant species in riparian zones attract a wide range of insects and other invertebrates, which serve as food sources for birds, bats, and other wildlife.

6.3 Assessment of riverside vegetation planting strategies

6.3.1 Selection of native plant species

Native plant species are those that naturally occur and have adapted to the local ecosystem and support local biodiversity by providing food, shelter, and habitat for native wildlife. They have co-evolved with local fauna, creating ecological relationships that are important for maintaining healthy ecosystems. Selecting native plant species helps to preserve and enhance local biodiversity and is well adapted to the local climate, soil types, and hydrological conditions. They have evolved mechanisms to cope with seasonal variations, droughts, floods, and other environmental factors. They are more likely to thrive and establish successfully in the riverside environment. Invasive plants can alter soil properties and microbial communities, leading to shifts in nutrient availability, soil moisture, and other factors that can influence plant growth (Suding et al., 2015).

Native plant species contribute to nutrient cycling, water filtration, soil stabilization, and other ecosystem services. By choosing native plant species, these critical ecosystem functions can be maintained and enhanced, D'Antonio et al. emphasize that exotic plant invasions can disrupt native plant communities, alter ecosystem processes, and reduce biodiversity. Native plant species have evolved within the local ecosystem and have developed defenses against invasive species. Planting native species can help reduce the risk of invasive plants outcompeting and displacing native vegetation, preserving the integrity of the riverside ecosystem.

Native plant species often have cultural and esthetic value, reflecting the local heritage and natural beauty of the region. Choosing native species for riverside vegetation planting can contribute to the cultural identity and esthetic appeal of the area. Hobbs et al. (2006) talk about the concept of novel ecosystems, which refers to ecosystems that arise from human activities and exhibit new combinations of species, ecological processes, and functions that differ from historical reference conditions. They highlighted that human-induced changes, such as land-use change, climate change, and species invasions, have increasingly led to the formation of novel ecosystems worldwide and discusses how these ecosystems are often characterized by high species turnover, assemblages of nonnative species, altered disturbance regimes, and novel ecological interactions. Selecting native plant species ensures that the plants are locally sourced and readily available. This helps to maintain the genetic diversity, adaptability, and resilience of the planted vegetation.

6.3.1.1 Importance of native species for ecosystem restoration

Native species have co-evolved with the local environment, including climate, soil conditions, and other organisms. As a result, they are well adapted to the specific ecological conditions of the area. Using native species ensures a better ecological fit, as they are suited to the local climate, hydrology, and nutrient cycles. They are more likely to establish successfully, grow vigorously, and contribute to the overall functioning of the ecosystem and contribute to nutrient cycling, energy flow, pollination, seed dispersal, and other ecological processes. The presence of native species helps maintain ecosystem stability and resilience, as they have adapted to local disturbances, climate fluctuations, and species interactions. Restoring native species enhances the overall functionality of the ecosystem and supports the provision of ecosystem services. Using native species in ecosystem restoration can help suppress the growth and spread of invasive species. Native species are often better competitors against invasives due to their co-evolutionary history and ecological adaptations. By reintroducing native species, they can outcompete invasives for resources, reducing their impacts and restoring the balance to the ecosystem.

6.3.1.2 Consideration of local climate and soil conditions

Native species have evolved and adapted to the specific climatic conditions of their native range. They are well suited to local temperature ranges, rainfall patterns, and seasonal variations. By selecting native species, restoration projects can increase the likelihood of successful establishment and long-term survival, as these species are already adapted to the local climate. Native vegetation is more resilient to climate change impacts, as they have already adapted to local conditions over evolutionary time. By restoring ecosystems with native species, there is a higher probability that the restored habitats will be more resilient to future climate change, including temperature increases, altered precipitation patterns, and extreme weather events.

6.3.2   Planting methodologies

When implementing riverside vegetation planting, the choice between direct seeding and transplanting depends on factors such as project scale and objectives. The timing and seasonality of planting should consider the optimal growing season and phenology of the target species. Vegetation spacing and density should be determined based on species characteristics, site conditions, and project goals. It is recommended to consult these sources directly to extract specific information and cite relevant findings or recommendations. Harris et al. (2002) explore that revegetation practices were implemented on landslides in the Queen Charlotte Islands with different techniques, such as seeding, planting, and natural regeneration, and evaluate their effectiveness in promoting vegetation establishment and slope stabilization.

6.3.2.1   Direct seeding versus transplanting

Direct seeding involves sowing seeds directly into the planting site. It can be a cost-effective method, particularly for large-scale projects. However, the success of direct seeding depends on factors such as seed quality, germination rates, and environmental conditions. It may require additional measures to protect seeds from predation or ensure adequate seed-to-soil contact. MacDonald et al. (2013) delve into the various techniques used in direct seeding and planting. They discuss the selection of suitable tree and shrub species, seed collection, and seedling production.

Transplanting involves the physical relocation of nursery-grown seedlings or established plants to the planting site. This method offers a higher likelihood of successful establishment since the plants are already developed. Transplanting is particularly useful for restoring specific plant species or when immediate vegetation cover is desired. However, it can be more labor-intensive and expensive compared to direct seeding. Lamb et al. explore different approaches and techniques for forest rehabilitation and restoration. They discuss the restoration of native tree species, reforestation, natural regeneration, agroforestry, and assisted natural regeneration.

6.3.2.2   Timing and seasonality of planting

The timing of planting is critical for successful establishment. Planting during the optimal growing season provides favorable conditions for root development and plant growth. It is important to consider factors such as frost dates, rainfall patterns, and temperature fluctuations when determining the ideal planting time. Early planting in the growing season allows plants to establish before summer heat or winter dormancy.

Native plants often have specific seasonal growth patterns and life cycles. Understanding the phenology of target species is important to align planting activities with their natural growth cycles. For example, deciduous species are typically planted during the dormant season when they are not actively growing, while evergreen species can be planted throughout the year.

6.3.2.3   Vegetation spacing and density

Spacing refers to the distance between individual plants within a planting area. Optimal spacing depends on various factors, including the growth habit of the species, site conditions, and desired vegetation density. Closer spacing can promote quicker canopy closure and provide better competition against weeds. However, overcrowding may lead to resource competition and reduced growth rates. Wider spacing allows more space for individual plant growth but may require additional canopy development and weed suppression time. vegetation to enhance infiltration and promote water storage in soils, thereby further reducing flood volumes (Kildisheva et al., 2016).

Vegetation density refers to the overall coverage of plants within a specific area. The desired density depends on the project objectives, site conditions, and the ecological function the vegetation is intended to provide. Higher densities may be appropriate for erosion control, habitat creation, or shading purposes. Lower densities may be suitable for more naturalistic or esthetic planting designs.

6.4   Case studies and best practices

An exploration into the impact of plant species and planting arrangements, coupled with nonstructural mechanical techniques, on reducing erosion along riverbanks presents a compelling avenue for potential case studies, particularly concerning the state's renowned rivers. Emphasize the importance of site assessment, native species selection, appropriate planting techniques, maintenance, monitoring, and community engagement for achieving positive outcomes in river restoration.

- Site Assessment: Conduct a thorough site assessment to understand the ecological conditions, hydrological dynamics, and soil characteristics of the planting area. This information will guide species selection and planting strategies.
- Native Species Selection: Select native plant species that are well-adapted to the local environment, including climate, soil, and hydrological conditions. Consider the ecological function and desired outcomes of the restoration project.
- Planting Techniques: Employ appropriate planting techniques, such as direct seeding or transplanting, depending on the project goals, scale, and site conditions. Follow best practices for seed preparation, planting depth, and spacing to maximize plant survival and establishment.
- Maintenance and Monitoring: Implement a maintenance and monitoring plan to ensure the long-term success of the planting project. This may include irrigation, weed control, and periodic monitoring of plant health, growth, and ecosystem responses.
- Collaboration and Community Engagement: Foster collaboration with stakeholders, local communities, and experts to garner support, share knowledge, and ensure project success. Engage the community in restoration activities and raise awareness about the importance of riverside vegetation planting for ecological and societal benefits.

6.4.1   Successful riverside vegetation planting projects case study 1

Restoration of River Yamuna floodplains- Yamuna Biodiversity ParkDelhi, an urban area blessed with unique natural endowments, has witnessed significant ecological transformations due to human endeavors and urban sprawl. While some green areas have been sacrificed to accommodate urban development, Delhi maintains its reputation as one Recognizing the alarming depletion of natural resources amid rapid economic growth, the Delhi Development Authority (DDA), in partnership with the Center for Environmental Management of Degraded Ecosystems (CEMDE), has initiated the establishment of the Yamuna Biodiversity Park (YBP). This endeavor aims to augment the city's natural wealth, ensuring the sustained provision of ecological goods and services for present and future generations. In addition to preserving natural heritage, the park will play a crucial role in raising public awareness about the significance and benefits of biodiversity conservation. Given the importance of the Yamuna River and the Aravali ranges to Delhi's heritage, joint efforts between DDA and CEMDE are underway to establish two biodiversity parks — the Yamuna Biodiversity Park and the Aravali Biodiversity Park. The Yamuna Biodiversity Park comprises two primary zones: the visitor zone and the nature reserve zone. The front section of the park, spanning 220 meters southward and 140 meters northward from the main entry gate, with a width of 20—30 meters, is designated as the Domesticated Biodiversity Zone. Enclosed by a hedge of poplar trees, this area showcases plants such as Ailanthus, Butea, and Bauhinia, which offer continuous seasonal interest due to their prolonged flower production throughout the season. Plant communities under various stages of development of Yamuna Biodiversity parks are Mixed deciduous with bamboo, Sal dominated mixed evergreen, Teak dominated mixed deciduous, Sal-dominated mixed deciduous, Grasslands, Thorn-scrub forests, Shallow wetland communities, Deep wetland Communities, The conservatory within the park boasts an impressive display of approximately 500 varieties spanning 80 species of fruit-yielding plants, showcasing the remarkable diversity found among these plants. Among them, both familiar and lesser-known fruits are presented, including Khirni (Mimusops hexandra), which was once locally extinct but now thrives alongside Kaith (Feronia limonia), pomegranate (Punica granatum), sapota (Achyrus sapota), jamun (Syzygium cumuni), guava (Psidium guajava), amla (Emblica officinalis), various Citrus species, grapes (Vitis sp), loquat (Eriobotrya japonica), and ber (Zizyphus mauritiana). Additionally, the conservatory provides habitat for a diverse array of birds such as parakeets, yellow-footed green pigeons, munias, babblers, coppersmith barbets, bulbuls, and peafowl, along with a variety of snakes, adding to the rich biodiversity of the park.

6.4.1.1   Cheonggyecheon Stream restoration, South Korea

The Cheonggyecheon Stream in Seoul was restored in 2005, transforming a concrete-lined urban stream into a thriving ecological corridor. The restoration involved the removal of concrete, reintroduction of natural water flow, and extensive riverside vegetation planting. Native plant species were selected to provide habitat, improve water quality, and enhance the esthetic appeal of the stream. The project has revitalized the area, attracted wildlife, and provided a recreational space for residents while mitigating flood risks (Lim et al., 2009).

6.4.1.2 Oostvaardersplassen Nature Reserve, the Netherlands

The Oostvaardersplassen Nature Reserve is an example of successful wetland restoration in the Netherlands. The restoration project involved the rewilding of a former polder, creating diverse wetland habitats. Riverside vegetation planting was a key component, with native reed beds and wetland species established to provide habitat for birds, mammals, and other wildlife. The restoration has been successful in creating a self-sustaining ecosystem and has become a haven for numerous bird species (Veeneklaas et al., 2010).

6.4.1.3 Spokane River restoration, the United States

The Spokane River restoration project in Washington State focused on enhancing riparian habitats and improving water quality. The restoration included extensive riverside vegetation planting with native species, such as willows and sedges, along the riverbanks. The planted vegetation helped stabilize the banks, reduce sedimentation, and create valuable habitats for fish and other wildlife. The project demonstrated the importance of integrating vegetation planting with other restoration measures to achieve comprehensive river restoration goals.

6.4.1.4 Best practices for successful riverside vegetation planting projects

Site assessment: A thorough site assessment is conducted to understand the ecological conditions, hydrological dynamics, and soil characteristics of the planting area. This information will guide species selection and planting strategies.

Native species selection: Native plant species that are well adapted to the local environment, including climate, soil, and hydrological conditions, are selected. The ecological function and desired outcomes of the restoration project are considered.

Planting techniques: Appropriate planting techniques, such as direct seeding or transplanting, are employed depending on the project goals, scale, and site conditions. Best practices are followed for seed preparation, planting depth, and spacing to maximize plant survival and establishment.

Maintenance and monitoring: A maintenance and monitoring plan is implemented to ensure the long-term success of the planting project. This may include irrigation, weed control, and periodic monitoring of plant health, growth, and ecosystem responses.

Collaboration and community engagement: A collaboration with stakeholders, local communities, and experts is fostered to garner support, share knowledge, and ensure project success. The community is engaged in restoration activities and raise awareness about the importance of riverside vegetation planting for ecological and societal benefits.

6.5 Conclusions

Effective management strategies are critical for the success of riverside vegetation planting projects. The selection of suitable native plant species, site preparation techniques, and planting methodologies are vital considerations. Additionally, long-term monitoring and maintenance of restored areas ensure the establishment and survival of the vegetation. By assessing various management approaches and lessons learned from past experiences, this chapter aims to provide guidance for practitioners and policymakers involved in river restoration initiatives.

Riverside vegetation planting represents an important and effective method for restoring and enhancing the ecological integrity of river ecosystems. Emphasis should be on considering the interactions between invasive plants, native species, and ecosystem processes when developing restoration strategies. This chapter will delve into the ecological benefits of such planting initiatives and present evidence-based management strategies for successful river restoration projects. By advancing our understanding of this practice, we can contribute to the conservation and sustainable management of rivers and their riparian zones, safeguarding these valuable ecosystems for future generations.

References

Bank Erosion Hazard Index (BEHI). (2001). *Natural Resources Conservation Service*. U.S. Department of Agriculture.

Carpenter, S. R., & Caraco, N. F. (1998). Nonpoint pollution of surface waters with phosphorus and nitrogen. *Ecological Applications, 8*(3), 559–568.

Crouzy, B., & Lane, S. N. (2017). Riparian vegetation and fluvial geomorphic processes. *Reviews of Geophysics, 55*(3), 788–822.

Darby, S. E., et al. (2016). Fluvial processes and vegetation encroachment in braided rivers: A review. *Earth Surface Processes and Landforms*, *41*(1), 38−62.

Flanagan, D. C., et al. (2017). Conservation practices to reduce runoff and sediment losses from rainfed agriculture. *Journal of Soil and Water Conservation*, *72*(4), 91A−95A.

Gurnell, A. M., et al. (2007). *Vegetation and the Fluvial Environment in River Restoration: Examples from European Projects*. Wiley.

Gurnell, A. M., et al. (2012). Vegetation and river channel dynamics: Insights from studies of hydroecological systems. *Hydrological Processes*, *26*(26), 3822−3829.

Harris, R., et al. (2002). Revegetation practices on landslides in the Queen Charlotte Islands, British Columbia, Canada. *Geomorphology*, *45*(1−2), 31−45.

Hobbs, R. J., et al. (2006). Novel ecosystems: Theoretical and management aspects of the new ecological world order. *Global Ecology and Biogeography*, *15*(1), 1−7.

Kildisheva, O. A., et al. (2016). The role of vegetation density in flood risk mitigation. *Environmental Research Letters*, *11*(12)124023.

Lim, M. H., et al. (2009). Restoration of the Cheonggyecheon Stream, Seoul, South Korea. *Landscape and Urban Planning*, *87*(2), 153−164.

Liu, B., et al. (2012). Root architecture alteration of *Zea mays* L. grown in soil amended with reclaimed coal-mine soil. *Plant and Soil*, *359*(1−2), 21−33.

MacDonald, S. E., et al. (2013). Direct seeding and planting of trees and shrubs in the boreal forest of Canada: A review. *Environmental Reviews*, *21*(3), 156−186.

Naiman, R. J., et al. (2005). Riparian corridors enhance biodiversity and provide critical ecosystem services in landscapes. *Ecology Letters*, *7*(6), 493−504.

Qiu, R., & Mitsch, W. J. (2015). Ecological engineering using vegetation for wastewater treatment: State of the art review. *Ecological Engineering*, *81*, 609−619.

Rowiński, P. M., Västilä, K., Aberle, J., Järvelä, J., & Kalinowska, M. B. (2018). How vegetation can aid in coping with river management challenges: A brief review. *Ecohydrology & Hydrobiology*, *18*, 345−354. Available from https://doi.org/10.1016/j.ecohyd.2018.07.003.

Salehi Gahrizsangi, H., Eslamian, S., Dalezios, N. R., Blanta, A., & Madadi, M. (2021). Vegetation advantages for water and soil conservation. In S. Eslamian, & F. Eslamian (Eds.), *Handbook of Water Harvesting and Conservation. vol. 1: Basic Concepts and Fundamentals* (pp. 323−336). New Jersey, USA: Wiley, Chapter 21.

Stokes, A., et al. (2008). Ecological benefits of vegetative filter strips in managing runoff from agricultural land. *Critical Reviews in Environmental Science and Technology*, *38*(2), 112−136.

Strobl, K., Wurfer, A.-L., & Kollmann, J. (2015). Ecological assessment of different riverbank revitalisation measures to restore riparian vegetation in a highly modified river. *Tuexenia*, *35*, 177−194. Available from https://doi.org/10.14471/2015.35.005.

Suding, K. N., et al. (2015). Consequences of plant−soil feedback in the invasion. *Journal of Ecology*, *103*(4), 1069−1080.

Tanimoto, S., Nakagoshi, N., & Nehira, K. (1999). Ecological evaluation on riverside vegetation with river restoration at early secondary successional stage. *Environmental Systems Research*, *27*, 315−321. Available from https://doi.org/10.2208/proer1988.27.315.

Veeneklaas, R. M., et al. (2010). The Oostvaardersplassen: History and future of a Dutch wetland. *Wetlands*, *30*(4), 699−709.

Wang, L., & Inoue, M. (2018). Bank erosion control using vegetation: A review. *Journal of Hydrology*, *556*, 899−912.

Yuan, J., Chen, L., Luo, J., Zhang, G., & You, F. (2020). An adaptive multi-layered ecological landscape: The ecological planting of herbaceous communities on river revetments in mountainous city. *Landscape Architecture Frontiers*, *8*, 44−57, https://doi.org/10.15302/J-LAF-1-020029 PAPERS.

Further reading

D'Antonio, C. M., et al. (2001). Exotic plant invasions in California: Challenges for restoration. *Ecological Restoration*, *19*(1), 55−67.

Lamb, D., et al. (2005). *Rehabilitation and Restoration of Degraded Forests*. CRC Press.

Mitsch, W. J., & Gosselink, J. G. (2015). *Wetlands* (5th ed.). Wiley.

Part III

River flow modeling

Chapter 7

Evaluating the reliability of open-source hydrodynamic models in flood inundation mapping: an exhaustive approach over a sensitive coastal catchment

Dev Anand Thakur[1], Vijay Suryawanshi[2], H. Ramesh[2] and Mohit Prakash Mohanty[1]

[1]Department of Water Resources Development and Management, Indian Institute of Technology Roorkee, Roorkee, Uttarakhand, India, [2]Department of Water Resources and Ocean Engineering, National Institute of Technology Karnataka, Mangaluru, Karnataka, India

7.1 Introduction

Urbanization, global warming, climate change, and uncontrolled exploitation of natural resources have led to many natural disasters over the entire globe in recent times (Thakur, Mohanty, Mishra, et al., 2024). Among all the natural disasters, flood stands out as one of the most enormously destructive and frequently occurring natural disasters (Mohanty et al., 2020). Floods causes loss of human life, derail the country's economic growth, decrease socioeconomic welfare, and damage ecosystems. Flooding is more prominent in urban areas (Piadeh et al., 2022), and this may be due to unplanned development, blockage of drainage channels, and inconsistent and altered rainfall. Due to global warming and climate change, rain has increased over the years, and the existing drainage networks are not capable of withstanding this change (Rangari et al., 2019). With the effects of floods on the life, economy, infrastructure, and growth of any nation being on a larger scale, it becomes imperative to understand and realize the necessity of proper and efficient flood management strategies. Floods in the Indian subcontinent have been devastating over the years, with their frequency of occurrence being higher than in most countries. The most recent floods that caused heavy destruction in the country are the Uttarakhand flood (2013), Srinagar flood (2014), Assam flood (2015), Mumbai flood (2017), Bihar flood (2017), Gujarat flood (2017), and Kerala flood (2018). Floods in India can be attributed to the Indian southwest monsoon and northeast monsoon. July, August, and September are the worst flood-affected months of the country, with Assam (Brahmaputra River basin) and Bihar (Ganga River basin) the worst affected states (Dhar & Nandargi, 2003; Prasad, 2016; Ray et al., 2019).

Flood inundation mapping serves a critical function in pinpointing key areas susceptible to current or potential flooding, thereby playing an indispensable role in flood risk management (Garg et al., 2022). This vital tool aids decision-makers in assessing necessary interventions and devising appropriate mitigation strategies. Through the detailed depiction of water depths and the extent of flooding, flood inundation maps offer invaluable data. This information is instrumental in enhancing the effectiveness of flood risk management strategies, facilitating comprehensive flood hazard assessments, and guiding the meticulous planning and execution of flood defense measures (Thakur & Mohanty, 2024). By providing a clear visual representation of flood-prone areas, these maps enable authorities and stakeholders to prioritize resources, improve emergency response actions, and implement proactive measures for community protection and infrastructure resilience.

Hydrological models and hydrodynamic models are two essential tools in the field of water resources management, each playing a crucial role in understanding and predicting flood events (Thakur & Mohanty, 2024). Hydrological models

Hydrosystem Restoration Handbook. DOI: https://doi.org/10.1016/B978-0-443-29802-8.00007-8

focus on the quantitative analysis of the water cycle, including precipitation, evaporation, infiltration, and runoff processes within a watershed. These models are primarily used to estimate the volume and timing of water flow, providing insights into how different rainfall events can lead to varying degrees of flood risk (Thakur & Mohanty, 2023). They are instrumental in predicting river discharge and identifying potential flood occurrence, enabling the early implementation of flood warnings and preparedness measures. On the other hand, hydrodynamic models delve into the detailed mechanics of water flow and its behavior in open channels, rivers, and floodplains. These models simulate the movement of water, taking into account its depth, velocity, and interaction with the surrounding environment, such as topography and man-made structures. Hydrodynamic models are particularly useful in creating flood inundation maps, which visually represent the extent of flooding and the depth of water in different areas (Mohanty & Karmakar, 2021). By integrating these models with geographic information system (GIS), researchers and decision-makers can obtain accurate, spatially explicit maps that highlight flood-prone zones, enabling targeted flood risk management and mitigation efforts. Together, hydrological and hydrodynamic models offer a comprehensive approach to flood mapping. They provide a foundation for developing effective flood management strategies, from forecasting and early warning systems to the design of flood defenses and land-use planning. By leveraging the strengths of both model types, water resource managers can enhance their understanding of flood dynamics, improve the accuracy of flood risk assessments, and implement more effective flood mitigation and adaptation measures (Abdelkrim et al., 2023).

Hydrological models include the Hydrologic Engineering Center's Hydrologic Modeling System and the Soil and Water Assessment Tool, which are pivotal in simulating precipitation-runoff processes and watershed management, respectively. In the domain of hydrodynamic modeling, notable examples are the Hydrologic Engineering Center's River Analysis System (HEC-RAS), used for floodplain analysis and river flow simulation, TELEMAC-2D for free-surface hydraulic computations, and MIKE + for comprehensive water dynamics in rivers, estuaries, and coastal areas. These models are instrumental in flood risk management and environmental impact assessments (Kumar et al., 2020; Liu et al., 2019; Pinos & Timbe, 2019).

The HEC-RAS stands out as an exceptionally apt software tool for crafting flood inundation maps tailored to a broad spectrum of applications. This versatile model is designed to conduct analyses across both steady and unsteady flow conditions, as well as subcritical and supercritical flow regimes, making it highly adaptable to various hydrological scenarios. The synergy between HEC-RAS and GIS significantly streamlines the process of constructing model geometries and enhances the efficiency of output postprocessing (Kumar et al., 2020; Pathan & Agnihotri, 2020). The simulation capabilities of HEC-RAS yield critical data on water depth, flow velocity, and temporal dynamics of floods at specific locations, enriching the comprehension and management of flood risks. Incorporating historical flood records, data from extreme rainfall events, and advanced GIS tools like ArcGIS and QGIS, alongside satellite imagery, researchers have been able to augment the precision and depth of their analyses. Such comprehensive data integration facilitates a more nuanced visualization and analysis of flood phenomena, significantly elevating the accuracy and applicability of research findings in the realm of flood risk management and mitigation planning (Quirogaa et al., 2016; Rangari et al., 2019).

In this research, a detailed analysis is carried out to construct accurate flood inundation maps and systematically identify areas at high risk of flooding within the Netravati−Gurupura River basin. This effort utilizes the advanced capabilities of the HEC-RAS two-dimensional model (2D) version 6.3, a sophisticated tool designed for comprehensive flood simulation and analysis. By integrating hydrological data and employing this model, the study aims to enhance our understanding of flood dynamics and vulnerability in the region, providing vital information for effective flood risk management and mitigation strategies. The paper is structured into five sections. Section 7.2 elaborates on the study area and data inventory and outlines the proposed methodology. Section 7.3 elucidates about data inventory and proposed methodology. Section 7.4 highlights the results and discussion. Lastly, Section 7.5 provides the conclusions and future research.

7.2   Study area

Netravati−Gurupura River basin originates in the Western Ghats and covers most of Dakshina Kannada district of Karnataka state, southern India. It lies between the latitude of $12°29'11''$ N and $13°11'11''$ N and longitude of $74°49'08''$ E and $75°47'53''$ E (details provided in Fig. 7.1). Netravati−Gurupura originate at an elevation of 1100 m above the mean sea level. This basin is an area of significant hydrological interest due to its complex river system, which plays a critical role in the region's water resource management, agriculture, and ecology. The Netravati River originates in the Western Ghats, a biodiversity hotspot, and flows westward toward the Arabian Sea, covering a substantial distance and encompassing a diverse range of ecosystems. The Gurupura River, also originating in the Western Ghats, merges with the Netravati River near Mangalore, before they collectively discharge into the Arabian Sea. This confluence area is

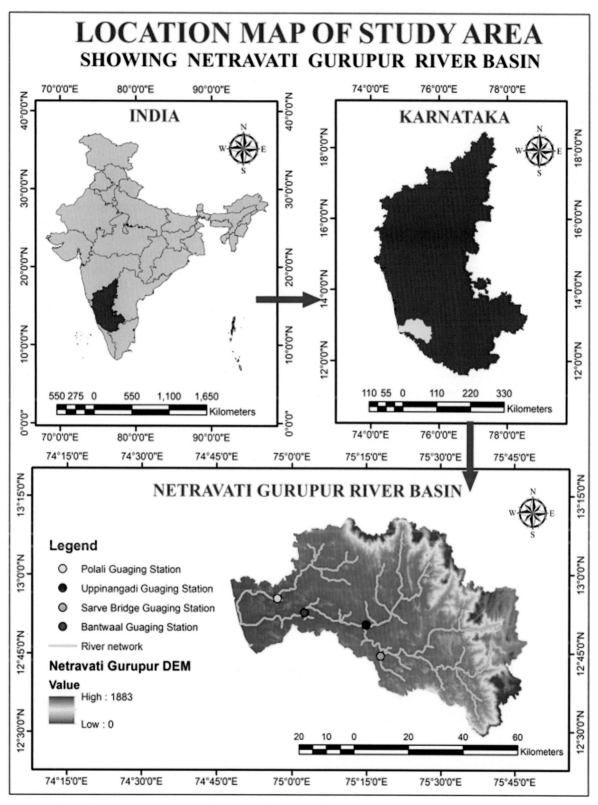

FIGURE 7.1 Details of the study area. Map lines delineate study areas and do not necessarily depict accepted national boundaries.

crucial for local fisheries, agriculture, and provides water for domestic and industrial use in the surrounding urban and rural areas. The basin's climate is tropical, with a significant portion of its rainfall occurring during the southwest monsoon season, leading to pronounced seasonal variations in river flow and water availability. The Netravati—Gurupura basin is also characterized by its varied topography, including highland regions with dense forest cover, midland zones with mixed agricultural use, and coastal plains that are highly productive but also vulnerable to flooding, especially during the monsoon season. These floods can affect agriculture, infrastructure, and livelihoods, making flood risk management a critical concern for the basin.

The total drainage area of the Netravati River basin is 3657 km^2 and that of the Gurupura river basin is 824 km^2. The major tributaries of the Netravati—Gurupura River are Kumaradhara, Shishila Hole, Gundiya Hole, and Neriya Hole. The climatic condition of the basin consists of wet months during June to September consisting of heavy showers, strong wind, and high humidity, October and November with little or no rains, cool months during December to February, followed by hot and humid climate with the rise in temperature from March to May. The annual rainfall in the basin is above 4000 mm, with the southwest monsoon being the most predominant, contributing 80% of annual rainfall. Northeast monsoon is not as dominant as the southwest monsoon, and most of the flow in the river basin is during June—September. Netravati river serves as the primary source of water supply for Mangaluru city, Bantwal, Beltangadi, Puttur, and other towns. The total population of the city is around 1.2 million. The city is in the confluence of Netravati and Gurupura rivers with an average elevation of 22 m. During monsoon, the Bantwal taluk of the Dakshina Kannada district is often submerged and affected by the flood. According to the residents, the major floods in Bantwal were in 1928 and 1974 (Sitharam, 2018).

7.3 Data inventory and methodology

7.3.1 Data inventory

For this comprehensive study, an array of diverse datasets sourced from various origins has been meticulously compiled and utilized. Essential for the development of detailed flood inundation maps, the required data encompasses a digital elevation model (DEM), measurements of river discharge, boundaries conditions at both upstream and downstream, and Manning's roughness coefficient "*n*." In the present study, 30-m spatial resolution Shuttle Radar Topography Mission DEM is utilized from the US Geological Survey. This DEM serves as the foundation for creating a digital terrain model (DTM), an essential component for the accurate execution of the HEC-RAS 2D model. The chosen DEM is carefully selected to ensure it captures critical geographical features of the study area, including rivers, dams, and other significant hydrological landmarks, thereby enhancing the model's precision in flood simulation. To construct a dynamic flow hydrograph, daily discharge data were obtained from the Central Water Commission, focusing on monitored gauging stations located at Bantwal across the Netravati River, and at Polali across the Gurupura river. Further augmenting the dataset, additional daily discharge data were acquired from gauging stations managed by the Karnataka Public Works Department at Uppinangady and Sarve Bridge along the Netravati River's upstream segment. These gauging stations are crucial for understanding the hydrological behavior of the river system and are meticulously detailed in Fig. 7.1.

7.3.2 Description of the Hydrologic Engineering Center's River Analysis System version 6.3 two-dimensional model setup

The HEC-RAS 2D version 6.3, developed by the US Army Corps of Engineers, is utilized in this analysis, showcasing its advanced capabilities in the realm of hydrodynamic modeling. This version introduces the HEC-RAS 2D module, specifically designed for intricate 2D flow routing. It adeptly handles both the comprehensive 2D Saint—Venant equations and the simplified 2D diffusive wave equations, providing flexibility based on analytical needs. The diffusive wave solver, by omitting the inertial terms from the momentum equations, simplifies the analysis without significantly compromising accuracy (Quirogaa et al., 2016; U.S. Army Corps of Engineers, 2021). HEC-RAS model uses Saint—Venant equation as governing one-dimensional mathematical equation given by Eqs. (7.1) and (7.2) (Chow et al., 1988; Timbadiya et al., 2011).

$$\frac{\partial A}{\partial t} + \frac{\partial Q}{\partial x} = \pm q \tag{7.1}$$

$$\frac{\partial Q}{\partial t} + \frac{\partial (Q^2/A)}{\partial x} + gA\frac{\partial H}{\partial x} + gA(S_o - S_f) = 0 \tag{7.2}$$

where A is the cross-sectional area normal to the flow (m^2), H is the elevation of the water surface above a specified datum (m), Q is the discharge (m^3s^{-1}), q is the lateral inflow/outflow in/from the channel (m^2s^{-1}), g is the acceleration due to gravity (ms^{-2}), S_o is the bed slope (dimensionless), S_f is the longitudinal boundary friction slope (dimensionless), t is the temporal coordinate (s), and x is longitudinal coordinate (m).

Two-dimensional Saint−Venant equations are given in Eqs. (7.3)−(7.5). While HEC-RAS is capable of processing both the Saint−Venant and diffusive wave equations, it has been observed that the outcomes of simulations are comparable in either scenario. However, simulations employing the diffusive wave equation tend to complete more faster than those utilizing the Saint−Venant equations (Quirogaa et al., 2016; U.S. Army Corps of Engineers, 2021).

$$\frac{\partial H}{\partial t} + \frac{\partial (hu)}{\partial x} + \frac{\partial (hv)}{\partial y} + q = 0 \tag{7.3}$$

$$\frac{\partial u}{\partial t} + u\frac{\partial u}{\partial x} + v\frac{\partial u}{\partial y} = -g\frac{\partial H}{\partial x} + v_t\left(\frac{\partial^2 u)}{(\partial x^2)} + \frac{\partial^2 u)}{(\partial y^2)}\right) - c_f u + fv \tag{7.4}$$

$$\frac{\partial v}{\partial t} + u\frac{\partial v}{\partial x} + v\frac{\partial v}{\partial y} = -g\frac{\partial H}{\partial y} + v_t\left(\frac{\partial^2 v)}{(\partial x^2)} + \frac{\partial^2 v)}{(\partial y^2)}\right) - c_f v + fu \tag{7.5}$$

where u and v are the velocity in the x- and y-directions, respectively ($m^2 s^{-1}$), q is a source/sink flux term ($m^2 s^{-1}$), g is the gravitational acceleration ($m s^{-2}$), v_t is the horizontal Eddy viscosity coefficient (dimensionless), c_f is the bottom friction coefficient (dimensionless), R is the hydraulic radius (m), and f is the Coriolis parameter (s^{-1}).

7.3.3 Proposed methodology

Fig. 7.2 provides a detail overview of the proposed methodology. To initiate the simulation using the HEC-RAS 2D model, a DTM of the study area is essential for accurately mapping the floodplain's geometry. This DTM is efficiently crafted by inputting the digital elevation model (DEM) into the RAS Mapper tool provided by HEC-RAS.

The creation of this DTM is crucial for the precise forecasting of flood inundation areas. In defining the flow area, care was taken to ensure the inclusion of the entire study region and its critical elements such as rivers, dams, canals, etc., into the model's framework. For the flow area, a mesh grid with dimensions of 100 m × 100 m was selected. Upon evaluating different mesh sizes (25 m, 50 m, 75 m, 100 m, and 200 m) it was determined that the model's accuracy remained relatively unchanged across these variations. However, a significant difference was noted in the simulation times; finer meshes led to longer simulation durations (Ongdas et al., 2020). This finding underscores the balance between mesh granularity and computational efficiency, highlighting the importance of selecting an optimal mesh size for effective and timely flood modeling. In this study, to create flood inundation maps for the Netravati−Gurupura River basin, historical flood events which caused severe flooding in the basin from six different years were analyzed for both river basins. Specifically, for the Netravati River basin, flood events from the years 1980, 1994, 2001, 2008, 2009, and 2013 were selected. Meanwhile, for the Gurupura river basin, the flood events chosen were from the years 2001, 2004, 2006, 2008, 2010, and 2014. For each selected year, daily discharge data from the respective gauging sites were utilized. The specifics of these data, including the source gauging sites and the methodology for data collection and analysis, are detailed in Table 7.1.

To start the simulation with the HEC-RAS 2D model, boundary conditions need to be defined for both the upstream and downstream areas of the study zone. These conditions can include normal depth, stage hydrograph, flow hydrograph, or a rating curve, depending on the specific requirements of the model setup. In this study, a flow hydrograph, constructed from daily discharge data, was applied as the boundary condition at the upstream end (as illustrated in Fig. 7.3). For the downstream boundary condition, a normal depth was chosen, with a channel slope specified at 0.0028, to facilitate the flow simulation process accurately. In conducting simulations with the HEC-RAS 2D model, it is crucial to define both the computation interval or time step and the interval for output mapping. These parameters play a significant role in determining the accuracy and resolution of the simulation results. A smaller computational interval is generally recommended as it allows water to flow through each computational cell more precisely, capturing the nuances of the hydrodynamic processes more effectively. This approach enhances the accuracy of the simulation by providing a more detailed picture of how water moves and accumulates across the modeled area. For this study, a computational interval of 15 min was selected. This choice reflects a balance between computational efficiency and the level of detail required to accurately simulate the flood dynamics within the Netravati−Gurupura River basin. By adopting this relatively short interval, the model is able to simulate the temporal variations in water flow and depth with high fidelity, ensuring that the resulting flood inundation maps are both reliable and informative.

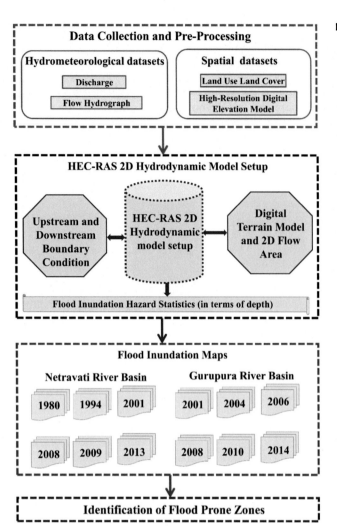

FIGURE 7.2 Proposed methodology.

TABLE 7.1 Daily discharge data of different years used in the study.

Basin	Year	Maximum discharge (m³/s)
Netravati Basin (Bantwal Gauging Station)	May 1, 1980 to October 1, 1980	6601
	May 1, 1994 to October 1, 1994	4846
	May 1, 2001 to October 1, 2001	2658
	May 1, 2008 to October 1, 2008	5610
	May 1, 2009 to October 1, 2009	4217
	May 1, 2013 to October 1, 2013	3949
Gurupura Basin (Polali Gauging Station)	May 1, 2001 to October 1, 2001	692
	May 1, 2004 to October 1, 2004	930
	May 1, 2006 to October 1, 2006	920
	May 1, 2008 to October 1, 2008	1042
	May 1, 2010 to October 1, 2010	1170
	May 1, 2014 to October 1, 2014	1083

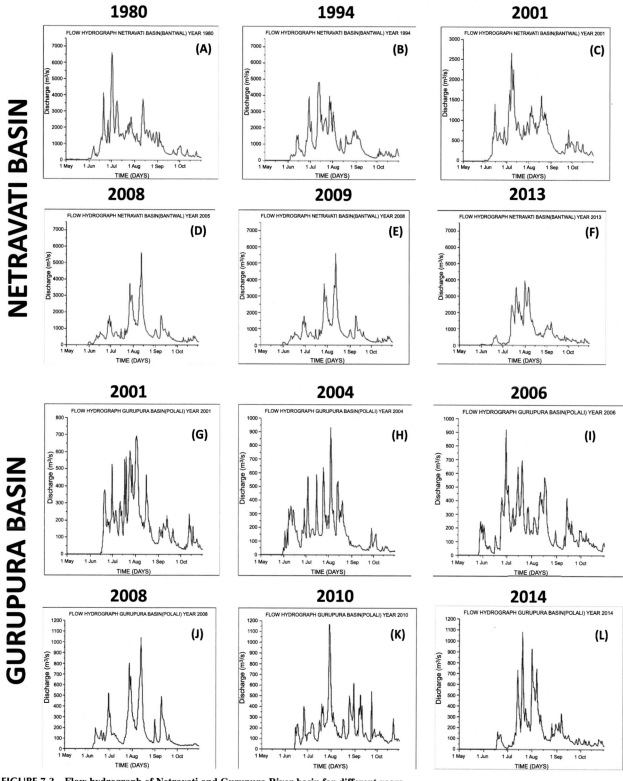

FIGURE 7.3 **Flow hydrograph of Netravati and Gurupura River basin for different years.**

7.3.3.1 Calibration of the model

In the context of hydrodynamic modeling with the HEC-RAS 2D model, setting the Manning's roughness coefficient "*n*" is a critical step, as it significantly influences the model's accuracy and reliability. The coefficient "*n*" reflects the channel's roughness, affecting how water flows and behaves within the model. Various values for Manning's roughness coefficient "*n*" are provided in Table 7.2, tailored to different types of channel conditions (Chow et al., 1988; Pathan & Agnihotri, 2020). For this study, particularly focused on the natural stream channels of the Netravati basin characterized by clean, straight streams, an initial "*n*" value of 0.030 was considered appropriate based on the guidelines in Table 7.2. However, recognizing the importance of precision in such simulations, the study did not settle for a fixed value without further scrutiny. Instead, Manning's roughness coefficient "*n*" was adjusted within a range from 0.025 to 0.035 to fine-tune the model's performance.

This adjustment process was meticulously carried out, with the model's output being evaluated against several statistical parameters to ensure the selection of an optimum "*n*" value that would yield the most accurate simulation results. To facilitate the model simulation, a computational mesh with a size of 100 m × 100 m was constructed for the flow area in the Netravati basin. The model utilized daily discharge data collected from the Uppinangady and Sarve bridge gauging sites during the period from June 1 to July 15, 2006, to develop a flow hydrograph. This period was chosen specifically for the simulation because the year 2006 experienced high flow levels in the river, making it a pertinent case study for flood modeling. The flow hydrograph derived from this data served as the upstream boundary condition for the simulation. At the downstream end, a normal depth condition was applied, with a channel slope of 0.0028, to accurately represent the physical conditions affecting water flow and to ensure the model's fidelity in predicting flood inundation and behavior.

The depth of water measured at the Bantwal gauging station was compared with the depth predicted by the model to evaluate its accuracy. This comparison involved assessing the variation between observed and simulated water depths through three statistical parameters as discussed below. Let O_i represent the observed depth at any given point, and P_i denote the simulated depth by the model at the same point. Further, let O_{bar} and P_{bar} be the mean values of the observed and simulated depths, respectively.

7.3.3.1.1 Coefficient of determination (R^2)

If O_i is the observed depth and P_i is the simulated depth from model simulation then, R^2 (shown in Eq. 7.6) is used to study the variation in simulated depth y when there is a variation in the observed depth x. The R^2 value ranges from 0 to 1, where a value closer to 1 indicates that the regression model closely fits the data. Conversely, a value near 0 suggests a poor fit, implying little to no correlation between the observed and simulated depths. This parameter is crucial for understanding the predictive power of the model and its ability to replicate real-world water depth variations accurately (Legates & McCabe, 1999; Waseem et al., 2008).

$$R^2 = \left(\frac{\sum_{i=1}^{n} (O_i - O_{bar})(P_i - P_{bar})}{\sqrt{\sum_{i=1}^{n} (O_i - O_{bar})^2} \sqrt{\sum_{i=1}^{n} (P_i - P_{bar})^2}} \right)^2 \tag{7.6}$$

TABLE 7.2 Manning's roughness coefficient "*n*" values for different channels.

Channel description	Manning's roughness coefficient values
Concrete material channel	0.012
Channel with bottom made up of gravel	0.020–0.033
Natural stream channels with clean and straight stream	0.030
Natural stream channels with clean and winding stream	0.040
Natural stream channels winding with weeds and pools	0.050
Flood plain consisting of field crops	0.040
Flood plain consisting of light brush and weeds	0.050
Flood plain consisting of dense brush	0.070
Flood plain consisting of dense trees	0.100

7.3.3.1.2 Nash−Sutcliffe efficiency

Nash−Sutcliffe efficiency (NSE) (shown in Eq. 7.7) serves as a widely recognized metric for evaluating the goodness of fit of hydrological and hydrodynamic models. It is a standard measure used to determine the accuracy of model simulations in comparison to observed data. NSE values range from $-\infty$ to $+1$, where a value of $+1$ indicates perfect agreement between model simulations and observations, suggesting an ideal model performance. Values closer to 0 imply that the model predictions are no better than a simple mean of the observed data, and negative values indicate that the model performs worse than the average observed data, highlighting a potentially biased or inaccurate model. This efficiency criterion is essential for assessing whether a model can reliably replicate observed conditions, thus guiding model selection and refinement in hydrological studies (McCuen et al., 2006).

$$NSE = 1 - \frac{\sum_{i=1}^{n}(O_i - P_i)^2}{\sum_{i=1}^{n}(O_i - O_{bar})^2} \tag{7.7}$$

7.3.3.1.3 Index of agreement (*d*)

The index of agreement (shown in Eq. 7.8) is a statistical tool used to quantify the degree to which model predictions deviate from error-free observations. It calculates the ratio between the mean square error observed in model predictions and the potential error, with its values spanning from 0 to 1. A value of 1 in the index of agreement indicates a perfect match between observed and simulated data, suggesting that the model predictions align precisely with real-world measurements. Conversely, a value of 0 signifies that there is no correlation between the observed and predicted data, pointing to significant discrepancies in the model's predictive accuracy. This measure is crucial for evaluating the reliability and precision of hydrological models, ensuring that they are capable of producing results that closely mirror the observed phenomena (Waseem et al., 2008; Willmott, 1981).

$$d = 1 - \frac{\sum_{i=1}^{n}(O_i - P_i)^2}{\sum_{i=1}^{n}(|P_i - O_{bar}| + |O_i - O_{bar}|)^2} \tag{7.8}$$

In light of the statistical parameters discussed, the selection of Manning's roughness coefficient is fine-tuned to ensure optimal outcomes for the coefficient of determination (R^2), NSE, and the index of agreement. The model's performance is deemed satisfactory if it meets specific benchmarks: an NSE value exceeding 0.50, an R^2 value above 0.6, and an index of agreement approaching 1. These criteria, as outlined by Moriasi et al. (2015) and Sowah et al. (2020), serve as a guideline to ascertain the reliability and accuracy of the model in simulating real-world water flow and flood inundation scenarios.

7.3.4 Flood inundation mapping and identification of flood-prone zones

Following the unsteady flow analysis, key outputs including water surface elevation, depth, and velocity were acquired from the model. These critical parameters facilitated the generation of detailed flood inundation maps that illustrate variations in depth, velocity, and water surface elevation. This was accomplished by exporting the simulation results from the HEC-RAS 2D model into ArcGIS, a process that enables the visual representation of flood dynamics over the landscape. Subsequent to the simulation exercises, the areas within the study region that experienced flooding during the specified years were accurately delineated. Particularly, areas in close proximity to the river's path that showed a recurring pattern of inundation were pinpointed and designated as potential flood-prone zones. This identification process is crucial for understanding the spatial distribution of flood risks within the river basin, allowing for targeted interventions and the development of effective flood risk management strategies. Through this analysis, stakeholders and decision-makers are equipped with actionable insights to mitigate flood impacts and enhance resilience against future flood events.

7.3.4.1 Validation of model

Upon creating the flood maps, their accuracy is evaluated to gauge the model's effectiveness in precisely predicting flood extent and depth. This validation process is critical for assessing the reliability of the flood simulation outcomes. It involves comparing the model-generated flood data with real-world flood information collected from a comprehensive flood validation survey conducted before the analysis. This survey gathers detailed data on the actual extent of flooding and the depth of floodwaters experienced in the area. Additionally, validation can be supplemented by leveraging high flood level data, when such data are accessible. This approach adds another layer of verification, ensuring that the model's predictions align closely with observed flood events. For this study, the model's performance and the accuracy of its flood inundation predictions were validated for the years 2007 and 2014. This step is indispensable for confirming the model's capability to serve as a reliable tool in flood risk management, planning, and mitigation efforts, enhancing the preparedness and response strategies for vulnerable regions.

7.4 Results and Discussion

7.4.1 Calibration and validation of the model setup

Using a mesh size of 100 m $\times$ 100 m resulted in the creation of 13,274 cells within the 2D flow area for the model's calibration process. Manning's roughness coefficient "n" was adjusted between 0.025 and 0.035, and for each value, the corresponding statistical parameters were computed and are presented in Table 7.3. The simulation revealed that a Manning's roughness coefficient of "$n = 0.032$" delivered the best performance, achieving a coefficient of determination (R^2) of 0.83, which surpasses the benchmark of 0.6, a NSE of 0.59, exceeding the minimum acceptable value of 0.5, and an index of agreement (d) of 0.87, approaching the ideal value of 1. Consequently, an "n" value of 0.032 was selected for the study area based on these optimal outcomes. Fig. 7.4 shows the variation in the observed and simulated depth for different values of "n."

The model's performance was critically assessed using the statistical parameters discussed in Section 7.3.3.1 of the chapter. The observed and simulated depth variations were graphically represented, as illustrated in Fig. 7.5. The statistical parameters' values for both years, as listed in Table 7.4, confirmed the model's satisfactory performance. This was evidenced by the coefficient of determination (R^2) exceeding 0.6, the NSE surpassing 0.5, and the index of agreement being close to 1, aligning with the thresholds established for acceptable model accuracy. These results indicate that the model is reliable for predicting flood conditions in the studied river basins. The satisfactory performance of the HEC-RAS 2D model in this study resonates with findings from several past research efforts. For instance, studies such as those by Costabile et al. (2021) and Ongdas et al. (2020) have demonstrated the model's robustness in simulating flood dynamics accurately across different geographical settings and flood events. Such studies, collectively, reinforce the confidence in HEC-RAS 2D as a reliable tool for flood risk assessment and align with the current study's findings, where the model's predictions fell within acceptable statistical ranges, indicating a high level of accuracy in simulating flood events for the Netravati−Gurupura River basin.

7.4.2 Flood inundation map and identification of flood-prone zones

Flood inundation maps were developed for the Netravati and Gurupura river basins, as illustrated in Figs. 7.6 and 7.7, respectively, for various years following the model simulation. The simulations also yielded data on the variation in

TABLE 7.3 Calibration results for variation of Manning's roughness coefficient "n."

Manning's roughness coefficient "n"	Nash−Sutcliffe efficiency	Coefficient of determination, R^2	Correlation coefficient, R	Index of agreement
0.025	0.451	0.828	0.910	0.835
0.026	0.362	0.746	0.834	0.811
0.027	0.461	0.799	0.894	0.833
0.028	0.356	0.772	0.879	0.805
0.029	0.462	0.808	0.899	0.831
0.030	0.434	0.799	0.894	0.828
0.031	0.471	0.801	0.895	0.829
0.032	0.588	0.827	0.909	0.873
0.033	0.469	0.793	0.891	0.834
0.034	0.515	0.808	0.899	0.843
0.035	0.463	0.786	0.887	0.824

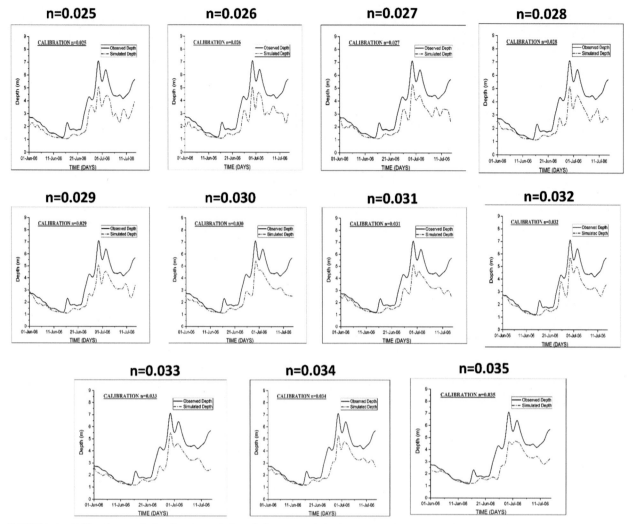

FIGURE 7.4   Calibration result showing the variation of observed and simulated depth for $n = 0.025$ to $n = 0.035$.

depth, velocity, and water surface elevation for both basins. Based on the depth variation, the inundated areas were categorized into three distinct regions: areas with a depth of less than 1 m, areas where the depth ranging from 1 to 5 m, and areas with a depth exceeding 5 m. These flood inundation maps are instrumental in identifying locations within the study area that have experienced flooding during extreme flood events in the past, providing crucial insights for future flood risk management and mitigation planning. These visualizations provide vital information for understanding the historical impact of flooding in these basins and assist in identifying areas that are repeatedly affected and are thus at higher risk. The maps serve as essential tools for local authorities and planners in developing targeted flood mitigation and adaptation strategies.

The flood inundation maps generated for the Netravati and Gurupura river basins have been instrumental in pinpointing areas that are vulnerable to flooding. The analysis delineated several potential flood-prone zones, as detailed in Fig. 7.8 for the Netravati River basin and Gurupura River basin.

Within the Netravati River basin, the localities of Amblamogru, Kannur, Adyar, and Arkula in Mangaluru taluk have been identified as high-risk zones. Additionally, the regions of Farangipete, Bantwal, Munnuru, Maninalkur, and Arikala in Bantwal taluk; Mogru and Niddle in Belthangady taluk; along with Bellipadi and Valal in Puttur taluk are also recognized as areas prone to flood threats. As for the Gurupura river basin, the zones marked as at an increased risk of flooding include Baikampady, Malavoor, Kolambe, Mulur, Badagaulipady, Tanilaulipady, Kenjar, and Ervuailu in Mangaluru taluk; Kariyangala and Karpe in Bantwal taluk; and Puchamogaru in Belthangady taluk. These identified zones are situated in close proximity to the river course, rendering them particularly susceptible to flooding. The mapping and identification of these areas are crucial for local authorities and communities, providing them with the

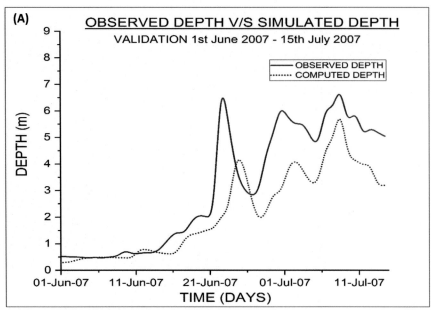

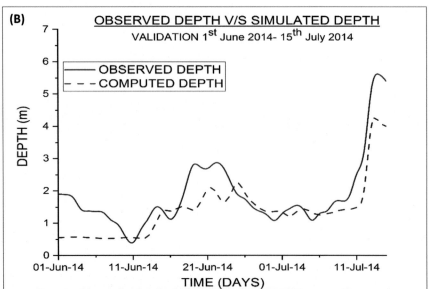

FIGURE 7.5 Validation result showing the variation of observed depth and simulated depth.

TABLE 7.4 Validation results.

Year	Nash–Sutcliffe efficiency	Coefficient of determination, R^2	Correlation coefficient, R	Index of agreement
2007	0.582	0.778	0.881	0.859
2014	0.566	0.742	0.861	0.865

necessary information to prepare and respond effectively to potential flood events. The highlighted zones should be a focal point for the development and implementation of flood prevention and mitigation strategies, aiming to minimize the impact of floods and safeguard the communities residing in these regions.

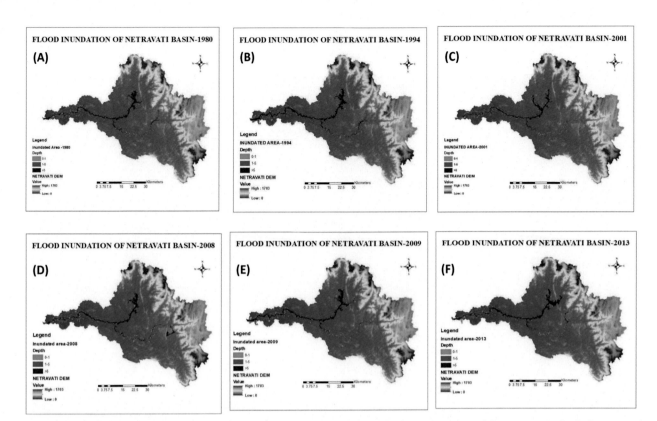

FIGURE 7.6 Flood inundation maps for the Netravati River basin. Map lines delineate study areas and do not necessarily depict accepted national boundaries.

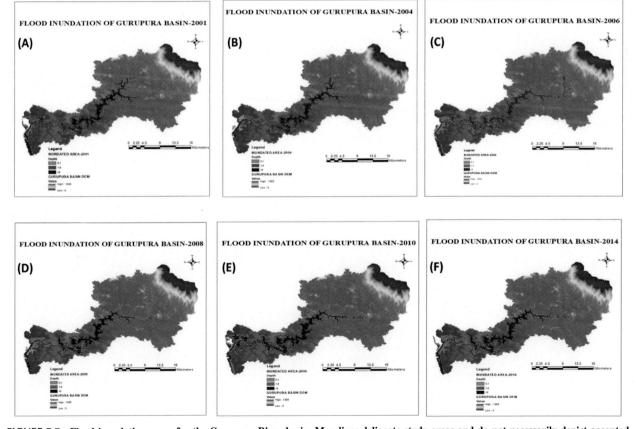

FIGURE 7.7 Flood inundation maps for the Gurupura River basin. Map lines delineate study areas and do not necessarily depict accepted national boundaries.

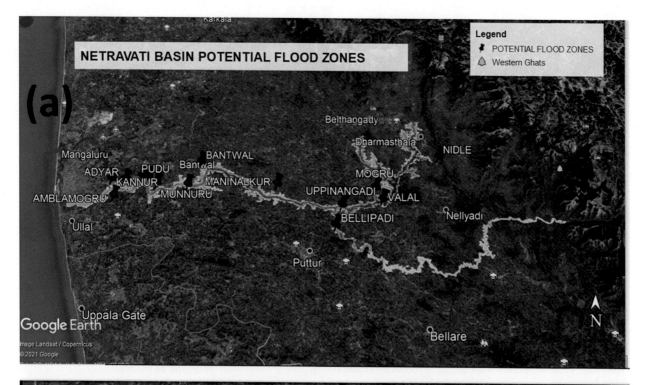

FIGURE 7.8 Flood-prone zones of Netravati and Gurupura River basin. Map lines delineate study areas and do not necessarily depict accepted national boundaries.

7.5 Conclusions

In the current study, the HEC-RAS 2D model was utilized to generate flood inundation maps and to pinpoint flood-prone zones for the Netravati and Gurupura river basins. This effort involved simulating various flood scenarios, characterized by high discharges, to assess the potential impact on these regions. Flow hydrographs were used to set the upstream boundary conditions, while a fixed normal depth, with a channel slope of 0.0028, served as the downstream boundary. The integration of the simulation outputs with geographic information systems like ArcGIS and visualization tools such as Google Earth Pro provided a clear and detailed representation of the flood extents. A key factor in the model's accuracy was the calibration of Manning's roughness coefficient "n." After careful evaluation, an optimal value of $n = 0.032$ was determined, which aligned with satisfactory model performance indicators such as R2, NSE, and the index of agreement.

This study, through the use of the HEC-RAS 2D model, offers a vital tool for disaster preparedness and mitigation in the Netravati and Gurupura river basins. By creating detailed flood inundation maps and identifying zones vulnerable to flooding, it equips local authorities and stakeholders with the necessary information to prioritize areas for intervention. The precise calibration of Manning's roughness coefficient ensures the reliability of the model, making the results a dependable basis for planning. The maps provide a visual assessment of potential flood extents, enabling emergency response teams to strategize evacuations and deploy resources effectively. Urban planners and engineers can use this data to design infrastructure resilient to predicted flood levels and to enhance existing flood defenses. Additionally, environmental agencies can better understand the impact of floods on local ecosystems and biodiversity, allowing for the development of conservation strategies.

Moreover, the study can inform the community about flood risks, fostering awareness and self-preparedness. The study recognizes the constraint posed by the scarcity of ground-based data for mapping flood inundation, suggesting that future research could benefit from improved data availability. Insurance companies might also use this information to adjust policies and premiums accurately. Overall, the study aids in creating a more flood-resilient society by laying the groundwork for comprehensive flood risk management and proactive planning to reduce future flood damage.

Acknowledgments

This study is a part of Science and Engineering Research Board of Department of Science and Technology, Govt of India funded project titled "Impounding of River flood waters along Dakshina Kannada Coast: A Sustainable strategy for water resource development" with project sanction No. IMP/2018/001298. The authors would like to thank Central Water Commission, India, and Karnataka Public Work Department, Karnataka, India, for providing the access to discharge data.

References

Abdelkrim, Z., Nouibat, B., & Eslamian, S. (2023). *Hydrological-hydraulic modeling of floodplain inundation: A case study in Bou Saada Wadi, a Subbasin, Algeria. Water Data Management Best Practices by Eslamian* (3, pp. 219–232). Elsevier.

Chow, V., Maidment, D. R., & Mays, L. W. (1988). *Development of Hydrology Applied Hydrology.* McGraw-Hill Book Company.

Costabile, P., Costanzo, C., Ferraro, D., & Barca, P. (2021). Is HEC-RAS 2D accurate enough for storm-event hazard assessment? Lessons learnt from a benchmarking study based on rain-on-grid modelling. *Journal of Hydrology, 603.* Available from https://doi.org/10.1016/j.jhydrol.2021.126962.

Dhar, O. N., & Nandargi, S. (2003). Hydrometeorological aspects of floods in India. *Natural Hazards, 28*(1), 1–33. Available from https://doi.org/10.1023/A:1021199714487.

Kumar, N., Kumar, M., Sherring, A., Suryavanshi, S., Ahmad, A., & Lal, D. (2020). Applicability of HEC-RAS 2D and GFMS for flood extent mapping: A case study of Sangam area, Prayagraj, India. *Modeling Earth Systems and Environment, 6*(1), 397–405. Available from https://doi.org/10.1007/s40808-019-00687-8.

Legates, D. R., & McCabe, G. J. (1999). Evaluating the use of 'goodness-of-fit' measures in hydrologic and hydroclimatic model validation. *Water Resources Research, 35*(1), 233–241. Available from https://doi.org/10.1029/1998WR900018, http://onlinelibrary.wiley.com/journal/10.1002/(ISSN)1944-7973.

Liu, Z., Merwade, V., & Jafarzadegan, K. (2019). Investigating the role of model structure and surface roughness in generating flood inundation extents using one- and two-dimensional hydraulic models. *Journal of Flood Risk Management, 12*(1). Available from https://doi.org/10.1111/jfr3.12347.

McCuen, R. H., Knight, Z., & Cutter, A. G. (2006). Evaluation of the Nash–Sutcliffe efficiency index. *Journal of Hydrologic Engineering, 11*(6), 597–602. Available from https://doi.org/10.1061/(ASCE)1084-0699(2006)11:6(597).

Mohanty, M. P., & Karmakar, S. (2021). WebFRIS: An efficient web-based decision support tool to disseminate end-to-end risk information for flood management. *Journal of Environmental Management, 288*, 112456.

Mohanty, M. P., Mudgil, S., & Karmakar, S. (2020). Flood management in India: A focussed review on the current status and future challenges. *International Journal of Disaster Risk Reduction, 49*, 101660.

Moriasi, D. N., Gitau, M. W., Pai, N., & Daggupati, P. (2015). Hydrologic and water quality models: Performance measures and evaluation criteria. *Transactions of the ASABE, 58*(6), 1763–1785.

Ongdas, N., Akiyanova, F., Karakulov, Y., Muratbayeva, A., & Zinabdin, N. (2020). Application of HEC-RAS (2D) for flood hazard maps generation for Yesil (Ishim) river in Kazakhstan. *Water (Switzerland), 12*(10), 1–20. Available from https://doi.org/10.3390/w12102672, http://www.mdpi.com/journal/water.

Pathan, A. K. I., & Agnihotri, P. G. (2020). 2-D unsteady flow modelling and inundation mapping for lower region of Purna basin using HEC-RAS. *Nature Environment and Pollution Technology, 19*(1), 277–285. Available from http://neptjournal.com/upload-images/(28)B-3622.pdf.

Piadeh, F., Behzadian, K., & Alani, A. M. (2022). A critical review of real-time modelling of flood forecasting in urban drainage systems. *Journal of Hydrology, 607*. Available from https://doi.org/10.1016/j.jhydrol.2022.127476.

Pinos, J., & Timbe, L. (2019). Performance assessment of two-dimensional hydraulic models for generation of flood inundation maps in mountain river basins. *Water Science and Engineering, 12*(1), 11–18. Available from https://doi.org/10.1016/j.wse.2019.03.001.

Prasad, R. K. Certified Organization Volume. In International Journal of Innovative Research in Science, Engineering and Technology An ISO. 3297 (2016), 2016.

Quirogaa, V. M., Kurea, S., Udoa, K., & Manoa, A. (2016). Application of 2D numerical simulation for the analysis of the February 2014 Bolivian Amazonia flood: Application of the new HEC-RAS version 5. *Ribagua, 3*(1), 25–33. Available from https://doi.org/10.1016/j.riba.2015.12.001.

Rangari, V. A., Sridhar, V., Umamahesh, N. V., & Patel, A. K. (2019). Floodplain mapping and management of urban catchment using HEC-RAS: A case study of Hyderabad city. *Journal of The Institution of Engineers (India): Series A, 100*(1), 49–63. Available from https://doi.org/10.1007/s40030-018-0345-0, http://www.springer.com/engineering/civil + engineering/journal/40030.

Ray, K., Pandey, P., Pandey, C., Dimri, A. P., & Kishore, K. (2019). On the recent floods in India. *Current Science, 117*(2), 204–218. Available from https://doi.org/10.18520/cs/v117/i2/204-218, https://www.currentscience.ac.in/Volumes/117/02/0204.pdf.

Sitharam, Feasibility Study on Coastal Reservoir Concept to Impound Netravati River Flood Waters Feasibility Study on Coastal Reservoir Concept to Impound Netravati River Flood Waters: A Sustainable Strategy for Water Resource Development for Mangaluru and Bengaluru. (2018), 2018.

Sowah, R. A., Bradshaw, K., Snyder, B., Spidle, D., & Molina, M. (2020). Evaluation of the soil and water assessment tool (SWAT) for simulating E. coli concentrations at the watershed-scale. *Science of the Total Environment, 746*, 140669.

Thakur, D. A., & Mohanty, M. P. (2023). A synergistic approach towards understanding flood risks over coastal multi-hazard environments: Appraisal of bivariate flood risk mapping through flood hazard, and socio-economic-cum-physical vulnerability dimensions. *Science of the Total Environment, 901*. Available from https://doi.org/10.1016/j.scitotenv.2023.166423, http://www.elsevier.com/locate/scitotenv.

Thakur, D. A., & Mohanty, M. P. (2024). Exploring the fidelity of satellite precipitation products in capturing flood risks: A novel framework incorporating hazard and vulnerability dimensions over a sensitive coastal multi-hazard catchment. *Science of the Total Environment, 920*. Available from https://doi.org/10.1016/j.scitotenv.2024.170884, https://www.sciencedirect.com/science/journal/00489697.

Thakur, D. A., Mohanty, M. P., Mishra, A., & Karmakar, S. (2024). Quantifying flood risks during monsoon and post-monsoon seasons: An integrated framework for resource-constrained coastal regions. *Journal of Hydrology, 630*. Available from https://doi.org/10.1016/j.jhydrol.2024.130683, https://www.sciencedirect.com/science/journal/00221694.

Timbadiya, P. V., Patel, P. L., & Porey, P. D. (2011). HEC-RAS based hydrodynamic model in prediction of stages of lower Tapi River. *ISH Journal of Hydraulic Engineering, 17*(2), 110–117. Available from https://doi.org/10.1080/09715010.2011.10515050.

U.S. Army Corps of Engineers (2021). HEC-RAS, Hydraulic Reference Manual, Version 6.0.

Waseem, M., Mani, N., Andiego, G., & Usman, M. (2008). A review of criteria of fit for hydrological models. *International Research Journal of Engineering and Technology, 9001*, 2008.

Willmott, C. J. (1981). On the validation of models. *Physical Geography, 2*(2), 184–194. Available from https://doi.org/10.1080/02723646.1981.10642213.

Chapter 8

Simulation of river flow (as a primary component for aquifer recharge) using deep learning approach

Mohammad Javad Zareian[1] and Fatemeh Salem[2]

[1]*Department of Water Resources Study and Research, Water Research Institute (WRI), Tehran, Iran*, [2]*Faculty of Computer Science and Engineering, Shahid Beheshtei University, Tehran, Iran*

8.1 Introduction

In recent decades, hydro-meteorological scientists have increasingly focused on the critical issue of climate change and its far-reaching impacts on the global ecosystem. These effects have triggered a spectrum of alterations in climatic, hydrological, and environmental variables worldwide, with projections indicating continued shifts in the coming decades (Berrang-Ford et al., 2011; Intergovernmental Panel on Climate Change [IPCC], 2013).

Climate change profoundly influences the hydro-climatic characteristics of the Earth, particularly the meteorological variables. Among them, key variations are in temperature, precipitation, evapotranspiration, and more (Georgescu et al., 2021; IPCC, 2013). Indirectly, climate change has affected noteworthy shifts in surface flows globally, leading to altered occurrences of floods or prolonged droughts (Nearing et al., 2004; Tucker & Slingerland, 1997). Furthermore, climate change has escalated demands for agricultural and domestic water (Eslamian et al., 2017; Zareian, 2021). This confluence of factors, coupled with population growth, positions climate change as a limiting factor for water supply in the forthcoming years. Consequently, water resources managers and policymakers are keen on obtaining predictive insights into future water resource conditions within the context of climate change, facilitating sustainable planning and management.

Understanding surface flow is crucial for both water allocation and groundwater studies (Green et al., 2011; Okhravi et al., 2017). In many regions, surface flows serve as the primary water supply for diverse demands, stored and distributed through various means, including dams. However, climate change, besides impacting surface flows directly, may also alter the temporal distribution pattern of these flows annually (Wilby & Keenan, 2012). Thus, it is imperative to scrutinize the effects of climate change on surface flows, especially in areas facing high water stress. Also, rivers play a crucial role in maintaining the overall hydrological balance of ecosystems, and their significance extends to influencing the groundwater recharge. The flow of rivers contributes significantly to the replenishment of underground aquifers, playing a vital role in sustaining groundwater levels. As rivers traverse landscapes, they collect rainwater and runoff, gradually percolating into the soil and infiltrating the groundwater reservoirs. This process not only ensures the availability of freshwater for various ecosystems but also directly impacts the quality and quantity of groundwater. Groundwater, in turn, serves as a vital source for drinking water, agricultural irrigation, and supporting ecosystems during dry periods. The interconnectedness of surface water flow and groundwater recharge highlights the critical importance of rivers in maintaining a sustainable water cycle, emphasizing the need for conservation and responsible management of river systems to safeguard both surface water and groundwater resources.

Attributed primarily to the escalating concentrations of greenhouse gases, climate change results from heightened human industrial activities, increased use of fossil fuels, and deforestation. The IPCC (2007) stands as the foremost reference in climate studies, disseminating outcomes through General Circulation Model (GCM) outputs, predicting Earth's future climate conditions under various emission scenarios (IPCC, 2013).

Hydrosystem Restoration Handbook. DOI: https://doi.org/10.1016/B978-0-443-29802-8.00008-X

Beyond forecasting the impacts of climate change on climate variables, such as temperature and precipitation, understanding its effects on meteorological variables necessitates examining their relationship with surface flows. Consequently, employing different statistical or mathematical methods becomes essential. While runoff modeling traditionally relies on complex and nonlinear conceptual models, recent years have witnessed the application of intelligent methods, including deep learning techniques, to model relationships between meteorological and hydrological variables. This approach, a subset of artificial neural networks, involves complex algorithms modeled and solved at multiple levels and layers, offering accelerated calculations and acceptable accuracy in hydrological modeling (Frame et al., 2022; Goodfellow et al., 2016; Voulodimos et al., 2018).

Numerous studies have explored the efficacy of deep learning methods in various hydrological modeling scenarios. For instance, Hussain et al. (2020) successfully employed a one-dimensional convolutional neural network (CNN) to predict the flow of the Gilgit River, Pakistan. Similarly, Liu et al. (2020) used a deep neural network to simulate the flow of the Yangtze River, China, combining Empirical Mode Decomposition and Encoder Decoder Long Short-Term Memory (En-De-LSTM) algorithms to enhance prediction accuracy. Lee et al. (2020) compared the predictive abilities of the LSTM model with the Soil and Water Assessment Tool (SWAT) model for the Mekong River flow, South Korea, under different emission scenarios, with LSTM providing superior estimates. Qian et al. (2019) demonstrated the effectiveness of a deep learning network in predicting floods in Austin, Texas, based on a solver based on Snow Water Equivalent. Eltner et al. (2021) utilized a deep learning network to improve river water level measurements' accuracy through photogrammetric techniques. Zhu et al. (2023) used a spatio-temporal runoff-rainfall forecasting model incorporating spatial information from high-resolution satellite precipitation products. The results showed that the accuracy of runoff prediction will increase significantly using this method.

Moreover, Wu et al. (2020) utilized the Gradient Boosting Decision Tree model to simulate floods in the Zhengzhou City urban basin, China, achieving a relative error (RE) of 11.5%. Barzegar et al. (2020) employed a hybrid deep learning model, combining LSTM and CNN models, to simulate water quality in Small Prespa Lake, Greece, with significantly enhanced accuracy.

The Eskandari Basin, situated in the arid and semiarid regions of central Iran within the Zayandeh-Rud Basin, plays a crucial role in surface water resource provision. The Eskandari Basin, a tributary of the Zayandeh-Rud River, significantly impacts the river's water flow. As a result, fluctuations in precipitation and temperature within the basin could greatly affect the flow of the Zayandeh-Rud River. Therefore implementing methods to link these changes in precipitation and temperature to variations in river flow within this region becomes highly important and practical. Given the region's significance in water resources management, there is a pressing need for a rapid and accurate tool to simulate streamflow in the Eskandari Basin and anticipate its changes over the next 18 years (2023−38) based on the Sixth Intergovernmental Panel on Climate Change assessment report projections. This study aims to evaluate the efficiency of CNN for this purpose, contributing valuable insights to the intersection of climate change and water resource dynamics.

8.2 Method

8.2.1 Study area

The research focuses on the Eskandari Basin, situated in the western part of the Zayandeh-Rud Basin, with coordinates ranging from $50°02'$ to $50°40'$ longitude and $32°11'$ to $32°45'$ latitude. Encompassing an area of 1649 km^2, this region is positioned in one of Iran's central and semiarid zones. The Pelasjan River, a significant tributary of the strategically important Zayandeh-Rud River, flows through this basin, and its discharge is regularly gauged at the Eskandari hydrometric station. With an average annual discharge of 131 million cubic meters, the Pelasjan River plays a pivotal role in the hydrology of the area. Fig. 8.1 provides an overview of the study area, emphasizing its geographical context. Monthly climate data from the Damneh synoptic station, including minimum and maximum air temperatures (T_{min} and T_{max}), precipitation (Pr), relative humidity (RH), and sunshine (N), along with the average monthly flow (Q) data from the Eskandari hydrometric station spanning the years 1981−2015, were utilized for the research analysis. Table 8.1 outlines the characteristics of the mentioned meteorological and hydrometric stations. Given the evident decrease in streamflow in recent years due to frequent droughts, there is a pressing need to forecast the impact of climate change on streamflow in the Eskandari Basin. Such predictions are crucial for sustainable water resources management, especially in the context of escalating water stress and existing water conflicts in the broader Zayandeh-Rud Basin (Gohari et al., 2014; Zareian et al., 2017).

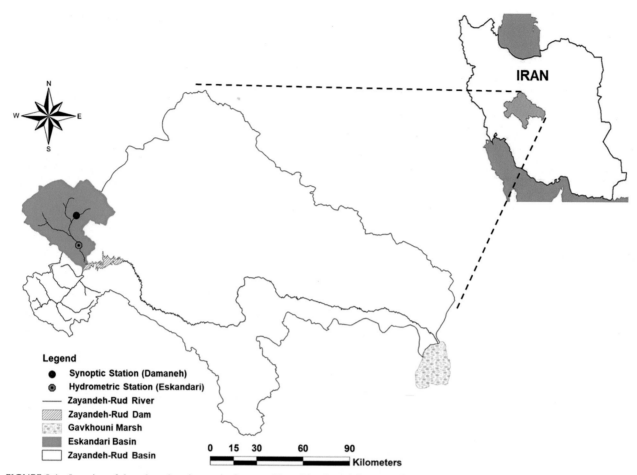

FIGURE 8.1 Location of the selected study area in the central Iran. Map lines delineate study areas and do not necessarily depict accepted national boundaries.

TABLE 8.1 Characteristics of the selected synoptic and hydrometric stations.

Latitude	Longitude	Type	Station
33°01′	50°29′	Synoptic station	Damaneh
32°49′	50°25′	Hydrometric station	Eskandari

8.2.2 Convolutional neural network

The CNN stands as a widely adopted architecture for enhancing various machine learning tasks (Cui & Fearn, 2018; Everingham et al., 2010). Functioning as a neural network integrated with deep learning operations, CNN utilizes advanced methods to train multiple information layers. Typically, a CNN comprises three primary layers: convolutional, pooling, and fully connected layers (Rawat & Wang, 2017). While physically based modeling offers detailed rainfall-runoff simulations, it requires extensive basin data (e.g., topography, land use, and soil moisture). These methods can be time-consuming, and inaccurate data may limit their effectiveness. Simpler approaches, like deep learning, can provide acceptable results for rapid applications, reducing modeling time.

This research entails the creation of seven distinct structures derived from the permutations of monthly minimum and maximum air temperatures (T_{min} and T_{max}), precipitation (Pr), relative humidity (RH), and sunshine (N) as predictor variables, alongside average monthly flow (Q) serving as the target variable (Table 8.2). For the calibration step, approximately 75% of the data were utilized, with the remaining 25% allocated for testing purposes. The efficacy of

CNN in simulating streamflow based on diverse structures was assessed using the following evaluation indices (Eqs. 8.1−8.3)

$$R^2 = \frac{\sum_{m=1}^{M} \left[\left(Q_m^o - \overline{Q}_m^o \right) \left(Q_m^s - \overline{Q}_m^s \right) \right]}{\sqrt{\sum_{m=1}^{M} \left(Q_m^o - \overline{Q}_m^o \right)^2 \left(Q_m^s - \overline{Q}_m^s \right)^2}} \tag{8.1}$$

$$MAE = \frac{1}{M} \sum_{m=1}^{M} \left| Q_m^o - Q_m^s \right| \tag{8.2}$$

$$RMSE = \sqrt{\frac{1}{M} \sum_{m=1}^{M} \left(Q_m^o - Q_m^s \right)^2} \tag{8.3}$$

where Q_m^o and Q_m^s are the observed and the predicted streamflow, respectively; $\overline{Q}_m^o$ and $\overline{Q}_m^s$ are the mean values of the observed and predicted streamflow, respectively; and M represents the number of data sets.

8.2.3 Assessing the effects of climate change in the study area

To assess the impact of climate change on temperature and precipitation in the upcoming period (2023−38) within the study area, the outputs of seven GCMs from the sixth IPCC assessment report were employed. The features of these models are detailed in Table 8.3 (O'Neill et al., 2016).

TABLE 8.2 Different combinations inputs for the CNN model (model structures).

Input structure	Name
M1	T_{min}, T_{max}, Pr, Q
M2	T_{min}, T_{max}, N, RH, Q
M3	T_{min}, T_{max}, Pr, N, Q
M4	T_{min}, T_{max}, Pr, RH, Q
M5	Pr, N, RH, Q
M6	T_{min}, T_{max}, RH, Q
M7	T_{min}, T_{max}, N, Q

TABLE 8.3 Characteristics of GCMs used in this study.

Developer	Resolution	Model
Beijing Climate Center, China Meteorological Administration	1.12° × 1.12°	BCC-CSM2-MR
Community Earth System Model Contributors	1.25° × 0.94°	CESM2
Centre National de Recherches Météorologiques Scientifique (CNRM)	1.40° × 1.40°	CNRM-CM6−1
Canadian Centre for Climate Modelling and Analysis-Canada	2.81° × 2.81°	CanESM5
National Institute for Environmental Studies, The University of Tokyo	1.40° × 1.40°	MIROC6
Meteorological Research Institute	1.12° × 1.12°	MRI-EMS2−0
Institute Pierre-Simon Laplace	2.50° × 1.26°	IPSL-CM6A-LR

Monthly alterations in temperature and precipitation for the future period were computed using Eqs. (8.4) and (8.5):

$$\Delta T_m = \frac{\sum_{G=1}^{7} \left[\overline{T}_{m(2023-2038)} - \overline{T}_{m(1981-2015)} \right]}{6} \tag{8.4}$$

$$\Delta P_m = \frac{\sum_{G=1}^{7} \left[\overline{P}_{m(2023-2038)} / \overline{P}_{m(1981-2015)} \right]}{6} \tag{8.5}$$

where G represents the GCMs; ΔT_m and ΔP_m are the average changes in temperature (°C) and precipitation (%) in each month (m) in the future (2023−38) compared to the base period (1981−2015), respectively; and $\overline{T}_m$ and $\overline{P}_m$ are the average observed or predicted values of temperature and precipitation, respectively. These changes were calculated for three emission scenarios: SSP126, SSP245, and SSP585. Each of these scenarios presents a certain amount of radiative forcing (W m^{-2}), with SSP585 being the highest and SSP126 the lowest (Shukla et al., 2019).

Following the determination of temperature and precipitation changes on an annual basis during the future period, monthly time series data for temperature and precipitation were generated. Subsequently, utilizing the optimal CNN model structure identified in Section 8.3, the streamflow time series were derived through the application of the selected deep training network architecture.

8.3 Results

8.3.1 Simulation of streamflow in the base period

Table 8.4 presents the CNN evaluation results for simulating streamflow during the base period (1981−2015) in both the calibration and test stages. Visual comparisons of observed and predicted streamflow values by the CNN model are illustrated in Figs. 8.2 and 8.3. The outcomes reveal that the most effective flow prediction structure in the study area is the M4 configuration, encompassing maximum air temperature ($T_{\min}$ and $T_{\max}$), precipitation (Pr), relative humidity (RH), and average monthly flow (Q) (calibration: $R^2 = 0.91$, MAE = 1.72, RMSE = 2.89; test: $R^2 = 0.82$, MAE = 2.63, RMSE = 5.19). Additionally, the M3 structure (comprising $T_{\min}$, $T_{\max}$, Pr, RH, N, and Q) demonstrated relatively good performance in simulating streamflow (training: $R^2 = 0.88$, MAE = 3.21, RMSE = 5.42; test: $R^2 = 0.73$, MAE = 4.23, RMSE = 6.56). Conversely, the M7 structure (including $T_{\min}$, $T_{\max}$, N, and Q) exhibited the lowest efficiency in flow simulation. Notably, the findings align with previous studies by Hussain et al. (2020) and Zhang et al. (2019), affirming that the CNN model can accurately predict surface runoff.

8.3.2 Investigating the effects of climate change in the study area

Fig. 8.4 illustrates the average monthly temperature changes anticipated for the study area during the future period (2023−38) under various emission scenarios. The findings indicate a consistent rise in air temperature across all months throughout the upcoming years. Temperature increases are projected to vary from 0.1°C (in January under SSP126) to 1.6°C (in September under SSP585) across different months. The annual temperature variations for the study area are detailed in

TABLE 8.4 Evaluation of the CNN model based on different performance indices (R^2, MAE, and RMSE) for modeling of the streamflow.

Model structure	Training			Test		
	R^2	MAE	RMSE	R^2	MAE	RMSE
M1	0.77	3.13	4.12	0.66	3.32	6.12
M2	0.36	9.12	12.11	0.32	9.25	14.21
M3	0.88	3.21	5.42	0.73	4.23	6.56
M4	0.91	1.72	2.89	0.82	2.63	5.19
M5	0.66	4.23	6.52	0.63	6.11	9.23
M6	0.36	6.12	9.12	0.34	9.56	11.88
M7	0.31	10.08	14.36	0.25	11.52	17.88

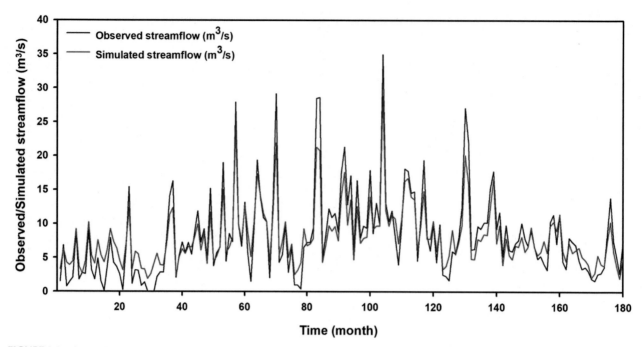

FIGURE 8.2   Comparison between the observed and simulated streamflow in the testing phase of the CNN.

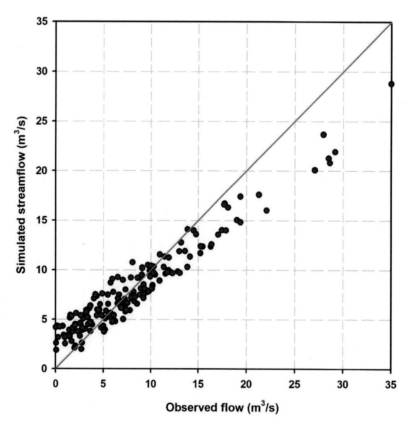

FIGURE 8.3   Scatter plots for the observed and predicted streamflow for the testing phase of CNN.

Table 8.5, revealing that the average annual temperature is expected to increase by 0.5°C, 0.7°C, and 1°C under SSP126, SSP245, and SSP585 emission scenarios, respectively. Overall, the SSP585 scenario suggests more severe temperature conditions, while the SSP126 scenario forecasts more favorable conditions. The average temperature changes, as indicated by various emission scenarios, suggest an overall temperature increase of 0.7°C in the future periods (Table 8.5).

Fig. 8.5 shows the changes in monthly precipitation for the study area during the future period. The corresponding annual variations are summarized in Table 8.5. According to the results, precipitation is anticipated to decrease in all

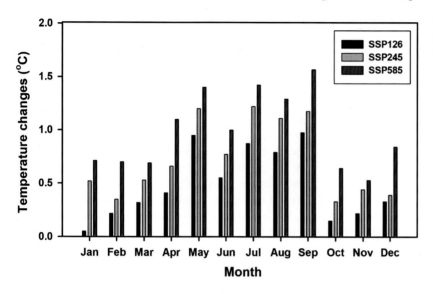

FIGURE 8.4 Monthly temperature changes in the study area under different climate change scenarios for the future period (2023−40).

TABLE 8.5 Average yearly changes in temperature and precipitation in the future (under different emission scenarios).

Emission scenarios	Temperature changes (°C)	Precipitation changes (%)
SSP2.6	0.49	−1.6
SSP4.5	0.72	−12.5
SSP8.5	0.99	−21.4
Average	0.73	−11.8

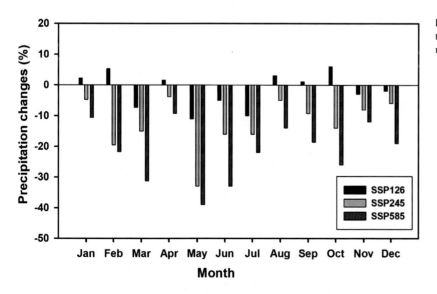

FIGURE 8.5 Monthly precipitation changes in the study area under different climate change scenarios for the future period (2023−40).

months and emission scenarios (with exceptions). The maximum projected reduction in precipitation is 39% in May under the SSP585 scenario, while the maximum increase is forecasted in October under the SSP126 scenario, reaching 6%. Annual precipitation changes indicate reductions of 1.6%, 12.5%, and 21.4% in the SSP126, SSP245, and SSP585 scenarios, respectively. On average, considering the SSP outputs, the study area is expected to experience an 11.8% reduction in precipitation (Table 8.5).

The outcomes of this study align with those of Zareian et al. (2021), who projected a temperature increase of 0.6°C–1.3°C and a precipitation decrease of 6.5%–31% in the Zayandeh-Rud Basin by 2044. Additionally, the findings are consistent with Javadinejad et al. (2021), who forecasted a 5%–10% decrease in precipitation in the same region by 2038.

8.3.3 Simulation of the effects of climate change on streamflow

Fig. 8.6 displays the simulated monthly streamflow in the study area using the CNN model under various emission scenarios. Additionally, Table 8.6 provides an overview of the annual streamflow changes anticipated during the future period. The outcomes suggest that streamflow may exhibit both increases and decreases in certain years of the upcoming period. Notably, the most substantial increases in streamflow are projected for the years 2023, 2026, 2029, and 2037. Conversely, a higher likelihood of streamflow reduction is indicated for 2028, 2030, 2032, and 2036 (Fig. 8.6).

Under SSP126, SSP245, and SSP585 emission scenarios, the annual streamflow is estimated at 4.46, 4.04, and 3.19 $m^3 s^{-1}$, respectively, reflecting corresponding reductions of 2.1%, 9.3%, and 21.1% compared to the base period. The average reduction across different emission scenarios is calculated to be 10.8%.

These findings align with prior research on the impact of climate change on river flow in the study area. Gohari et al. (2017) asserted that climate change could potentially reduce the Zayandeh-Rud River flow by 8%–43% by 2044. Similarly, Khalilian et al. (2021) identified an increased likelihood of a 10% reduction in the Zayandeh-Rud River flow due to climate change by 2050.

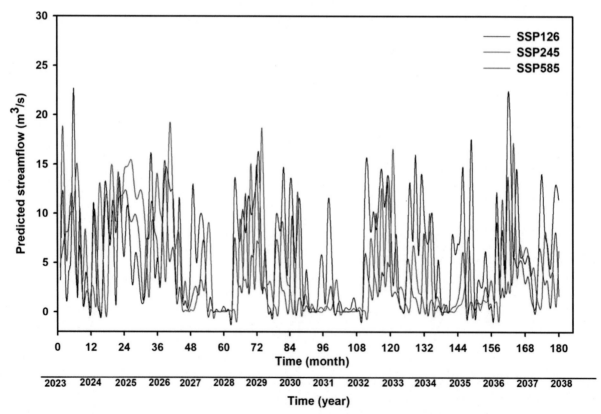

FIGURE 8.6 Predicted future stream flow by CNN model in the study area under different climate change scenarios (2023–40).

TABLE 8.6 Average yearly changes in streamflow in the future (2023−40) under different emission scenarios.

Emission scenarios	Predicted streamflow (m^3 s^{-1})	Changes in streamflow (%)
SSP2.6	4.46	−2.1
SSP4.5	4.04	−9.3
SSP8.5	3.19	−21.1
Average	3.89	−10.8

8.4   Conclusions

Climate change is a global phenomenon that causes changes in the Earth's climate in different regions. Necessary preparation for climate change management in the future is very important, especially in the field of water resource management, which is one of the most sensitive sectors of climate change. Necessary forecasting and modeling in the field of the effects of climate change on water resources is an important prerequisite for this issue. In this research, a practical and efficient tool based on deep learning model, named CNN, was used to estimate the change in the streamflow in the Eskandari Basin, Iran. The results showed that the increase in temperature and decrease in precipitation in the future will be very possible in this area. Simulation of streamflow using CNN also showed that this model can accurately predict the streamflow of the area and can be used as a practical and reliable model to simulate the streamflow changes in the region affected by climate change. The results of future flow simulations show that the streamflow will decrease in the future. Rising temperatures and declining precipitation, along with declining streamflow in the Eskandari Basin, which is an important region in the semiarid region of central Iran, is a serious threat to the sustainability of water resources in this region, where a significant population lives and depends on surface flow. Therefore, water resources managers and policymakers must plan properly to adapt this region to climate change.

The intricate interplay between climate change, river flows, groundwater recharge, and water resources management underscores the complexity of sustaining water ecosystems in the face of environmental shifts. Climate change, marked by altered precipitation patterns, rising temperatures, and extreme weather events, directly impacts river flows by influencing the timing and intensity of runoff. These changes, in turn, affect the recharge of groundwater aquifers, as altered river flows can impact the natural replenishment of subsurface water sources. The interaction becomes crucial in the context of water resource management, where the availability and distribution of water resources are pivotal for human, agricultural, and industrial needs. As climate change accelerates, managing river flows and groundwater recharge becomes increasingly challenging, requiring adaptive strategies to ensure sustainable water supply and mitigate the impacts on ecosystems and communities. Implementing integrated water resource management practices that consider the interconnected dynamics of climate, rivers, and groundwater is essential for building resilience in the face of a changing environment. It is important to note that the rainfall-runoff model developed in this research is specific to the Eskandari Basin. For application in other Iranian regions, similar tailored models would be necessary.

References

Barzegar, R., Aalami, M. T., & Adamowski, J. (2020). Short-term water quality variable prediction using a hybrid CNN−LSTM deep learning model. *Stochastic Environmental Research and Risk Assessment, 34*, 415−433.

Berrang-Ford, L., Ford, J. D., & Paterson, J. (2011). Are we adapting to climate change? *Global Environmental Change, 21*, 25−33.

Cui, C., & Fearn, T. (2018). Modern practical convolutional neural networks for multivariate regression: Applications to NIR calibration. *Chemometrics and Intelligent Laboratory Systems, 182*, 9−20.

Eltner, A., Bressan, P. O., Akiyama, T., Gonçalves, W. N., & Marcato Junior, J. (2021). Using deep learning for automatic water stage measurements. *Water Resources Research, 57*, e2020WR027608.

Eslamian, S., Safavi, H. R., Gohari, A., Sajjadi, M., Raghibi, V., & Zareian, M. J. (2017). *Climate change impacts on some hydrological variables in the Zayandeh-Rud River basin. Reviving the dying giant.* Cham, Iran: Springer.

Everingham, M., Van Gool, L., Williams, C. K., Winn, J., & Zisserman, A. (2010). The pascal visual object classes (voc) challenge. *International Journal of Computer Vision, 88*, 303−338.

Frame, J. M., Kratzert, F., Klotz, D., Gauch, M., Shalev, G., Gilon, O., Qualls, L. M., Gupta, H. V., & Nearing, G. S. (2022). Deep learning rainfall-runoff predictions of extreme events. *Hydrology and Earth System Sciences, 26*(13), 3377−3392.

Georgescu, M., Broadbent, A. M., Wang, M., Krayenhoff, E. S., & Moustaoui, M. (2021). Precipitation response to climate change and urban development over the continental United States. *Environmental Research Letters, 16*, 044001.

Gohari, A., Bozorgi, A., Madani, K., Elledge, J., & Berndtsson, R. (2014). Adaptation of surface water supply to climate change in central Iran. *Journal of Water and Climate Change, 5*, 391.

Gohari A., Zareian M.J., Eslamian S., Nazari R. (2017) Interbasin transfers of water: Zayandeh-Rud River basin. In Handbook of drought and water scarcity. CRC Press.

Goodfellow I., Bengio Y., Courville A. (2016) Deep learning (p. 599). MIT Press.

Green, T. R., Taniguchi, M., Kooi, H., Gurdak, J. J., Allen, D. M., Hiscock, K. M., Treidel, H., & Aureli, A. (2011). Beneath the surface of global change: Impacts of climate change on groundwater. *Journal of Hydrology, 405*, 532−560.

Hussain, D., Hussain, T., Khan, A. A., Naqvi, S. A. A., & Jamil, A. (2020). A deep learning approach for hydrological time-series prediction: A case study of Gilgit river basin. *Earth Science Informatics, 13*, 915−927.

IPCC (2013) Climate Change 2013: The Physical Science Basis. In Stocker T.F., Qin D., Plattner G.K., Tignor M.M., Allen S.K., Boschung J., Nauels A., Xia Y., Bex V., Midgley P.M. (Eds.). Contribution of Working Group I to the Fifth Assessment Report of IPCC the Intergovernmental Panel on Climate Change.

IPCC. (2007). The physical science basis. In S. Solomon, et al. (Eds.), *Contribution of working group I to the fourth assessment report of the intergovernmental panel on climate change*. New York: Cambridge University Press.

Javadinejad, S., Dara, R., & Jafary, F. (2021). Climate change simulation and impacts on extreme events of rainfall and storm water in the Zayandeh Rud Catchment. *Resources Environment and Information Engineering, 3*, 100−110.

Khalilian, S., Sarai Tabrizi, M., Babazadeh, H., & Saremi, A. (2021). Assessing the impact of climate change on the inflow on Zayandehrood Dam. *Water and Soil (JWSS), 24*, 255−271, In Persian].

Lee, D., Lee, G., Kim, S., & Jung, S. (2020). Future runoff analysis in the Mekong River Basin under a climate change scenario using deep learning. *Water, 12*, 1556.

Liu, D., Jiang, W., Mu, L., & Wang, S. (2020). Streamflow prediction using deep learning neural network: Case study of Yangtze River. *IEEE Access, 8*, 90069−90086.

Nearing, M. A., Pruski, F. F., & O'neal, M. R. (2004). Expected 331 climate change impacts on soil erosion rates: A review. *Journal of Soil and Water Conservation, 59*, 43−50.

O'Neill, B. C., Tebaldi, C., Vuuren, D. P. V., Eyring, V., Friedlingstein, P., Hurtt, G., Knutti, R., Kriegler, E., Lamarque, J. F., Lowe, J., & Meehl, G. A. (2016). The scenario model intercomparison project (ScenarioMIP) for CMIP6. *Geoscientific Model Development, 9*, 3461−3482.

Okhravi, S., Eslamian, S., & Esfahany, S. T. (2017). *Drought in Lake Urmia. Handbook of drought and water scarcity*. CRC Press.

Qian K., Mohamed A., Claudel C. (2019) Physics Informed Data Driven Model for Flood Prediction: Application of Deep Learning in Prediction of Urban Flood Development. arXiv preprint arXiv: 1908.10312.

Rawat, W., & Wang, Z. (2017). Deep convolutional neural networks for image classification: A comprehensive review. *Neural Computation, 29*, 2352−2449.

Shukla P.R., Skea J., Calvo Buendia E., Masson-Delmotte V., Pörtner H.O., Roberts D.C., Zhai P., Slade R., Connors S., Van Diemen R., Ferrat M. (2019) IPCC: Climate Change and Land: An IPCC Special Report on Climate Change, Desertification, Land Degradation, Sustainable Land Management, Food Security, and Greenhouse Gas Fluxes in Terrestrial Ecosystems.

Tucker, G. E., & Slingerland, R. (1997). Drainage basin responses to climate change. *Water Resources Research, 33*, 2031−2047.

Voulodimos, A., Doulamis, N., Doulamis, A., & Protopapadakis, E. (2018). Deep learning for computer vision: A brief review. *Computational Intelligence and Neuroscience.*.

Wilby, R. L., & Keenan, R. (2012). Adapting to flood risk under climate change. *Progress in Physical Geography: Earth and Environment, 36*, 348−378.

Wu, Z., Zhou, Y., Wang, H., & Jiang, Z. (2020). Depth prediction of urban flood under different rainfall return periods based on deep learning and data warehouse. *The Science of the Total Environment, 716*, 137077.

Zareian, M. J. (2021). Optimal water allocation at different levels 355 of climate change to minimize water shortage in arid regions (Case Study: Zayandeh-Rud River Basin, Iran). *Journal of Hydro-environment Research, 35*, 13−30.

Zareian M.J., Eslamian S., Gohari A., Adamowski J.F. (2017) The effect of climate change on watershed water balance. In Mathematical Advances Towards Sustainable Environmental Systems (p. 215-238). Springer.

Zhang, W., Yu, Y., Qi, Y., Shu, F., & Wang, Y. (2019). Short-term traffic flow prediction based on spatio-temporal analysis and CNN deep learning. *Transportmetrica A: Transport Science, 15*, 1688−1711.

Zhu, S., Wei, J., Zhang, H., Xu, Y., & Qin, H. (2023). Spatiotemporal deep learning rainfall-runoff forecasting combined with remote sensing precipitation products in large scale basins. *Journal of Hydrology, 616*, 128727.

Part IV

Climate changes impacts and adaptation

Chapter 9

Accuracy of climate and weather early warnings for sustainable crop water and river basin management

Punnoli Dhanya[1], Vellingiri Geethalakshmi[2], Subbiah Ramanathan[1], Kandasamy Senthilraja[3], Manickam Dhasarathan[3], Punnoli Sreeraj[4], Ganesan Dheebakaran[1], Chinnasamy Pradipa[1], Kulanthaisamy Bhuvaneshwari[1], N.S. Vidhya Priya[5], Sasirekha Sivasubramaniam[5], Prasad Arul[6,7] and S. Vigneswaran[8]

[1]Agro Climatic Research Centre, Tamil Nadu Agricultural University, Coimbatore, Tamil Nadu, India, [2]Tamil Nadu Agriculture University, Coimbatore, Tamil Nadu, India, [3]Tamil Nadu Agricultural University, Coimbatore, Tamil Nadu, India, [4]Thangal Kunju Musaliar College of Engineering, Kollam, Kerala, India, [5]ECE, Coimbatore, Tamil Nadu, India, [6]Krishi Vigyan Kendra, Tamil Nadu Agriculture University, Tirur, Tamil Nadu, India, [7]District Agrometeorology Unit, Chennai, Tamil Nadu, India, [8]Institute of Forest Genetics and Tree Breeding, Coimbatore, Tamil Nadu, India

9.1 Introduction

Anthropogenic warming due to rising greenhouse gas emissions is amplifying extreme weather events across the planet. Being a climate-sensitive sector, agriculture faces the heavy brunt of weather-related disasters. The world has been alerted to the fact that there is not enough time to prevent the worst effects of climate crises by the UN Intergovernmental Panel on Climate Change during the recent COP 28 conference (UN, 2023). Due to its booming economy, a large portion of India's working population still depends on agriculture for a living. A rise in the number and length of heat wave days are projected to rise under 1.5°C warming for India by 2050. Based on a study conducted by Chakraborty et al. (2018), exceptionally warm day temperatures (ExWD) pose a serious threat to the southern plateau, northeastern India, and the east and west coast plains (IPCC, 2021). The likelihood of a 5% or greater yield loss for kharif cereals might increase from 17% to 53% and that for Rabi cereals from 11% to 43% if ExWD increases from 20% to 60%. Efficient management and conservation of availability water resources is essential under the purview of rising evapotranspiration due to warming (Manda et al., 2014). Upsurges in moisture levels on the low-level monsoon westerlies heading toward the Indian subcontinent are linked to warming in the western Indian Ocean, and this might potentially result in a rise in the frequency of precipitation extremes during crop-growing seasons (Roxy et al., 2017). With this background, India needs to enhance its institutional capacity to tackle impacts from climate extremities at local level.

For climate change adaptation, early warning systems are one of the climate-smart options for farmers, as they are relatively cheaper and more effective in protecting people and assets from torrential rains, storms, floods, droughts, heatwaves, and so on (FAO, 2010). Farmers' sensitivity, which is based on their exposure to climate change as well as their coping capability and resilience, increases their ability to respond to catastrophic events and climate variability (Afkhami, Zahraie and Ghorbani, 1047; Duncan et al., 2015; Khichar & Bishnoi, 2003).

The Intergovernmental Panel on Climate Change (IPCC) sixth assessment report foresees a general increase in irrigation water demand by 2080, which will lead to a further reduction in rain-fed agriculture. Water for agriculture is becoming increasingly limited due to rising water demands from many sectors, with serious water shortages emerging in several nations, especially India, if there is no adaptation (IWMI Report accessed on 23rd March, n.d.). With 18% of the world's population but only 4% of its water resources, India is among the nations with the greatest water stress in the world. Apart from that, teleconnections also exacerbate the challenges of climate extremes in several years in India.

Hydrosystem Restoration Handbook. DOI: https://doi.org/10.1016/B978-0-443-29802-8.00009-1

El Niño years cause crop productivity to be negatively impacted by warmer, dry spells, longer-lasting heatwaves, and a delay in the onset of the Indian summer monsoon (Goswami et al., 2010; Murari et al., 2016). It is now imperative to increase farm productivity by opting for crops with high land suitability, especially less-water-intensive crops that are weather-smart for arid and semiarid regions.

Agroecosystem adaptations in water-stressed areas must focus on climate-resilient food crops that need lesser inputs and produce sustainable yields even under biotic and abiotic stresses. Hence, drought heat- and flood-tolerant varieties are the crops for future under climate change. Early warning and agromet advisories provide weather-smart, water-smart, and knowledge-smart options to the farmers. In order to provide early warning, weather research and forecasting uses a regional climate model to forecast/downscale micrometeorological weather variables (Skamarock et al., 2008). Being a peninsular region, India has a complex tropical monsoon type of climate. Crops and water supplies are completely dependent on monsoonal rains; hence, weather forecasts are extremely beneficial especially during extreme events. Due to its proximity to Arabian Sea on the west and Bay of Bengal on the east, the state of Tamil Nadu is more severely affected by these kinds of catastrophic weather phenomena. Apart from that, many parts of the state still lack adaptive capacity to manage risks (SAPCC, 2014). According to Aggarwal et al. (2019) and Goswami et al. (2010), in order to address climate variability at the regional level and to enable the adoption and scaling out of such methods, however, enormous investment, policy, and institutional support will be required. In India, agromet advisories are disseminated biweekly, considering the prevailing crop conditions under the Gramin Krishi Mausam Sewa Project run by the India Meteorological Department (IMD) in collaboration with the Indian Institute of Tropical Meteorology (IITM), Pune, All India Coordinated Re-search Project on Agrometeorology, Central Research Institute for Dryland Agriculture (CRIDA), and Indian Council of Agri-cultural Research. IITM has developed the Ensemble Prediction Systems on Climate Forecast System Models. Rathore et al. (2001) discussed the weather forecasting program of the National Centre for Medium Range Weather Forecasting that aims to provide location-specific weather forecasts to the Agromet Advisory Services units spread across India three days in advance. The potential to lessen the susceptibility of vulnerable economic sectors to weather fluctuations exists with the growing ability to give accurate and skillful weather forecasts (Hansen, 2002).

Information and Communications Technology (ICT)-based solutions support farmers with accessibility to weather updates, crop management, and market services (Pongnumkul et al., 2015). It is stated that a synergy between autonomous and planned adaptation at farm level will play a crucial role in minimizing agriculture risks. Hence, the present study was undertaken. Moreover, there are very limited studies reported on the impact of early warning disseminations among sorghum farmers, especially on the west agroecological zones of Tamil Nadu. Further, no research information is available in the existing literatures on the impacts of early warning on farming communities under anticipated climate change, except for Cauvery Delta Zones of the state of Tamil Nadu. Climate-smart and less-water-intensive crops such as sorghum needs to be promoted at a larger scale, especially in water-insecure basins. Moreover, a knowledge gap also exists on the performance of subseasonal climate, weather forecasts, and agro advisories for climate risk aversion at farmers' fields. Hence, this chapter provides the background and context of the usage and significance of seasonal and subseasonal climate and weather updates to minimize weather risk in a given case study.

9.2 Materials and methods

9.2.1 Crop production in the study area

Groundnut and sorghum are mainly cultivated in the rainfed areas of the basin, providing many essential nutrients for human beings, animals, and birds alike. Groundnut and sorghum farmers of Anamalai, Pollachi, and Udumalpet blocks in the Parambikulam Aliyar (PAP) Basin, Coimbatore district, are chosen for the study. The study area lies in the rain shadow regions of Western Ghats and hence is semiarid in nature. The southern side is bordered with forested hills of PAP Basin (Fig. 9.1). It lies in the west agroclimatic zone of the state (Fig. 9.1). The major crops cultivated by the rainfed farmers of Anamalai are groundnut, maize, and sorghum. This chapter is about the impacts of weather warnings and seasonal and subseasonal climate information on the sorghum and beetroot farmers in the basin (Fig. 9.1).

In this research, an assessment is carried out to evaluate the significance of technological adaptations using ICTs to minimize climate and weather vagaries among the sorghum farmers on one side of the Parambikulam Aliyar (PAP) Basin. Parambikulam Aliyar project provides water for both Kerala and Tamil Nadu states. The PAP Basin area comes under the Coimbatore district of the state and lies between $10°10'00''$ N to $10°57'20''$ N latitude, $76°43'00''$ E to $77°12'30''$ E longitudes. It has irrigated lands and the other part is completely rain-fed. In the command areas of the basin, water-intensive crops such as rice and maize are cultivated apart from groundnuts and plantation crops such as

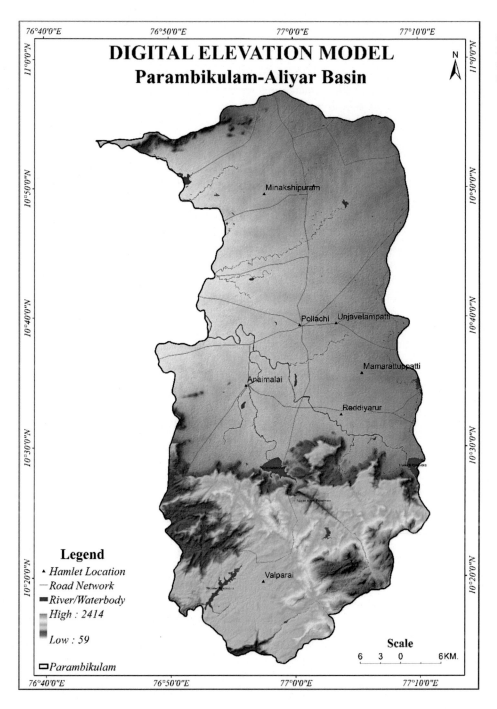

FIGURE 9.1 Digital elevation model map of the PAP Basin. Map lines delineate study areas and do not necessarily depict accepted national boundaries.

coconuts and cocoa and Nutmeg and Tamarind. Mainly large farmers have plantation agriculture. Also, small and marginal farmers mainly focus on sorghum (*Sorghum bicolor*) in the rain-fed tracks (locally known as *Vaanam Partha Bhoomi**). Farmers in these regions rely primarily on rainfall to grow sorghum. Based on the rainfall time series analysis, this interstate basin has experienced severe droughts in the years 1976, 1983, 2002, 2003, 2012, and 2016. Intense dry spells and droughts are frequent in the basin, affecting the water supply and irrigation severely. The middle and northern regions of the basin receive precipitation in the range of 700−1100 mm and are more susceptible to droughts. Southern side of the basin receives comparatively better rainfall in the range of 1200−4000 mm.

Farmers in these areas are heavily dependent on rainfall for groundnut cultivation. Groundnut is cultivated during *Chithirai pattam** (April) and *Vaikasi pattam* (May) in the rain-fed areas of the river basin. With respect to groundnut crop production, the major cultivating seasons found in Anamalai block was *summer/Zaid season/Chithirai*

Pattam (sowed in April/May and harvested in August using varieties such as TMV7, CO-2, CO-3, VRI-2, VRI-8, and BSR8) and during *Khariff/Ani Pattam* (sowed in June and harvested in September using varieties such as TMV7, CO-2, and VRI-2).

The area under groundnut crop in Coimbatore has decreased from 59,401 ha in 1992 to 4877 ha in 2015−16. The groundnut yield in Coimbatore district has shown a slight increasing trend during the period 1987 to 2015−16 (Fig. 9.2). During the drought year, 2002−03, the yield of groundnut was seen to be reduced to 942 kg ha^{-1} in the district. However, the share of groundnut yield of the state during the drought year was only 30%. The study area achieved its maximum share during 2011−12 (Fig. 9.3).

Sorghum (*Sorghum bicolor* L.) is the fifth most important staple grain crop globally (Pongnumkul et al., 2015). It is one of the hardy C$_4$ crops. Sorghum is known as *Cholam* in the vernacular language Tamil. PAP Basin located in the Coimbatore district of Tamil Nadu is known for its sorghum production since way back in 1980s and 1990s. This could be due to its less crop water requirement of less than 500 mm, hence this area suits very well for its cultivation. However, as the area under majority of crops has declined over the past two decades, even the area under sorghum crop has reduced from 1,36,421 ha in 1987−88 to 27,441 ha in 2015−16 (Fig. 9.4). Fig. 9.4 The share of the net sown area of the state was the lowest during 2012−13. Based on the *Agriculture Statistical Hand Book* for the year 2020−21, the area under sorghum cultivation in Coimbatore was 26,462 ha, contributing 6.53% to the state's total area sown. The average productivity of sorghum in Coimbatore also shows a fluctuating trend, well below the state's average (1054 kg ha^{-1}) in most of the recent years. Especially, during the drought year 2002−03, the mean yield fell to 284 kg ha^{-1} (Fig. 9.5). Sorghum is used by the rain-fed farmers for meeting local demands of food, energy, and fodder for livestock and birds. The common varieties are CO (S) 28, TNAU Sorghum variety CO 30, BSR 1, and so on. Multicut fodder sorghum is also cultivated in these areas for fodder purposes.

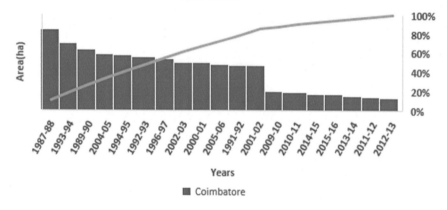

FIGURE 9.2 Pareto chart. The trend in the relative share of groundnut-growing area in Coimbatore district to the overall mean yield in the state of Tamil Nadu from 1987 to 2016 (source: Department of Agriculture Economics and statistics, DoAES, GoTN).

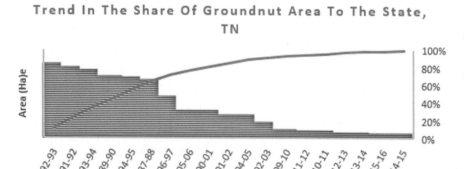

FIGURE 9.3 Pareto chart. The trend in the relative share of groundnut-productivity in Coimbatore district to the overall mean yield in the state of Tamil Nadu from 1987 to 2016 (source: Department of Agriculture Economics and statistics, DoAES, GoTN).

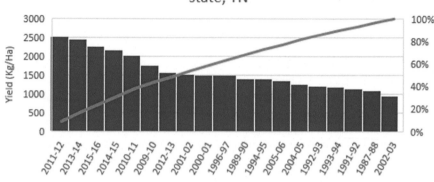

FIGURE 9.4 Pareto chart. The trend in the relative share of sorghum area in Coimbatore district to the total area in the state of Tamil Nadu from 1987 to 2016 (source: Department of Agriculture Economics and statistics, DoAES, GoTN).

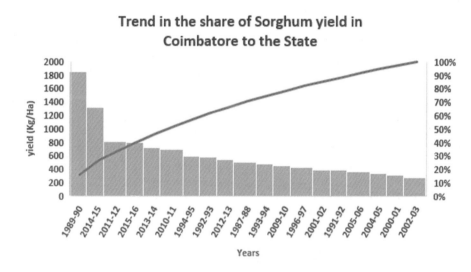

FIGURE 9.5 Pareto chart showing the temporal trend in the relative share of sorghum productivity in Coimbatore district to the total percentage yield in the state of Tamil Nadu from 1987 to 2016.

9.2.2 Farmer's participatory survey

Focus groups and household surveys were the techniques employed for gathering primary data. Statistical verification was carried out to understand the accuracy of the weather forecast. An extensive survey was conducted to understand the farmer's challenges at various crop-growing periods during 2021. An Automatic Weather Station (AWS) was used to receive real-time weather data. Block-level forecasts provided by the IMD were used to disseminate biweekly weather updates and agro advisories to the farmers. Standardized precipitation index is utilized by IMD to evaluate the status of monthly, seasonal, and weekly meteorology and identify the regions with prevailing or beginning/ending of the extremely/severely/moderately dry or wet conditions. IMD provides medium- and extended-range forecasts specific to states, districts, and blocks. Forecast verification for a given place can improve the weather forecast even more in future.

9.2.3 Verification of the weather forecasts

Assessment of forecast accuracy was done using qualitative skill scores. Verification of the forecast was done on observed and forecasted weather variables to evaluate its reliability across growing seasons. IMD's SOP is followed for verification purpose. Detailed methodology is adopted from Mani and Mukherjee (2016). The standard methodology proposed by IMD is used for the verification.

Here, Hanssen and Kuiper's score can also be interpreted as (accuracy for events) + (accuracy for nonevents) − 1, and the reliability of forecast is the average agreement between the forecast values and the observed values and bias is the correspondence between the mean forecast and mean observation.

The forecasts and outputs from climate change impact assessment using crop simulation studies were communicated to the stakeholders, especially extension workers and farmers during awareness meetings and discussions.

9.3   Results

9.3.1   Perspectives from groundnut and sorghum farmers

Socioeconomic profile of the surveyed farmers shows that the majority of them are small and marginal farmers owning below 1 acre (0.404 ha) of farm lands. There are exemptions, a few farmers owning more than 10 acres were also seen in Pollachi blocks. The field visits started during the seedling emergence stage till harvest.

Field visits and survey were conducted to understand each specific growth stages in the study area (annexure 1 and 2; Fig. 9.6). There incidence of stem rot, leaf minor, leaf minor, late tikka, and leaf spot noted in the farmers' field (Fig. 9.7). The COVID-19 pandemic had impacted logistics and availability of labor for a short time period in the study area during 2020. Later, family and friends themselves supported each other in completing the farm operations (Dhanya et al., 2020). During the land preparations and sowing, the local weather was influenced by the effects of cyclone Tauktae and Yaas. It was the first cyclone to form in the Arabian Sea in 2021. On May 17, 2021, it made landfall in southern Gujarat and was categorized as a very severe cyclonic storm. Cyclonic impacts cannot always be treated as destructive to the farming communities. In the summer of 2021, the early warning disseminated helped the farmers to a significant extent as a boon. This area got good summer showers during this period, which helped the farmers to complete farm operations and in good germination rates. However, there was a break period of rainfall during the month of June. Premonsoon cyclone impact was also felt in the year 2020, when cyclones Amphan and Nisarga also showed their impacts in the study area during crop land preparations.

It was an active cyclone period. Even just months after cyclone Yaas left, and a depression (BOB-3 September-1st week) and another cyclone Gulaab (September 25) were also formed. The IMD issued weather warnings during that time. The agromet advisories played a critical role during the harvesting period in September 2021. Heavy precipitation events had forecasted for the area during the first week of September by IMD as there was a cyclonic depression formed in Bay of Bengal. Majority of the registered farmers harvested the groundnut crop early due to the forecast of heavy precipitation events during the end of kharif/*Kuruvai* season in the first week of September 2021. This is because

FIGURE 9.6   Annexure 1. Photographs from the groundnut field visits.

FIGURE 9.7   Annexure 2. Photographs from the sorghum field visits.

the matured groundnut kernels will germinate again in the soil if not harvested on time and when soil becomes saturated once hit by rainfall. Apart from that, sale of wet groundnut stubbles as fodder may pose indigestion issues for the live-stock. Farmers received $1100-1300 \, \text{kg ha}^{-1}$ as they harvested their produce on time. Survey revealed that early warning and agromet advisories through ICTs helped the rain-fed groundnut farmers of Anamalai significantly during the harvest of kharif in 2021 to achieve local food and livelihood security and sustain income even during weather extremities. The precipitation values forecasted were fairly accurate for the season. Refinement of the model forecast can improve precision and dependability.

The areas covered under the survey were Vettaikaran Pudur, Sethumadai, Udayakulam, Jellipatti, Tensithur, Pethanaickanur and Aliyar, Pollachi, and Poosaripatti in the Coimbatore district.

Farmers were informed about the seasonal weather forecasts of IMD through SMS and social media apps such as WhatsApp. Furthermore, a website was launched to provide hourly and daily weather updates in vernacular language and agromet advisory services to enrolled farmers. IMD forecasted a above normal NE monsoon rainfall for Tamil Nadu. Special weather updates (bulletins) received from Regional Meteorological Centre in Chennai and agromet advisories prepared by the team were circulated to the farmers. Agromet advisories were issued to the farmers on a biweekly basis from sowing to harvest. Initial surveys during the early stages of the research revealed that almost half of the surveyed households in the basin had no access to seasonal or daily weather forecast (scientific) and almost the other half of the respondents received it through radios or TV. However, in later stages of the research project in 2022, it was seen that the access and availability improved in the study region. Comparing smartphone apps to other ICT tools, they are more user friendly especially among the young farmers; however, lack of affordability was noted among the aged small and marginal farmers. Certain villages in the foothills of Anamalai block lack network connectivity.

Annexure 1 Photographs taken during the groundnut field visits.

Annexure 2 Photographs of the sorghum field visits.

9.3.2   Analyzing the climatological trends

The data on historical observed climate for the study area is shown in Fig. 9.8. The observed data shows that the month of January receives the lowest rainfall (Fig. 9.6). However, occasionally, unexpected rainfall occurred during January

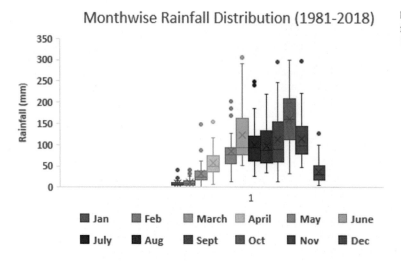

FIGURE 9.8 Box plot showing month wise rainfall distribution in the study area for the period 1981−2018.

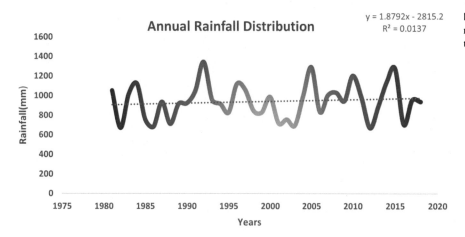

FIGURE 9.9 Line graph showing annual rainfall distribution in the study area for the period 1981−2018.

2021, posing havoc to harvest ready rabi crops. Many unexpected above-normal rainfalls were noted during May and July months also. The maximum rainfall was received during October (160 mm). Southwest monsoon and northeast monsoon together contribute about 78.9% of the yearly rainfall. The annual mean rainfall received in the study area for the period of 1981−2018 is 940 mm (Fig. 9.9). Rainfall deficit years noted are 1982, 1986, 1988, 2002, 2003, 2012, 2016, and 2018. The excess rainfall years in the study area were 2005, 2010, and 2015 during the period 1981−2018. These extremities shows the necessity to harvest and conserve available water resources in the basin consistently. The mean annual temperature is 28.2°C in the plains and 15.2°C in the hills of the basin. The temperature is lowest in the month of January and maximum in May. On the basis of temperature, northern part of the basin is isohyperthermic.

For disseminating anticipated climate change information about the basin to the stakeholders, regional climate model output of MIROC5 under representative concentration pathways (RCPs), 4.5 and 8.5 climate scenarios were used and shared with the stakeholders during awareness meeting. The rise in maximum temperature is projected to be 1.1°C and 1.3°C by 2040s and 3.1°C and 3.2°C by 2100 for RCP 4.5 and RCP 8.5, respectively. The minimum temperature is expected to increase in the basin by 1.2°C and 1.3° C by 2040s and 3.2°C and 3.3°C for RCP 4.5 and RCP 8.5, respectively. Due to the rise in temperatures, soil moisture stress in the farms may be exacerbated in future. As far as rainfall deviation is concerned, there is a decrease noted for nearly a century and there after a rise of 4% and 6% by the end of the century for RCP 4.5 and RCP 8.5, respectively, from the base period 1971−2000. This region-specific information on the anticipated scenarios may support the stakeholders with the knowledge of local conditions to make more informed decisions related to how to respond and adapt to climate-related risks in future such as extreme temperatures and heavy precipitation events.

The mean meteorological values recorded at the AWS located at the farmers field during the crop-growing period is given in Table 9.1. It was an above-normal rainfall period.

TABLE 9.1 The climatological values for the growing season, recorded from Automatic Weather Station and IMD normal for the period 2021–22.

S. no.	Variable	AWS values	IMD normals
1	Total rainfall (mm)	723.5	434.78
2	Mean maximum temperature (°C)	28.52	27.83
3	Mean minimum temperature (°C)	21.33	21.88
4	Mean relative humidity (%)	74.78	67.2
5	Wind speed (km h^{-1})	5.1	8
6	Average sunshine (hours per day)	5.2	5.8

AWS, Automatic Weather Station; *IMD*, India Meteorological Department.

TABLE 9.2 Farmer's perception on early warnings and agromet advisories.

Significance of forecast	*f*	%
Not required	17	22.37
Required	26	34.21
Very essential	33	43.42
Total (*N*)	**76**	**100**
Is it beneficial		
Yes	67	88.2
No/do not know	9	11.8
Total (*N*)	**76**	**100.0**

No permission required.

The mean RH (relative humidity) was maximum during November (82%), followed by October (79.7%) and March being the driest month with mean RH of only 60%.

Some of the sorghum farmers were planning to take a second crop (late rabi) during *Thai or Masipattam* as the soil moisture was good after heavy downpour in the month of November and December. As shown in Fig. 9.8, even the observed data also showed that the forecasted rainfall for the season was around 315 mm; however, the actual observed was around 724 mm for the crop-growing season from the second half of September 2021 till the end of February 2022. Famers' perceptions on the significance of early warning is presented in Table 9.2. The data revealed that the majority of the farmers (88%) suggested that early warning should be provided to them consistently, especially for utilizing the rainwater efficiently. Farmers wanted to safeguard their crops from bad weather, benefit from the favorable weather conditions, reduce input cost and ensure and enhance farm income.

There were heavy precipitation events in the November month, especially on November 3 and 17, 2021 due to cyclonic depressions in Bay of Bengal. IMD had issued a yellow alert for the district. Sorghum crops were in the heading and flowering stage at that time. Farmers were informed about the rainfall event; however, it was extreme rainfall in a day more than 70–95 mm downpour had happened in and around the study area. It was interesting to note that some of the farmers who had farms in the low-lying/flood-prone lands did not want to take risk and hence they preharvested their crops due to timely Agromet Advisory Service (AAS) and sold them in the local area as fodder. This reveals that the risk due to weather aberrations vary from crop to crop. It is time, location, crop, and farmer specific. It was sad to note that the total revenue for the season was well below US$33.3 ha^{-1}, after all expenses.

9.3.3   Demonstration of uses of weather alerts and mobile apps

It was very clear from the field visits that the sorghum crop was able to withstand the torrential rains to a great extent in many locations. Even though AAS and weather forecast have helped a small portion of the farming communities; however, farmers were not bothered much and perceive that the crops can sustain due to their natural resistance capacity. Farmers had complained about the shoot flies, stem borer, and shoot borer in-fluxes during the vegetative and flowering stages. Remedies and solutions such as foliar spray of cow urine and panchgavya and neem seed kernel extract were suggested to the farmers, consulting with experts like entomologists and agronomists. The weather information that came in the form of SMS was used by the farmers. Additionally, the farmers were shown how to download and register for mobile apps like Meghadoot, Uzhavan, and others in order to get government updates on farming. Timely information through SMS has helped the farmers to make decisions on the timing of land preparation, sowing, irrigation scheduling, fertilizer and insecticide application, and harvesting, particularly in water-stressed areas of the basin.

9.3.4   Seasonal comparison of verification of forecast

The actual weather data recorded by the AWS deployed in the farmer's field was compared with the weather forecasts obtained from IMD to ascertain the validity and accuracy of weather forecast during the crop growth period in 2021 using statistical procedures and skill scores. Quantitative verification and usability analyses for weather parameters were carried out using skill scores and critical values for crop growing seasons (Table 9.3).

9.3.5   For groundnut-growing season

Skill score for rainfall for kharif the major groundnut-growing season 2021 (Table 9.3).

With an accuracy score of (HI) of 0.57, Hanssen and Kuipers score (HK) of 0.13, probability of detection of hits of 0.94, root mean square error (RMSE) of 4.17, and correlation coefficient (R^2) of 0.27, the rainfall reflected a rather excellent forecast for the season, albeit irregular. Forecast of day-time temperature has a correlation coefficient of 0.39. However, the forecast of night-time temperature and relative humidity showed poor correlation between the observed and forecasted data (Table 9.3). Survey with the farmers revealed that forecast helped them during harvest, followed by land preparation and sowing.

It was noteworthy to understand that the probability of false detection (PoFD or false alarm rate) is 0.81 for kharif season (June−September), indicating that for 81% of the observed "no-rain" events, the forecasts were incorrect, similarly, for rabi (September−February) season, PoFD was 0.30, indicating that for only 30% of the observed "no-rain" events, the forecasts were incorrect. The study area gets a major share of its rainfall from northeast monsoon/retreating monsoon during rabi/samba season, unlike other parts of India. Hence, the forecast is more reliable during rabi season than the *kharif* season. The threat score (critical success index) for both the season indicates that slightly more than half of the "rain" events (observed and/or predicted) were correctly forecast for the rabi season (0.60). The odds ratio for the rabi season is 22.94, indicating that the odds of a "yes" prediction being correct are over 23 times higher than the likelihood that a "yes" forecast being incorrect. The reliability of maximum and minimum temperature forecast was relatively very low during both the seasons. The correlation coefficient is very less for night-time temperatures and relative

TABLE 9.3 Skill score for rainfall for kharif the major groundnut-growing season, 2021.	
Accuracy ratio	**0.56**
Hit rate	**0.93**
False alarm ratio	0.45
Threat score (critical success rate)	0.52
H/K score	0.13
RMSE	**4.17**
Correlation coefficient	**0.27**
No permission required.	

humidity in the rabi season. This could be due to the uniqueness in the topography of the study area and its proximity to thickly forested hills in the Anamalai Parambikulam region. Survey revealed that the sorghum farmers who had harvested their crops at maturity had received only yield below the mean yield of the area of 600 kg ha^{-1} for the season.

9.4 Discussion

During an adverse seasonal climate forecast situation, contingent agricultural practices and livelihood diversification are suggested to the farmers. A review by Ricart (2022) discussed the necessity of generic meteorological forecasts as opposed to predictions focused to provide tailored ones based on specific impacts. To reduce the impact of weather and climate hazards on farmers' livelihoods, the provision of an impact-based prediction enables them to take action proactively prior to disasters. The medium of forecast dissemination is also important, as it ensures farming communities have access to the required information and process it effectively to make the right decisions at the right time (Bacci et al., 2020; Guido et al., 2021; Hlophe-Ginindza & Mpandeli, 2021; Panda, 2016; Perera et al., 2022). Voice messages sent through "Kissan call centers" in India provide such valuable information. This study found that radio and television are more effective than mobile phones at disseminating weather and climate information, mainly due to farmer's poor economic status. Android phones are not owned by the majority of the small and marginal farmers. They depend on friends and other family members to access weather updates. This is in agreement with the findings of Mani and Mukherjee (2016), who had an opinion that accurate forecasting of weather variables remains a major challenge for the scientific community, especially in tropical regions. An economic impact assessment showed that data quality and reliability of agromet advisories issued from IMD could reduce cost of farming operations by 5%−10% on an average and increase the crop yield from 10% to 25% (Khichar & Bishnoi, 2003). Khichar and Bishnoi (2003) reported on the accuracy of weather forecasts for the western agroclimatic zone of Haryana during the kharif season (Rathore et al., 2001). They reported that more than 60% farmers realized the weather prediction and agromet advisories to be useful for scheduling irrigation time, pest/diseases management, and harvesting of crops. Currently, Satellite Rainfall Products are also in widespread use around the world as a better alternative for scarcely observed rain gauge data (White et al., 2015). In remote places with fewer real-time observational capacities, satellite products are an alternative. River basin management has been explored by various researchers across the globe in various geographical, hydrological, and socioeconomic contexts (Eslamian et al., 2024).

Due to climate change, 4.5%−9% loss in crop productivity is anticipated in major cereal crops. In this chapter, groundnut and sorghum crops are promising crops for future, provided farmers adopt adaptations (Baul & McDonald, 2015; Dhanya & Ramachandran, 2020; Dhanya et al., 2022; Sultan et al., 2013). Adopting a new crop calendar or changing the existing crop calendar that suits the changed weather situation based on extended weather forecast can save the available water resources in many places. This will minimize the overexploitation of surface and ground water resources and help in hydrological restorations to a great extent. Knowing its significance, the year 2023 was declared as the International Year of Millets. These climate-resilient crops can not only withstand sudden droughts and dry spells and suit the dry land areas of the world but also provide nutritional and livelihood securities to the local communities. Sultan et al. reported that simulations show that the photoperiod-sensitive traditional cultivars of millet and sorghum used by local farmers for centuries seem more resilient to future climate conditions than modern cultivars bred for their high yield potential (Baul & McDonald, 2015; Prasad et al., 2008; Singh et al., 2014; Srivastava et al., 2010; Sultan et al., 2013). Hence, traditional knowledge also plays a significant role in tackling climate and weather aberrations, which are local and crop-specific. Baul and McDonald have also advocated an integration of indigenous knowledge in addressing climate change (Dhanya & Ramachandran, 2020; Dhanya et al., 2022). These outcomes are valuable climate information for all stakeholders especially, the modelers, breeders, and extensions scientists. Cultivating crops that are hardy facilitates to augment better water resource management in water-insecure areas.

9.5 Conclusions and way forward

It has been demonstrated that seasonal, subseasonal, and biweekly forecasts are gaining momentum in the basin. Through this research specifically, it was clear that the level of damage may be decreased with better impact-based warnings. It improves at-risk populations' resilience and strengthens their capacity to handle extreme weather occurrences. There is only limited real-time data available currently from the AWS, so collecting more data would support better and accurate validation of forecasts. Upscaling climate- and weather-smart technologies to other regions is the need of the hour.

Investing in adaptation and resilience is crucial at this moment to achieve sustainable development goals and Intended Nationally Determined Contributions for a country like India. Analysis of real-time weather data and forecast dissemination creates opportunities for better management of resources for crop production. Researchers working in agrometeorology and agroclimatology play a bridging job among the climate and weather forecasters, agronomists, and grass root-level stakeholders. Research studies help to enhance climate services in a region through action-oriented research. Regional studies support to derive insightful and customized solutions from real-time weather data, assessing risks, improving methodologies and validating forecasts for enhancing the modeling capacities, and interpreting existing climate knowledge within their local communities and among scientific communities. There are other remote regions with similar climates and water stress requiring such support for sustainable crop production. Creating more awareness on the adoption of agromet advisories among the farmers and concerned officials is crucial for these regions.

- There is scope for improving the content and access to customized weather forecast and agromet advisories, catering to the needs of diverse farmers. At present, lack of income and knowledge on the use of technology are found to be the limiting factors.
- Resilience is more crop- and farmer-specific. It is highly significant to make sure that dissemination of the existing climate and weather forecasts should be tailor made to the farmers' specific needs in each location.

Funding

The corresponding author would like to acknowledge the research fellowship provided by KIRAN-WISE division, Department of Science and Technology, Government of India.

References

Afkhami, M., Zahraie, B., & Ghorbani. M. (2022). Quantitative and qualitative analysis of the dimensions of farmers' adaptive capacity in the face of water scarcity. Journal of Arid Environment, 199, 104715.

Aggarwal, P., Vyas, S., Thornton, P., & Campbell, B. M. (2019). How much does climate change add to the challenge of feeding the planet this century? *Environmental Research Letters, 14*(4), 043001. Available from https://doi.org/10.1088/1748-9326/aafa3e.

Bacci, M., Baoua, Y. O., & Tarchiani, V. (2020). Agrometeorological forecast for smallholder farmers: A powerful tool for weather-informed crops management in the Sahel. *Sustainability (Switzerland), 12*(8). Available from https://doi.org/10.3390/SU12083246, https://www.mdpi.com/2071-1050/12/8/3246.

Baul, T. K., & McDonald, M. (2015). Integration of indigenous knowledge in addressing climate change. National Institute of Science Communication and Information Resources (NISCAIR). *Journal of Traditional Knowledge, 14*(1), 20–27. Available from http://nopr.niscair.res.in/bitstream/123456789/32021/1/IJTK%201%281%29%2020-27.pdf.

Chakraborty, D., Sehgal, V. K., Dhakar, R., Varghese, E., Das, D. K., & Ray, M. (2018). Changes in daily maximum temperature extremes across India over 1951–2014 and their relation with cereal crop productivity. *Stochastic Environmental Research and Risk Assessment, 32*(11), 3067–3081. Available from https://doi.org/10.1007/s00477-018-1604-3, http://link.springer-ny.com/link/service/journals/00477/index.htm.

Dhanya, P., Geethalakshmi, V., Ramanathan, S., Senthilraja, K., Sreeraj, P., Pradipa, C., Bhuvaneshwari, K., Vengateswari, M., Dheebakaran, G., Kokilavani, S., Karthikeyan, R., & Sathyamoorthy, N. K. (2022). Impacts and climate change adaptation of agrometeorological services among the maize farmers of west Tamil Nadu. *AgriEngineering, 4*(4), 1030–1053. Available from https://doi.org/10.3390/agriengineering4040065.

Dhanya, P., & Ramachandran, A. (2020). The current policies and practices behind scaling up climate-smart agriculture in India, In: *Global Climate Change: Resilient and Smart Agriculture*, Springer.

Duncan, J. M. A., Dash, J., & Atkinson, P. M. (2015). Elucidating the impact of temperature variability and extremes on cereal croplands through remote sensing. *Global Change Biology, 21*(4), 1541–1551. Available from https://doi.org/10.1111/gcb.12660UnitedKingdom, http://onlinelibrary.wiley.com/journal/10.1111/(ISSN)1365-2486.

Eslamian, S., Huda, B., Rather, M., A., N., & Eslamian, F. (2024). *Handbook of Climate Change Impacts on River Basin Management, . Fundamentals and Impacts* (Vol. 1). USA: Taylor and Francis, CRC Group.

FAO, Climate Smart Agriculture: Policies, Practices and Financing for Food Security, Adaptation and Mitigation. FAO (2010), 2010.

Goswami, B. N., Kulkarni, J. R., Mujumdar, V. R., & Chattopadhyay, R. (2010). On factors responsible for recent secular trend in the onset phase of monsoon intraseasonal oscillations. *International Journal of Climatology, 30*(14), 2240–2246. Available from https://doi.org/10.1002/joc.2041.

Guido, Z., Lopus, S., Waldman, K., Hannah, C., Zimmer, A., Krell, N., Knudson, C., Estes, L., Caylor, K., & Evans, T. (2021). Perceived links between climate change and weather forecast accuracy: New barriers to tools for agricultural decision-making. *Climatic Change, 168*(1-2), 1–20. Available from https://doi.org/10.1007/s10584-021-03207-9.

Hansen, J. W. (2002). Realizing the potential benefits of climate prediction to agriculture: Issues, approaches, challenges. *Agricultural Systems, 74*(3), 309–330. Available from https://doi.org/10.1016/S0308-521X(02)00043-4.

Hlophe-Ginindza, S. N., & Mpandeli, N. S. (2021). *The Role of Small-Scale Farmers in Ensuring Food Security in Africa*. IntechOpen. Available from 10.5772/intechopen.91694.

IPCC, (2021) Accessed from https://www.ipcc.ch/2021/08/09/ar6-wg1-20210809-pr/ on March,12th, 2023.

IWMI Report accessed on 23rd March (n.d.).

Khichar, M. L., & Bishnoi, O. P. (2003). Accuracy of weather forecast for western agroclimatic zone of Haryana during kharif season. *Haryana Agricultural University Journal of Research*, *33*(2), 2003.

Manda, A., Nakamura, H., Asano, N., Iizuka, S., Miyama, T., Moteki, Q., Yoshioka, M. K., Nishii, K., & Miyasaka, T. (2014). Impacts of a warming marginal sea on torrential rainfall organized under the Asian summer monsoon. *Scientific Reports*, *4*(1), 5741. Available from https://doi.org/10.1038/srep05741.

Mani, J. K., & Mukherjee, D. (2016). Accuracy of weather forecast for hill zone of West Bengal for better agriculture management practices. *Indian Journal of Research.*, *5*, 325–328, 2016.

Murari, K. K., Sahana, A. S., Daly, E., & Ghosh, S. (2016). The influence of the El Niño Southern Oscillation on heat waves in India. *Meteorological Applications*, *23*(4), 705–713. Available from https://doi.org/10.1002/met.1594, http://onlinelibrary.wiley.com/journal/10.1002/(ISSN)1469-8080.

Panda, A. (2016). Exploring climate change perceptions, rainfall trends and perceived barriers to adaptation in a drought affected region in India. *Natural Hazards*, *84*(2), 777–796. Available from https://doi.org/10.1007/s11069-016-2456-0.

Perera, H., Senaratne, N., Gunathilake, M. B., Mutill, N., & Rathnayake, U. (2022). Appraisal of satellite rainfall products for Malwathu, Deduru, and Kalu River Basins, Sri Lanka. *Climate*, *10*(10). Available from https://doi.org/10.3390/cli10100156.

Pongnumkul, S., Chaovalit, P., & Surasvadi, N. (2015). Applications of smartphone-based sensors in agriculture: A systematic review of research. *Journal of Sensors*, *2015*, 1–18. Available from https://doi.org/10.1155/2015/195308.

Prasad, P. V. V., Pisipati, S. R., Mutava, R. N., & Tuinstra, M. R. (2008). Sensitivity of grain sorghum to high temperature stress during reproductive development. *Crop Science*, *48*(5), 1911–1917. Available from https://doi.org/10.2135/cropsci2008.01.0036.

Rathore, L. S., Gupta, A., & Singh, K. K. (2001). Medium range weather forecasting and agricultural production. *Journal of Agricultural Physics*, *1*, 43–47, 2001.

Ricart, S. (2022). On farmers' perceptions of climate change and its nexus with climate data and adaptive capacity. A comprehensive review Environmental Research Letters, 17, 083002.

Roxy, M. K., Ghosh, S., Pathak, A., Athulya, R., Mujumdar, M., Murtugudde, R., Terray, P., & Rajeevan, M. (2017). A threefold rise in widespread extreme rain events over central India. *Nature Communications*, *8*(1). Available from https://doi.org/10.1038/s41467-017-00744-9.

SAPCC, T.N. (2014) accessed on 16th December, 2021 from https://moef.gov.in/wp-content/uploads/2017/09/Tamilnadu-Final-report.pdf.

Singh, P., Nedumaran, S., Traore, P. C. S., Boote, K. J., Rattunde, H. F. W., Prasad, P. V. V., Singh, N. P., Srinivas, K., & Bantilan, M. C. S. (2014). Quantifying potential benefits of drought and heat tolerance in rainy season sorghum for adapting to climate change. *Agricultural and Forest Meteorology*, *185*, 37–48. Available from https://doi.org/10.1016/j.agrformet.2013.10.012.

Skamarock, W., Klemp, J., Dudhia, J., Gill, D., Barker, D., Duda, M., Huang, X., Wang, W., & Powers, J. (2008). A Description of the Advanced Research WRF Version 3. *NCAR/TN-475 STR; NCAR Technical Note; Mesoscale and Microscale Meteorology Division, 2008.*

Srivastava, A., Naresh Kumar, S., & Aggarwal, P. K. (2010). Assessment on vulnerability of sorghum to climate change in India. *Agriculture, Ecosystems and Environment*, *138*(3-4), 160–169. Available from https://doi.org/10.1016/j.agee.2010.04.012.

Sultan, B., Roudier, P., Quirion, P., Alhassane, A., Muller, B., Dingkuhn, M., Ciais, P., Guimberteau, M., Traore, S., & Baron, C. (2013). Assessing climate change impacts on sorghum and millet yields in the Sudanian and Sahelian savannas of West Africa. *Environmental Research Letters*, *8*(1). Available from https://doi.org/10.1088/1748-9326/8/1/014040.

UN, (2023) https://www.un.org/en/climatechange/early-warnings-for-all accessed on November 17th, 2023.

White, J. W., Alagarswamy, G., Ottman, M. J., Porter, C. H., Singh, U., & Hoogenboom, G. (2015). An overview of CERES−sorghum as implemented in the cropping system model version 4.5. *Agronomy Journal*, *107*(6), 1987–2002. Available from https://doi.org/10.2134/agronj15.0102.

Chapter 10

Flood risk assessment for the integrated disaster risk management and climate change adaptation action plan: La Mojana Region, Colombia

Omar Dario Cardona[1,2], Gabriel Andres Bernal[2,3] and Mabel Cristina Marulanda[2]

[1]*Universidad Nacional de Colombia, IDEA, Manizales, Colombia,* [2]*INGENIAR: Risk Intelligence, Bogota, Colombia,* [3]*Universidad Nacional de Colombia, Ingenieria Civil y Agricola, Bogota, Colombia*

10.1 Introduction

The expansive delta of the La Mojana region, nestled in the northwest of Colombia, spans over a million hectares. Here, the convergence of three mighty rivers—the Cauca, Magdalena, and San Jorge—creates a landscape characterized by its strikingly flat terrain, enhanced by an intricate network of water channels and permanent and temporary wetlands (Ministerio del Medio Ambiente, 1999, 2002). This region, a subset of the broader Depresión Momposina, was once inhabited by the pre-Columbian Zenu community. Utilizing advanced hydraulic infrastructure 2000 years ago, it plays crucial ecological roles essential to the area's equilibrium. Acting as a reservoir for floodwaters, it retains excess water during inundations, sustains surface water levels during dry seasons, and accumulates vital sediments. These functions are the cornerstone of the region's environmental stability and resilience (Aguilera, 2004). Inhabited by over 400,000 people across 11 municipalities spanning four departments, La Mojana is home to communities primarily engaged in subsistence livelihoods (DANE, 2005). Fig. 10.1 shows the region, the municipality names, and some of the archaeological remains of the works created by the Zenu community.

Dependent on livestock rearing, subsistence agriculture, and traditional fishing, these communities confront significant vulnerability to the cycles of droughts and floods. Poverty affects 83.3% of the population. The devastating impact of the 2010−11 rainy season, exacerbated by the impacts of La Niña, inflicted widespread havoc across the region. Homes and commercial establishments were obliterated, croplands submerged, and grasslands inundated, dealing a severe blow to the agriculture sector (Olaya, 2015). The ramifications reverberated, profoundly affecting the lives and sustenance of thousands within the region.

In response to these challenges, a comprehensive study funded by the Colombian government through the Adaptation Fund set out to undertake a meticulous flood risk assessment for La Mojana (Cardona et al., 2017; Cardona, 2014a, 2014b, 2014c, 2015a, 2015b, 2016a, 2016b). Employing a fully probabilistic approach, the findings of the study played a pivotal role in informing the strategy and action plan of the Adaptation Fund for disaster risk management and climate adaptation in the area. Furthermore, the assessment helped prioritize housing rehabilitation initiatives following the catastrophic flood event of 2010−11. Leveraging the insights garnered, the region secured funding for the ambitious "Scaling Up Climate-Resilient Management Practices for the Vulnerable Communities of La Mojana" project, valued at US$117 million. This endeavor received partial backing from the Green Climate Fund and aimed to bolster the community's resilience against climate-related adversities (PNUD, 2019).

Hydrosystem Restoration Handbook. DOI: https://doi.org/10.1016/B978-0-443-29802-8.00010-8

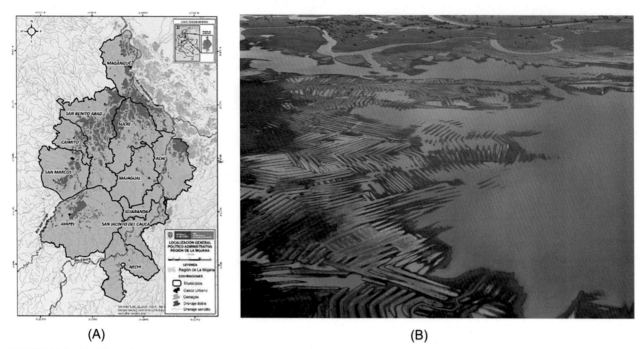

(A)　　　　　　　　　　　　　　　　　(B)

FIGURE 10.1 (A) The Mojana region in northwest Colombia is the hydraulic buffer zone for the Cauca, San Jorge, and Magdalena rivers. (B) Ancient indigenous communities built infrastructure to coexist with the floods.

10.2 Probabilistic risk assessment methodology

The probabilistic risk assessment model for flood hazard in the La Mojana region evaluates risk across 11 municipalities. This model considers various exposed elements, including residences, medical facilities, educational institutions, government structures, lifelines, roads, and agricultural fields. Given the unpredictable nature of hazardous events, the model employs a set of scenarios derived from the hazard model. These scenarios encompass all potential manifestations of the hazard, accounting for both their likelihood and severity. The process of computing the probabilistic risk model is iterated for all hazard events, covering the complete spectrum of flood hazards. Each event triggers specific impacts to the exposed elements, outlining a theoretical scope of potential consequences. By employing suitable mathematical calculations, individual losses are consolidated for each event, offering a comprehensive evaluation of the impact of each hazard scenario on the entire exposure portfolio.

The primary components of the probabilistic catastrophe risk analysis for La Mojana involve the following:

- *Hazard assessment*: Identifying a comprehensive array of events and their respective frequencies for floods. This incorporates spatial probability parameters to model intensities as random variables.

While this area commonly faces floods, historical records lacked adequate details to understand the causes behind past extreme events. The hazard assessment team obtained 38 years' worth of incomplete precipitation and flow data. From this limited information, only five instances of extreme flooding could be accurately identified. As a result, it became crucial to simulate stochastic flood events in the region. This aimed to evaluate potential flood impacts that had not occurred yet or were poorly documented. Over 150 stochastic events were simulated, considering the hydrological and hydraulic characteristics of the entire basin. This encompassed the intricate network of river streams and swamps that contribute to the region's susceptibility to floods. Additionally, these simulations considered the potential failure of the existing flood defense, accounting for its fragility and probability of malfunction. Fig. 10.2 depicts flood hazard assessment footprints for various return periods, considering the current conditions of the "Marginal Dike," a highly vulnerable embankment defense.

- *Exposure assessment*: Creating an inventory of exposed assets, specifying their geographic locations, replacement values or fiscal liabilities, and building classifications. The building of the exposure database involved analyzing Landsat images for urban centers and scattered rural settlements. With the support of local authorities and nongovernment organizations, data related to building types, number of floors, terrain elevation, and occupancy were collected. Likewise, constructing the database required collaboration with authorities in the education, health, and infrastructure sectors, at the local and national levels.

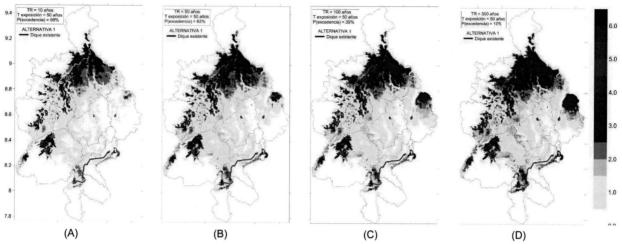

FIGURE 10.2 Integrated flood hazard maps (A–D) for 10-, 50-, 100-, and 500-year return periods (left to right) under current conditions in the La Mojana region. The "Marginal Dike" is shown in red, spanning 70 km.

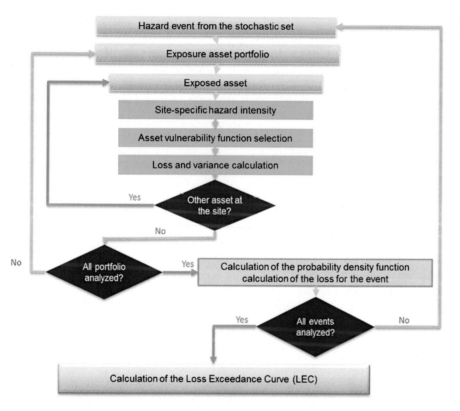

FIGURE 10.3 Flowchart of probabilistic risk assessment process.

- *Vulnerability assessment*: Establishing vulnerability functions for various building classes and infrastructure elements. These functions characterize asset behavior during hazardous events. Vulnerability functions for crop portfolios also consider the four phenological phases. These functions provide probability distributions of losses based on increasing hazard intensity.

The process of computing the loss exceedance curve (LEC) follows the outlined sequence of steps illustrated in Fig. 10.3.

The LEC provides essential details for understanding the occurrence of loss events. Various risk metrics are derived from the LEC, including the Average Annual Loss (AAL), Probable Maximum Loss (PML), probability of ruin, interevent times, number of events, and the next event (Bernal & Cardona, 2018; Bernal et al., 2017, 2021; Cardona et al., 2023a, 2023b; Marulanda et al., 2013; Ordaz, 2000; Torres et al., 2014). Among these metrics, the PML and AAL are the most widely used and discussed.

- *PML:* This signifies a relatively uncommon loss event, usually linked to extended intervals between occurrences or a low frequency of incidence. The return period stands as the reciprocal of the exceedance annual rate.
- *AAL:* This metric consolidates the impact of various hazardous scenarios affecting vulnerable elements into a single numerical representation. It stands out as a reliable risk indicator due to its ability to summarize the entire loss-time process in a singular value while demonstrating minimal sensitivity to uncertainties. The AAL reflects the anticipated value of annual losses, portraying the annual compensation needed to cover all future losses over the long term. In a simplified insurance context, the AAL would equate to the annual pure premium. Its computation involves the integration of the LEC. Even in situations where hazards are depicted not as specific scenarios but as a compilation of uniform hazard maps, albeit lacking a comprehensive risk evaluation, the AAL can still be calculated.

10.3 Disaster risk results

The region faces increasing risk due to the construction of inappropriate drainage and protective infrastructure that provides the population with a false sense of security. To mitigate current risk conditions in the La Mojana region and decrease the vulnerability of its inhabitants, a series of alternative interventions focusing on flood defense infrastructure was developed. The probabilistic risk assessment was carried out for the current conditions of the Marginal Dike as well as for four different types of interventions on that defense and supplementary engineering works, aiming to estimate the potential reduction in losses in the event of flooding occurrences in the region.

The costs and benefits of the intervention alternatives were assessed to reduce risk, ranging from no intervention at all of current conditions of the existing 70-km dike (ALT 1), enhancement and reinforcement of the existing dike (ALT 2), enhancement, reinforcement, and 80-km extension of the dike (ALT 3), reinforcing the existing dike but with bypass structures that allowed water to flow from one water body to another (ALT 4), and constructing a parallel dike and floodgates in both dikes (ALT 5).

This approach allows for the comparison of potential future losses associated with each alternative and the implications for the 11 municipalities within La Mojana. It also assesses the level of impact on each of the portfolios considered in the evaluation. Crop evaluation differs slightly from construction evaluation, as the risk is calculated for each of the four growth phases of the plants.

Fig. 10.4 shows maps of the probability of exceeding 0.5 m of flooding for a 50-year return period for the various alternative interventions considered on the Marginal Dike.

Table 10.1 shows the expected annual loss and PML for various return periods for the entire buildings' portfolio, and Fig. 10.5 shows the comparison of total relative AAL for the entire buildings' portfolio; both for current conditions, as alternative 1, and for the other four structural alternatives considered. Surprisingly, the findings indicate that the enhancement and reinforcement of the existing Marginal Dike leads to an increase in expected annual loss from 16‰ in the current state to 39.7‰ if the intervention is implemented. Conversely, reinforcing the existing dike and constructing a new longer one results in a decrease in expected annual losses to 13.7‰. However, caution is warranted as the impacts may not follow a linear pattern, potentially causing harmful effects in specific municipalities that might not offset the overall reduction in losses. Regarding the alternative of reinforcing the existing and constructing floodgates, the

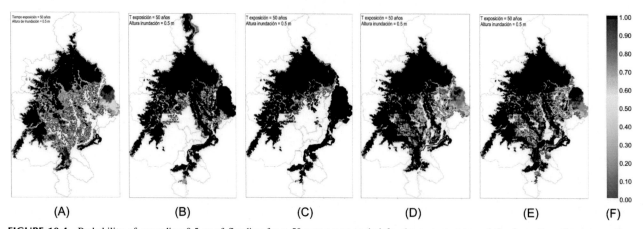

FIGURE 10.4 Probability of exceeding 0.5 m of flooding for a 50-year return period for the current state and the four alternative interventions considered on the Marginal Dike.

TABLE 10.1 Risk results for the entire building portfolio, comparing Average Annual Loss and Probable Maximum Loss for various return periods.

Exposed Value		ALT 1		ALT 2		ALT 3		ALT 4		ALT 5	
		COP$ x10^6		COP$ x10^6		COP$ x10^6		COP$ x10^6		COP$ x10^6	
		10,462,308		10,462,308		10,462,308		10,462,308		10,462,308	
AAL		COP$ x10^6	‰	COP$ x10^6	‰	COP$ x10^6	‰	COP$ x10^6	‰	COP$ x10^6	‰
		166,601	15.9	415,090	39.7	143,509	13.7	175,590	16.8	177,972	17.0
PML	years	COP$ x10^6	%	COP$ x10^6	%	COP$ x10^6	%	COP$ x10^6	%	COP$ x10^6	%
	100	590,000	5.6	1,036,986	9.9	389,037	3.7	511,854	4.9	431,960	4.1
	250	666,000	6.4	1,211,883	11.6	452,646	4.3	584,593	5.6	524,489	5.0
	500	788,000	7.5	1,318,429	12.6	489,878	4.7	632,098	6.0	581,013	5.6
	1000	887,000	8.5	1,413,181	13.5	524,736	5.0	677,100	6.5	632,708	6.0

The comparison includes the current stage or alternative 1 and four other alternative interventions for the Marginal Dike. AAL, Average Annual Loss; PML, Probable Maximum Loss.

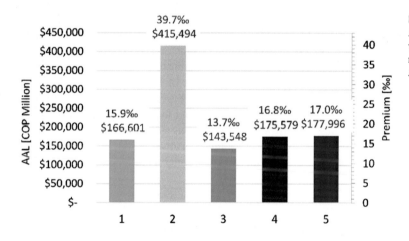

FIGURE 10.5 Comparison of total and relative Average Annual Loss for the entire building portfolio. The ALT 2 was the highest resulting risk while ALT 3, 4, or 5 did not offer any significant advantages to the current state or ALT 1.

expected annual loss is 16.8‰ with the intervention. Furthermore, constructing a parallel dike with floodgates in both dikes yields an expected AAL of 17‰. These latter two intervention alternatives demonstrate no significant risk reduction compared to the "current state."

10.4 Cost−benefit analysis

For the cost−benefit analysis, consideration was given to the five alternatives concerning the Marginal Dike, structural intervention options for rural housing (stilt houses), and the construction of walls or defenses as perimeter protections for municipal urban centers. The analytical methodology involved combining different interventions on the dike with interventions on buildings, namely, elevated rural houses and protective defenses for urban centers.

Each intervention incurred implementation costs. With 11 municipalities to apply these interventions to, or not, countless combinations needed to be considered. An optimization methodology was developed to evaluate these combinations, determining the optimal combination of interventions based on a predefined scoring system. This methodology involves a genetic algorithm for optimization, using artificial or computational intelligence, where individuals correspond to different intervention alternatives in La Mojana. The genotype of these individuals represents the combination of intervention measures on the dike and in the municipalities. The evolutionary process begins with a population of randomly generated individuals, undergoing an iterative process in which individuals are crossed and mutated (i.e., their genotype is altered) to create the next generation. In each generation, the strength of each individual in the population according to various evaluation criteria is assessed.

Three criteria were defined to evaluate the conditions of individuals: highest cost—benefit ratio, maximum benefit, and minimum cost. Additionally, two conditions were set: to include (or not) walls or defenses surrounding the municipal centers and to elevate rural houses with a minimum height value. A Forced Evolution approach was employed, where the fittest individual (referred to as the "champion") was crossed with all other individuals in its generation, creating the subsequent generation. This method ensured that the conditions achieved in each generation were at least as good as those of the previous one. The new generation of individuals was used in the next iteration of the genetic algorithm, which terminates upon reaching a maximum number of generations.

During each optimization process, populations of 100,000 individuals and a total of 10 generations were defined, resulting in the evaluation of approximately 200 million alternatives in total. After obtaining a set of 202 optimal intervention options, five classification types were defined, creating five "Top 10" lists based on the above-mentioned criteria and conditions, namely, (1) best cost—benefit ratio, (2) maximum benefit, (3) minimum cost, (4) best cost—benefit ratio while minimizing the number of protective defenses in municipal centers to be built, (5) maximum average benefit—cost value of the municipalities, ensuring all municipalities have a cost—benefit ratio greater than zero. Each intervention option within each "Top 10" list was assigned a score based on its ranking, ranging from 10 (highest score) to 1 (lowest score). Subsequently, the total score of each intervention option was calculated by summing the scores obtained in each of the rankings per criterion. After organizing the scores and removing redundant alternatives, the six best intervention options were identified, as shown in Table 10.2 and Fig. 10.6. Additionally, four more options were selected within a specific cost range (400—600 billion Colombian pesos) to address project resource limitations. Green shades in the table indicate favorable values, while red shades represent less favorable conditions.

The findings reveal that most options align with alternative 1 (maintain current conditions), except for option 264. These interventions vary in effectiveness and costs; for instance, constructing protective walls is costly but effective in reducing risk compared to raising rural homes. Some options show high benefits at moderate costs, while others display varying premiums relative to their effectiveness in reducing risk. Despite implementing structural measures, the risk is not reduced to zero, highlighting the need for complementary nonstructural measures due to the inherent impossibility of eliminating expected losses.

Fig. 10.7 illustrates the LECs for the different selected options, showcasing a range of losses between option 132 maximum and option 414 minimum. The options were categorized into three groups based on their loss outcomes: options 414, 264, and 426 incurred the lowest losses, potentially indicating that reinforcing the existing dike (alternative 4) could be as effective as building defenses in urban areas, despite cost variations. Options 97 and 98 showed comparatively low losses, followed by options 82, 2, and 5, which resulted in moderate losses. However, options 132 and 326 demonstrated the highest losses with minimal benefits, despite significant differences in intervention costs.

TABLE 10.2 Final scores based on the top 10 classifications of intervention options.

ID option	Dike ALT	Benefit COP $x10^6$/ year	Cost COP $x10^6$	B/C	Time Equiv. years	AAL [‰]	Walls [UN]	Houses on Stilts	Municipal B/C Average	Score
264	Alt4	138,019	1,163,915	0.12	8.43	2.21	0	24,242	0.19	19
132	Alt1	13,497	59,785	0.23	4.43	14.30	0	1,163	0.05	19
2	Alt1	87,370	136,032	0.64	1.56	7.30	2	0	0.14	16
5	Alt1	52,634	89,068	0.59	1.69	10.61	1	0	0.07	16
414	Alt1	163,386	2,137,857	0.08	13.08	0.37	9	56,336	0.08	10
426	Alt1	161,354	1,207,070	0.13	7.48	0.90	9	27,608	0.12	10
82	Alt1	112,939	576,219	0.20	5.10	4.74	2	13,437	0.15	Cost criterion
326	Alt1	9,895	514,649	0.02	52.01	14.69	0	15,053	0.02	Cost criterion
97	Alt1	126,458	468,683	0.27	3.71	3.40	2	10,118	0.24	Cost criterion
98	Alt1	142,506	542,814	0.26	3.81	1.90	2	12,406	0.20	Cost criterion

AAL, Average Annual Loss.

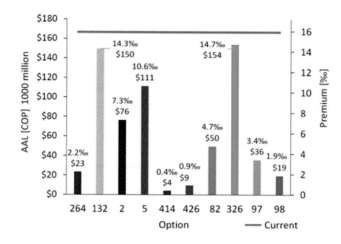

FIGURE 10.6 Average Annual Loss, absolute and relative (the premium), for the optimal options of structural interventions in La Mojana.

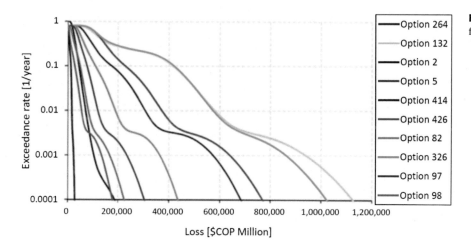

FIGURE 10.7 Loss Exceedance Curve for the 10 different selected options.

10.5 Holistic risk evaluation

The holistic evaluation offers a simplified view of a multidimensional concept, facilitating interpretation across various fields and promoting integrated efforts among social, economic, environmental, and cultural aspects. This holistic assessment enables the incorporation of underlying social, environmental, and institutional factors that are the root causes of vulnerability and risk in the region. By breaking down the results, it becomes possible to identify the factors toward which risk reduction actions should be directed and to assess their effectiveness.

The methodology used is based on the conceptual framework of holistic risk estimation proposed in the late 1990s by Cardona (2001), expanded by Carreño (2006), and further improved jointly by Carreño et al. (2007, 2022). Holistic evaluation is an indicator applied within the socioeconomic context, based on probabilistic modeling of physical risk. This indicator results from the combination of direct physical risk, RF, and the aggravating coefficient, F, i.e., the underlying conditions that exacerbate it. RF is determined by normalizing the expected annual losses relative to the exposed value of each municipality with the minimum and maximum values of the entire portfolio. On the other hand, the aggravating coefficient, F, is derived from the weighted sum of aggravating factors representing social fragility and lack of resilience. These aggravating factors are calculated by normalizing the value of each respective descriptor using a transformation function for each descriptor, obtaining values between 0 and 1. Each factor has a weight or relative contribution. These weights sum up to 1 and are obtained using the Analytic Hierarchy Process. Table 10.3 presents the selected descriptors representing social fragility and lack of resilience, along with their respective weights for the holistic evaluation study of the La Mojana region. Hence, the total risk, RT, is defined as $RT = RF(1 + F)$, where $(1 + F)$ is the factor representing the impact of physical damage in the socioeconomic context (Carreño, 2006; Carreño et al., 2007, 2022; Eslamian & Eslamian, 2023; Marulanda et al., 2009, 2020, 2022; Cardona, 2001). This expression explicitly incorporates the natural, socio-natural, and anthropic characteristics of the various aspects that control disaster risk into a single indicator (Adaptación, 2016). Fig. 10.8 depicts the composition of RT for all municipalities in the current state.

TABLE 10.3 Descriptors representing the aggravating factors for social fragility and lack of resilience and their weights.

Aspect		Descriptor	Definition/units	Weight dimension	Weight
Social fragility	X_{SF1}	People in extreme poverty	Unsatisfied basic needs index, component people in extreme poverty/[%]	55%	0.16
	X_{SF2}	Social disparity	GINI index/[%]		0.17
	X_{SF3}	Mortality rate	Crude mortality rate in children under 1 year/ [per 1000 live births]		0.142
	X_{SF4}	Rural displacement	Percentage relative to the total number of displaced persons.		0.104
	X_{SF5}	Housing	Unsatisfied basic needs index, housing component/[%]		0.047
	X_{SF6}	Drinking water and basic sanitation services	Unsatisfied basic needs index, services component (drinking water and sanitation)/[%]		0.094
	X_{SF7}	People insured to health services	Proportion of household members, over 5 years old, insured by Social Security in Health		0.094
	X_{SF8}	Dependent population	Unsatisfied basic needs index, dependency component/[%]		0.094
	X_{SF9}	Unemployment	Percentage of the population.		0.094
Lack of resilience	X_{LR1}	Quality of life	Quality of life index/[%]	20%	0.132
	X_{LR2}	Roads	Density of tertiary roads by municipal area.		0.043
	X_{LR3}	Health centers	Number of health centers per 10,000 inhabitants.		0.043
	X_{LR4}	Governance	Comprehensive performance index/[%]		0.236
	X_{LR5}	Legal coherence of territorial ordering planning	Lack of coherence in the land-use plans with the risk management plan and legislation.		0.124
	X_{LR6}	Land tenure	Land GINI Index/[%]		0.132
	X_{LR7}	Education	Illiterate population/[%]		0.039
	X_{LR8}	Archaeological influence area	Percentage relative to the total area of the municipality.		0.039
	X_{LR9}	Wetland influence area	Percentage relative to the total area of the municipality.		0.213

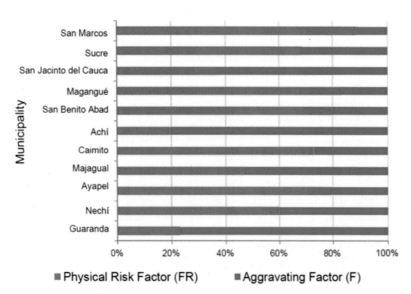

FIGURE 10.8 Composition of total risk for the 11 municipalities at the current state.

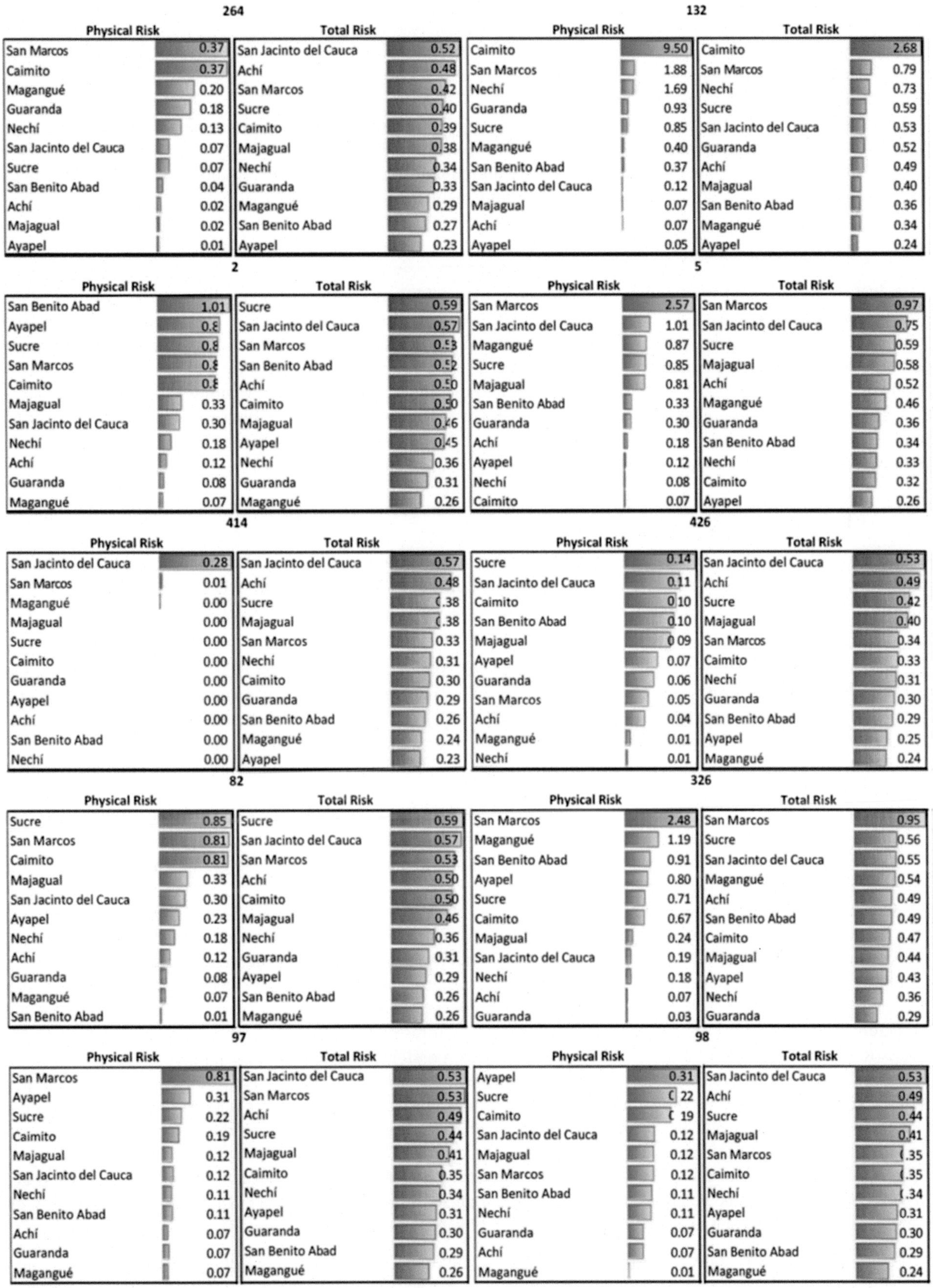

264

Physical Risk		Total Risk	
San Marcos	0.37	San Jacinto del Cauca	0.52
Caimito	0.37	Achí	0.48
Magangué	0.20	San Marcos	0.42
Guaranda	0.18	Sucre	0.40
Nechí	0.13	Caimito	0.39
San Jacinto del Cauca	0.07	Majagual	0.38
Sucre	0.07	Nechí	0.34
San Benito Abad	0.04	Guaranda	0.33
Achí	0.02	Magangué	0.29
Majagual	0.02	San Benito Abad	0.27
Ayapel	0.01	Ayapel	0.23

132

Physical Risk		Total Risk	
Caimito	9.50	Caimito	2.68
San Marcos	1.88	San Marcos	0.79
Nechí	1.69	Nechí	0.73
Guaranda	0.93	Sucre	0.59
Sucre	0.85	San Jacinto del Cauca	0.53
Magangué	0.40	Guaranda	0.52
San Benito Abad	0.37	Achí	0.49
San Jacinto del Cauca	0.12	Majagual	0.40
Majagual	0.07	San Benito Abad	0.36
Achí	0.07	Magangué	0.34
Ayapel	0.05	Ayapel	0.24

2

Physical Risk		Total Risk	
San Benito Abad	1.01	Sucre	0.59
Ayapel	0.8	San Jacinto del Cauca	0.57
Sucre	0.8	San Marcos	0.53
San Marcos	0.8	San Benito Abad	0.52
Caimito	0.8	Achí	0.50
Majagual	0.33	Caimito	0.50
San Jacinto del Cauca	0.30	Majagual	0.46
Nechí	0.18	Ayapel	0.45
Achí	0.12	Nechí	0.36
Guaranda	0.08	Guaranda	0.31
Magangué	0.07	Magangué	0.26

5

Physical Risk		Total Risk	
San Marcos	2.57	San Marcos	0.97
San Jacinto del Cauca	1.01	San Jacinto del Cauca	0.75
Magangué	0.87	Sucre	0.59
Sucre	0.85	Majagual	0.58
Majagual	0.81	Achí	0.52
San Benito Abad	0.33	Magangué	0.46
Guaranda	0.30	Guaranda	0.36
Achí	0.18	San Benito Abad	0.34
Ayapel	0.12	Nechí	0.33
Nechí	0.08	Caimito	0.32
Caimito	0.07	Ayapel	0.26

414

Physical Risk		Total Risk	
San Jacinto del Cauca	0.28	San Jacinto del Cauca	0.57
San Marcos	0.01	Achí	0.48
Magangué	0.00	Sucre	0.38
Majagual	0.00	Majagual	0.38
Sucre	0.00	San Marcos	0.33
Caimito	0.00	Nechí	0.31
Guaranda	0.00	Caimito	0.30
Ayapel	0.00	Guaranda	0.29
Achí	0.00	San Benito Abad	0.26
San Benito Abad	0.00	Magangué	0.24
Nechí	0.00	Ayapel	0.23

426

Physical Risk		Total Risk	
Sucre	0.14	San Jacinto del Cauca	0.53
San Jacinto del Cauca	0.11	Achí	0.49
Caimito	0.10	Sucre	0.42
San Benito Abad	0.10	Majagual	0.40
Majagual	0.09	San Marcos	0.34
Ayapel	0.07	Caimito	0.33
Guaranda	0.06	Nechí	0.31
San Marcos	0.05	Guaranda	0.30
Achí	0.04	San Benito Abad	0.29
Magangué	0.01	Ayapel	0.25
Nechí	0.01	Magangué	0.24

82

Physical Risk		Total Risk	
Sucre	0.85	Sucre	0.59
San Marcos	0.81	San Jacinto del Cauca	0.57
Caimito	0.81	San Marcos	0.53
Majagual	0.33	Achí	0.50
San Jacinto del Cauca	0.30	Caimito	0.50
Ayapel	0.23	Majagual	0.46
Nechí	0.18	Nechí	0.36
Achí	0.12	Guaranda	0.31
Guaranda	0.08	Ayapel	0.29
Magangué	0.07	San Benito Abad	0.26
San Benito Abad	0.01	Magangué	0.26

326

Physical Risk		Total Risk	
San Marcos	2.48	San Marcos	0.95
Magangué	1.19	Sucre	0.56
San Benito Abad	0.91	San Jacinto del Cauca	0.55
Ayapel	0.80	Magangué	0.54
Sucre	0.71	Achí	0.49
Caimito	0.67	San Benito Abad	0.49
Majagual	0.24	Caimito	0.47
San Jacinto del Cauca	0.19	Majagual	0.44
Nechí	0.18	Ayapel	0.43
Achí	0.07	Nechí	0.36
Guaranda	0.03	Guaranda	0.29

97

Physical Risk		Total Risk	
San Marcos	0.81	San Jacinto del Cauca	0.53
Ayapel	0.31	San Marcos	0.53
Sucre	0.22	Achí	0.49
Caimito	0.19	Sucre	0.44
Majagual	0.12	Majagual	0.41
San Jacinto del Cauca	0.12	Caimito	0.35
Nechí	0.11	Nechí	0.34
San Benito Abad	0.11	Ayapel	0.31
Achí	0.07	Guaranda	0.30
Guaranda	0.07	San Benito Abad	0.29
Magangué	0.07	Magangué	0.26

98

Physical Risk		Total Risk	
Ayapel	0.31	San Jacinto del Cauca	0.53
Sucre	0.22	Achí	0.49
Caimito	0.19	Sucre	0.44
San Jacinto del Cauca	0.12	Majagual	0.41
Majagual	0.12	San Marcos	0.35
San Marcos	0.12	Caimito	0.35
San Benito Abad	0.11	Nechí	0.34
Nechí	0.11	Ayapel	0.31
Guaranda	0.07	Guaranda	0.30
Achí	0.07	San Benito Abad	0.29
Magangué	0.01	Magangué	0.24

FIGURE 10.9 Physical risk and total risk for the 10 intervention options.

Fig. 10.9 displays the outcomes for physical risk (FR) and total risk (RT) for the 10 intervention options. It notes that the hierarchy of total risk does not align with the hierarchy based on physical risk, indicating the impact of social vulnerability in altering the positions of the municipalities. Additionally, it observes that physical risk outcomes vary for each intervention option, despite similar aggravating factors across municipalities. This leads to differences in total risk levels among the various intervention options.

These rankings demonstrate substantial differences among intervention options, highlighting how each option affects physical risk, altering the proportions of physical risk involvement and social vulnerability conditions. Notably, option 414 significantly minimizes physical risk across nearly all municipalities. Conversely, options 5 and 132 elevate physical risk to the highest proportion of total risk in most municipalities.

10.6 Discussion

The project for the probabilistic flood risk assessment in the La Mojana region was divided into three parts: probabilistic risk assessment to estimate potential losses in exposed elements, cost−benefit analysis to evaluate various structural intervention alternatives in the Marginal Dike, urban municipalities, and rural homes, and comprehensive risk assessment to consider multidimensional aspects contributing to the construction of vulnerability and risk. This study lays the groundwork for understanding flood risk in the region and the underlying factors contributing to risk in La Mojana. It presents structural and nonstructural intervention alternatives to improve the region's physical and socioeconomic conditions (Leibovich et al., 2015). The results presented in the previous section summarize the total project outcomes, covering different exposed sectors, various flood return periods, and potential losses.

10.6.1 Challenges and limitations

This section presents some considerations regarding other aspects not considered in the risk assessment but are relevant during decision-making processes as they are crucial in defining more suitable and feasible actions and interventions:

- Any structural intervention poses challenges: for instance, (1) the response of the community, as it may be perceived as an interference in their way of life, (2) associated costs like purchasing land, for example, to build embankments or other protective measures, (3) in other flood protection interventions, drought conditions may arise, impacting ecosystems; although floodgates management could be a solution, if not properly managed and efficiently operated, it might lead to legal responsibility issues and social, economic, and environmental implications.
- Interventions might be concentrated in only a few places, resulting in significant benefits for only some municipalities. While specific engineering interventions may benefit the entire region, this perception might not be evident, as municipalities may prioritize investments for themselves. Therefore a joint and integrated effort involving all municipalities within the region is essential, requiring significant administrative representation, community participation, and collaboration in project development. Involvement is a key strategy for empowering beneficiaries to take action.
- In such an extensive area as the La Mojana region, the impact of selected structural and nonstructural measures may not be very noticeable. As mentioned, this might give the impression of specific benefits, leading to rejection and dissatisfaction from local administrations and the community. Prioritization of interventions should consider their visibility, achieving a balance with risk reduction. For instance, interventions like constructing stilt houses reduce loss, but in the case of La Mojana, unless extensively implemented in urban areas, it may not be as effective as other structural measures. However, the construction of housing, infrastructure, sanitation, and public services, in general, tends to be more noticeable, especially if executed in most possible municipalities. Social investments can also garner significant visibility and therefore greater acceptance from the community. Finally, the decision of which alternative to select, besides considering highly visible structural measures, should weigh actions that partly compensate for the investment imbalance between municipalities, seeking visibility of actions before the community, media, and local and regional political authorities.
- The flood risk assessment estimates potential losses in exposed elements, but it does not evaluate the impact on ecosystem connectivity or the potential effects on the flood pulse. It is crucial to preserve the states, characteristics, and services of ecosystems and assess how they are affected by each alternative.

Some of the environmental assessment aspects that are important to consider in selecting the alternative to implement were outlined by García (2016) in his document for the action plan formulation. This document emphasizes connectivity as a key element for the vitality of La Mojana's ecosystems. It also exposes the different impacts that each dike intervention alternative considered in this risk assessment has on connectivity, flood pulses, and, therefore, on ecosystem services (García, 2014, 2015). In addition to estimating the risk assessment of the La Mojana region, efforts must be made to protect hydric connectivity by restoring the vegetative cover of streams and creeks that connect rivers to the floodplain and marshes. This will help improve the regulation of managing extreme events.

10.6.2 Findings and implications

The significant progress in this study lies in having the probabilistic modeling of flood risk in the La Mojana region, which is fundamental today for well-founded decision-making based on the cost−benefit analysis of physical interventions in the area, whether these are defense or water regulation structures or adaptive interventions for housing in a region fundamentally composed of interconnected aquatic ecosystems and wetlands. However, it is inadequate or insufficient for adaptation to consider only the factors that determine or increase flood hazard because, to be effective, it is necessary also to address the causes of vulnerability. Effective adaptation or risk management processes involve comprehensive actions within the framework of development and the pursuit of sustainability through management at the socioeconomic, environmental, and cultural levels, accounting for the various dimensions of vulnerability (Cardona et al., 2012; Cardona, 2012). The complexity posed by the La Mojana region goes beyond obtaining this probabilistic risk modeling; therefore other relevant aspects associated with economic, social, and institutional factors, and historical aspects, which have determined what this territory is today over decades, must be involved.

With the holistic risk assessment, within a single conceptual framework, hazard and vulnerability (including physical vulnerability, understood as the susceptibility to damage of exposed elements, and contextual vulnerability, expressed through socioeconomic factors) are considered. This approach aims to reflect risk from a comprehensive perspective. On one hand, it considers physical risk, or potential physical damage directly related to the occurrence of events. On the other hand, it captures the underlying drivers or risk amplifiers—social, economic, and institutional factors—that create conditions exacerbating or adding to the physical risk. Within this holistic approach, it is understood that to reduce existing risk or prevent the generation of new risk, a control system (institutional structure) and action systems (public policies, actions, and strategies) are required. These systems must intervene not only in vulnerable elements but also in societal conditions that favor the creation or increase of risk.

10.7 Conclusions

It is imperative that long-term planning in La Mojana and, generally, in flood-exposed territories in Colombia, to build their climate resilience, starts with understanding the territory as a socioecological system. Its characteristics are heterogeneous, complex, adaptive, and multiscalar, essentially composed of the permanent interaction between the population (sociocultural system) and nature (biodiversity and its ecosystem services). In this sense, the development planning in La Mojana requires actions on landscapes (territories) in a holistic and systemic way considering their environmental and cultural particularities, overcoming the error of trying to solve territorial problems with fragmented and disjointed approaches, increasing the fallacy that conservation and production are irreconcilable, while social gaps widen.

Thus a policy is justified from the perspective of integrated disaster risk management developed with joint and articulated actions of the state, associated with (1) reducing vulnerability and adapting exposed elements, (2) improving conditions, livelihoods, and sources of income for the population, and (3) strengthening institutional capacities for planning, territorial ordering, and responsiveness to unavoidable events. This policy will apply a socioecosystem logic for risk management so that actors influencing La Mojana carry out their activities in an articulated, corrective, and forward-thinking manner, based on an appropriate risk assessment, reducing vulnerability and derived risk, effectively, and promptly managing any disaster that may occur; these principles are also supported by the National Policy for Disaster Risk Management.

This work presents an unprecedented study concerning hazard and risk assessment in the La Mojana region. Not only does this work serve as a national reference, but it is also internationally recognized, with studies of this level of detail and rigor being scarce. Consequently, this study stands as a significant contribution to a process involving multiple stakeholders. It represents a valuable knowledge asset that the Adaptation Fund and the country, in general, should aim to leverage. By ensuring that future investments in hazard and risk reduction are based on careful and rigorous studies addressing risk assessment in all its complexity, this approach guarantees that decisions are informed and grounded in the best available knowledge.

References

Adaptación, F. (2016). *Plan de Acción Integral para la reducción del riesgo de inundaciones y adaptación al cambio climático en la región de La Mojana.*

Aguilera, M. (2004). La Mojana: Riqueza natural y potencial económico. *Documentos de Trabajo Sobre Economía Regional, 48.*

Ambiente, M. del M. (1999). *Identificación de Prioridades de Gestión Ambiental en Ecosistemas de Páramos.* Geoingeniería Ltda, Geoingeniería Ltda.

Ambiente, M. del M. (2002). *Política Nacional para Humedales Interiores de Colombia.*

Bernal, G. A., & Cardona, O. D. (2018). Next Generation CAPRA Software. In Proceedings of the 16th European Conference on Earthquake Engineering.

Bernal, G. A., Cardona, O. D., Marulanda, M. C., & Carreño, M. L. (2021). *Dealing with uncertainty using fully probabilistic risk assessment for decision-making. Handbook of disaster risk reduction for resilience: New frameworks for building resilience to disasters* (pp. 299–340). Colombia: Springer International Publishing. Available from https://link.springer.com/book/10.1007/978-3-030-61278-8, http://doi.org/10.1007/978-3-030-61278-8_14.

Bernal, G. A., Salgado-Gálvez, M. A., Zuloaga, D., Tristancho, J., González, D., & Cardona, O. D. (2017). Integration of probabilistic and multi-hazard risk assessment within urban development planning and emergency preparedness and response: Application to Manizales, Colombia. *International Journal of Disaster Risk Science, 8*(3), 270–283. Available from https://doi.org/10.1007/s13753-017-0135-8, http://www.springer.com/earth + sciences + and + geography/natural + hazards/journal/13753.

Cardona, O. D. (2001). Evaluación holística del riesgo utilizando sistemas dinámicos complejos. Tesis doctoral.

Cardona, O. D. (2012). Un marco conceptual común para la gestión del riesgo y la adaptación al cambio climático: Encuentros y desencuentros de una iniciativa insoslayable. In F. Briones (Coordinador), *Perspectivas de investigación y acción frente al cambio climático en Latinoamérica* (pp. 13–38). La Red, CIGIR.

Cardona, O. D. (2014a). Documento de análisis de la información de los elementos expuestos e información mínima necesaria para la articulación con la evaluación probabilista. Producto B de consultoría. *Informe para el Fondo Adaptación.*

Cardona, O. D. (2014b). Documento con las recomendaciones sobre la metodología y el procedimiento general de análisis de la amenaza por inundación para efectos de la modelación del riesgo por inundaciones. Producto A de consultoría. *Informe para el Fondo Adaptación.*

Cardona, O. D. (2014c). Documento que recoja el resultado de la modelación de vulnerabilidad ante inundaciones y la definición de curvas de vulnerabilidad para los diferentes elementos expuestos existentes en la región de La Mojana. Producto C de consultoría. *Informe para el Fondo Adaptación.*

Cardona, O. D. (2015a). Documento con la evaluación probabilista del riesgo de inundación en la Región de La Mojana para su estado actual (Sin intervenciones). Producto E de consultoría. *Informe para el Fondo Adaptación.*

Cardona, O. D. (2015b). Documento con la evaluación probabilista del riesgo de inundación basado en escenarios históricos de amenaza en la Región de La Mojana. Producto D de consultoría. *Informe para el Fondo Adaptación.*

Cardona, O. D. (2016a). Documento con el análisis costo-beneficio de las alternativas seleccionadas por el Fondo Adaptación y Documento con las recomendaciones al Fondo Adaptación sobre las alternativas de intervención para reducción de la amenaza y/o de la vulnerabilidad con su respectiva combinación de medidas estructurales y no estructurales, conforme a los lineamientos suministrados. Productos G y H de consultoría. *Informe para el Fondo Adaptación.*

Cardona, O. D. (2016b). Documento con la evaluación probabilista del riesgo de inundación en la Región de La Mojana con alternativas de intervención al dique marginal. *Producto F de consultoría. Informe para el Fondo Adaptación.*

Cardona, O. D., Van Aalst, M. K., Birkmann, J., Fordham, M., Mc Gregor, G., Rosa, P., Pulwarty, R. S., Schipper, E. L. F., Sinh, B. T., Décamps, H., Keim, M., Davis, I., Ebi, K. L., Lavell, A., Mechler, R., Murray, V., Pelling, M., Pohl, J., Smith, A. O., & Thomalla, F. (2012). Determinants of risk: Exposure and vulnerability. *Managing the Risks of Extreme Events and Disasters to Advance Climate Change Adaptation: Special Report of the Intergovernmental Panel on Climate Change.* Available from https://doi.org/10.1017/CBO9781139177245.005.

Cardona, O. D., Bernal, G., & Escovar, M. A. (2023a). *Flood and drought risk assessment, climate change, and resilience. Disaster risk reduction for resilience: Climate change and disaster risk adaptation* ((pp. 191–214). Colombia: Springer International Publishing. Available from https://link.springer.com/book/10.1007/978-3-031-22112-5, http://doi.org/10.1007/978-3-031-22112-5_9.

Cardona, O. D., Bernal, G., & Escovar, M. A. (2023b). *Climate change, food security, and resilience: hydrologic excess and deficit measurement. Disaster risk reduction for resilience: Climate change and disaster risk adaptation* (pp. 333–359). Colombia: Springer International Publishing. Available from https://link.springer.com/book/10.1007/978-3-031-22112-5, http://doi.org/10.1007/978-3-031-22112-5_15.

Cardona, O. D., Bernal, G. A., Zuloaga, D., Escovar, M. A., & Olaya, J. C. (2017). Modelación probabilista de inundaciones en La Mojana. *Informe para el Fondo Adaptación.*

Carreño, M. L. (2006). Técnicas innovadoras para la evaluación del riesgo sísmico y su gestión en centros urbanos: Acciones ex ante y ex post. *Tesis doctoral.*

Carreño, M. L., Cardona, O. D., & Barbat, A. H. (2007). Urban seismic risk evaluation: A holistic approach. *Natural Hazards, 40*(1), 137–172. Available from https://doi.org/10.1007/s11069-006-0008-8.

Carreño, M. L., Cardona, O. D., & Eslamian, S. (2022). *Index of resilience and effectiveness of disaster risk management. In. Handbook of hydroInformatics: Vol. II: Advanced machine learning techniques* (pp. 305–314). Spain: Elsevier. Available from https://www.sciencedirect.com/book/9780128219614, http://doi.org/10.1016/B978-0-12-821961-4.00020-8.

DANE. (2005). Bogotá: Departamento Administrativo Nacional de Estadística-DANE. DDTS-DANE (Dirección de Desarrollo Territorial Sostenible – Departamento Nacional de Planeación). (s.f.) Fichas de caracterización territorial.

Eslamian, S., & Eslamian, F. (2023). *Disaster risk reduction for resilience: Climate change and disaster risk adaptation.* Springer Nature.

García, L. (2014). Funcionamiento del Sistema Natural y Características de los Ecosistemas del Núcleo de Once Municipios de la Región de La Mojana.

García, L. (2015). Valoración integral de la biodiversidad y los servicios ecosistémicos en el núcleo de once municipios de la región de La Mojana.

García, L. (2016). Environmental assessment of intervention alternatives for flood risk mitigation in the core of eleven municipalities in the La Mojana region as part of the action plan formulation. *Proyecto Modelación Hidrodinámica de La Mojana.*

Leibovich, J., Guerrero, P., Llinás, G., Morales, A., & Pereira, M. (2015). Documento de diagnóstico de la problemática con el manejo del recurso hídrico de La Mojana.

Marulanda, M. C., Cardona, O. D., & Barbat, A. H. (2009). Robustness of the holistic seismic risk evaluation in urban centers using the USRi. *Natural Hazards*, *49*(3), 501−516. Available from https://doi.org/10.1007/s11069-008-9301-z.

Marulanda, M. C., Cardona, O. D., Marulanda, P., Carreño, M. L., & Barbat, A. H. (2020). Evaluating risk from a holistic perspective to improve resilience: The United Nations evaluation at global level. *Safety Science*, *127*. Available from https://doi.org/10.1016/j.ssci.2020.104739.

Marulanda, M. C., Carreño, M. L., Cardona, O. D., Ordaz, M. G., & Barbat, A. H. (2013). Probabilistic earthquake risk assessment using CAPRA: Application to the city of Barcelona, Spain. *Natural Hazards*, *69*(1), 59−84. Available from https://doi.org/10.1007/s11069-013-0685-z, http://www.wkap.nl/journalhome.htm/0921-030X.

Marulanda, P., Cardona, O. D., Marulanda, M. C., & Carreño, M. L. (2022). *Unveiling the latent disasters from a holistic and probabilistic view: Development of a national risk atlas. Disaster risk reduction for resilience: Disaster economic vulnerability and recovery programs* (pp. 313−336). Colombia: Springer International Publishing. Available from http://doi.org/10.1007/978-3-031-08325-9_15.

Olaya, J.C. (2015). Flood vulnerability assessment for agricultural cultivated plants: Methodology and testing in La Mojana region (Colombia).

Ordaz, M. (2000). Metodología para la evaluación del riesgo sísmico enfocada a la gerencia de seguros por terremoto.

PNUD. (2019). USD117 Million will be invested in environmental project Mojana: Climate and Life.

Torres, M. A., Jaimes, M. A., Reinoso, E., & Ordaz, M. (2014). Event-based approach for probabilistic flood risk assessment. *International Journal of River Basin Management*, *12*(4), 377−389. Available from https://doi.org/10.1080/15715124.2013.847844, http://www.tandfonline.com/loi/trbm20.

Chapter 11

Impact of climate changes on stream flow discharge of some African rivers

A. Adediji[1], M.O. Ibitoye[2], J.J. Idolor[3] and Saeid Eslamian[4]

[1]Department of Geography, Obafemi Awolowo University, Ile-Ife, Nigeria, [2]Department of Remote Sensing and Geoinformation Science, Federal University of Technology, Akure, Nigeria, [3]Institute of Ecology & Environmental Management, Obafemi Awolowo University, Ile-Ife, Nigeria, [4]Department of Water Sciences and Engineering, College of Agriculture, Isfahan University of Technology, Isfahan, Iran

11.1 Introduction

Stream flow discharge (Q) can be described as the volume of water being moved or transported through the stream channel per unit time. The value of stream flow discharge is usually expressed in $cm^3 \ s^{-1}$ (US) for small stream or $m^3 \ s^{-1}$ for large stream (river). It is expressed quantitatively as the product of stream flow velocity (ms^{-1}) and cross-sectional area (m^2) of the stream channels. Information on stream flow discharge of river is important for determining stream flow sediment discharge (i.e., volume of load/sediments being transported through the river channel) as well as flood monitoring. In fact, it is the major component for computing sediment transport or yield of major rivers at both global and regional levels. Also, the most important climatic elements/parameters that influence stream flow discharge of rivers in drainage basins globally are precipitation and temperature. The long-term effects or impacts of these parameters/variables have been measured in many parts of the world. These two elements (rainfall and temperatures) and their associated characteristics have been found to exert great influence on river discharge of many large rivers. This has been done through the use of Global Climate Change Models (Mujere & Eslamian, 2014).

At this juncture, this chapter attempts to examine the concept of climate change before exploring its impact on river discharge of some African rivers especially large drainage basins in the tropical Africa. Further, change in climate (especially change in precipitation and temperature over a long period of time) has been found to have notable effects on global run off regime and stream flow discharge in many drainage basins of the world (Arora & Boer, 2001; Sperna Weiland et al., 2012). Also, several studies exist on the hydrological consequences of change in climate at a global level/scale (IPCC, 2007; Mechl et al., 2009; Murphy et al., 2018).

A major effect of global climate change is a shift in the frequency, duration, and intensity of extreme rainfall and temperature events (Mishra et al., 2010). According to Merino et al. (2018), impacts of consecutive rainfall and temperature extremes for prolong duration are more intense. Persistent dry periods with little or no rainfall in addition to increased temperature extremes for prolong periods could result in drought conditions. On the other hand, rainfall extremes due to prolonged and continuous increase in intensities of extreme rainfall events such as very heavy rainy days, extremely heavy rainy days, and rainfall events above a threshold (such as rainfall events above 10 mm/day) can result in severe flooding events in drainage basins (Li et al., 2017). Therefore, assessment of the interrelationships between rainfall extremes, temperature extremes, antecedent precipitation index, and streamflow (discharge) is invaluable for water resource management, disaster management, and climate change mitigation and adaptation (Idolor, 2023; Singh et al., 2014). However, this chapter focuses on the review of studies on the impacts of climate change on stream flow discharge of some African rivers. Examples of some of studies on African rivers include those by Idolor (2023), Sperna Weiland et al. (2012), Viviroli et al. (2011), and Qaisrani et al. (2021).

11.2 Literature review

River drainage is a natural land management and environment unit, which can be described as the area contributing water and sediment to the main stream channel and its tributaries. It is also commonly referred to as river catchment or watershed.

Hydrosystem Restoration Handbook. DOI: https://doi.org/10.1016/B978-0-443-29802-8.00011-X

Streams and rivers are dynamic freshwater components of the hydrological cycle and contains about 25% of the world's freshwater. Estimating freshwater supply from streams and rivers involves streamflow discharge/river discharge of drainage basin. However, recent developments in numerical simulation modeling, computers, and geo-information technology have become important and efficient tools for finding solution to diverse water resources and environmental management challenges (Duan et al., 2019). Also, the information on the world's larger river by discharge and length is shown in Table 11.1.

Information on river stage/level is very significant for forecasting and monitoring of river or stream flow discharge. it is also important in establishing and developing effective robust policies and managerial schemes for mitigating the adverse effects of hydrological events such as flood and drought, which are often accompanied by scarcity of water and increased demands for water.

Further, the water level of inland river channel within the watershed is of significance to inland water transport and navigation of vessels and their loading. In this regard, their precise prediction of the average, minimum, and maximum water levels (stage) of rivers across the season is necessary for the maintenance and management of the waterway with the goal of improving the safety, volume, and traffic of humans and other commodities (Seo et al., 2016). Also, Pan et al. (2020) combined hybrid deep neural networks to predict water levels of Yangtze River in China. The Yangtze River is the third longest river in the world and in fact the longest river in the continent of Asia.

Moreover, according to Pan et al. (2020), the hybrid convolutional neural networks and gate recurrent unit (CNN−GRU) model showed higher production accuracy and better performance than wavelet artificial neural networks (ANNs) and Autoregressive Integrated Moving Average models. The study by Pan et al. (2020) showed a higher forecast precision for the GRU model in the dry season and wet season. However, Nguyea et al. (2015) used Randon Forests and support vector regression (SVR) to forecast time series for water level at Thakhek station at the lower Mekong drainage basin situated within the boundaries of Cambodia, Laos, Thailand, and Vietnam in the continent of Asia. The SVR indicated lower prediction error for a 5 lead day scenario, with a mean absolute error of 0.486 m, which was found to be below the Mekong River commission limit of between 0.5 and 0.75 m.

Also, Khan and Coulibaly (2006) asserted the prediction performance of two Machine Learning Models (MLP) and Support Vector Machine (SVM) for modeling the water level/stage of a lake and reported that SVM indicated better performance than MLP. However, Imami et al. (2018) and Sulaiman et al. (2011) used Artififial Neural Network (ANN), extreme learning machine, and relevance vector machine for river water level forecasting. For instance, Seo et al. (2016) used hybrid wavelet packet adaptive neuro fuzzy infertile system, hybrid wavelet packet ANN and hybrid wavelet packet support vector machine to predict/forecast daily water level at Gam station, South Korea. In this regard, it was reported that a combination of the machine learning models and the wavelet packet decomposition can be adopted to monitor and forecast daily water level/stage in drainage basins more accurately than conventional hydrological and hydraulic models.

It is also worth of noting that natural disasters such as flooding is increasingly reoccurring across the world within the last five decades and are becoming an environmental concern. For example, according to reports by the United Nations (2007), hydroclimatic extreme events constitute over 60% of disasters. Floods are the most common

TABLE 11.1 The world's largest rivers by discharge and length.

Name	Rank by discharge volume	Rank by length	Approximate discharge ($m^3\,s^{-1}$)	Approximate length (km)	Approximate discharge area (km^2)	Continent
Amazon	1	2	210,000	6,400	5,800,000	South America
Congo	2	9	40,000	4700	400,000	Africa
Ganges-Brahma	3	23	39,000	2900	1,7730,000	Eurasia
Yangtze	4	3	21,000	4900	1,900,000	Eurasia
Parana-La Plata	5	8	19,000	4900	2,000,000	South America
Yenissei	6	5	17,000	5500	880,000	South America
Mississippi-Missouri	7	4	17,000	6000	880,000	North America
Orinoco	8	27	17,000	2700	880,000	South America
Nile	29	1	5000	6650	2,870,000	Africa

reoccurring natural disasters across the globe and are often associated with displacement of people from their homes, loss of livelihood, loss of life, and environmental stress. The severe floods result from extremely large stream flow or river discharge. This could also be associated with extreme rainfall during monsoon as well as rain failures especially in the semiarid regions in the tropics. For instance, communities along the Inland Niger Delta in Mali have not been spared from the menace of severe flood events that reoccur every year, subjecting the inhabitants of the region to constant fear during the monsoon rainy season; therefore, food vulnerability assessment is increasingly being adopted in most regions of the world for emergency services. For instance, countries in Asia, Pacific region, and West Africa have not been spared by the natural disaster of flooding. In light of the above review, it is evidently clear that climate change indicators such as global warming (increasing temperature), decrease in precipitation amounts, increased frequency of extreme hydroclimatic events (e.g., droughts and floods), sea level rise, melting of ice, exponential population growth (which African countries are well known for), increased industrial growth (typical of industrialized countries of the West), urbanization, and rapid development in flood plains have raised a number of key questions regarding the safety of riverine areas and coastal regions of the world. This chapter, therefore, attempts to review some of the existing studies on the impact of climate change especially extreme climate events on the stream flow discharge and water level of some African rivers.

11.3 Characterization of precipitation/rainfall parameters and temperature as well as stream flow discharge (Q) variability

Some of the parameters or variables of precipitation considered by various studies as having impact or influence on river discharge as documented by Sperna Weiland et al. (2012), Redelsperger et al. (2006), and Arora and Boer (2001) as well as Idolor (2023) using re-analysis data of precipitation on both daily and monthly basis. They all observed that high-frequency variability comes from the re-analyses and not from direct observations. Some of the characteristics of selected basins in Africa are shown in Table 11.2.

Also, in a recent study by Idolor (2023), rainfall and temperature parameters such as extreme rainfall, wet days (R95p), extremely wet days (R99p), rainfall greater than 10 mm (R20mm), maximum 5-day consecutive rainfall (RX5day), maximum 1-day rainfall (Rx1day), simple daily intensity index or average rainfall in rain days (RX5days), minimum temperature, summer days or number of days when maximum temperature is greater than 25°C (SU25), cool nights or minimum temperature less than 10th percentile (TN10p), warm nights or minimum temperature greater than 90th percentile (TN90p), and warm spell duration index (WSDI) in addition to the daily rainfall amounts and antecedent precipitation index (API) were determined and related to the stream flow discharge (Q) and stage/level (H) variables in the Niger Drainage Basin. The details of the procedures involved in the determination/estimation of the selected rainfall/precipitation and temperature parameters enumerated in this section are documented elsewhere (see Idolor, 2023).

Idolor (2023) assessed the interrelationship between rainfall extremes and temperature extremes for four major flood events (Benue, Guinean, Sahelian, and Sudan) within the selected river gauging stations (Jiderebode, Lokoja, and Niamey) of the Niger River Basin (see Fig. 11.1). According to Idolor (2023) the Benue flood event occurs at Lokoja subcatchment of the Lower Niger Basin and the Sahelian flood events occurs at Niamey subcatchment of the Middle Niger River Basin (Aich et al., 2016; Idolor, 2023). The Sudan flood event occurs at the Jiderebode subcatchment of the Lower Niger River Basin (Idolor, 2023), while the Guinean flood event occurs at the both Jiderebode subcatchments of the Lower Niger River Basin and Niamey subcatchments of the Middle Niger River Basin (Idolor, 2023).

TABLE 11.2 Selected drainage basins characteristics such as basin area (A), average observed discharge (Q_{avg}), and location of gauges of some African rivers (Sperna Weiland et al., 2012).

Basin	Area (km^2)	Q_{avg} (m^3 s^{-1})	Gauge's location
Niger	2,117,700	5589	Dire
Orange	973,000	365	Aliwal North
Congo	4,014,500	41,000	Kinshasa
Zambezi	1,390,000	3400	Katoma Mulilo

Source: Adapted from Sperna Weiland F.C., van Beek L.P.H., Kwadijk J.C.J., Bierkens M.F.P., Global patterns of change in discharge regimes for 2100. Hydrology and Earth System Sciences. 16 (4) (2012), 1047–1062, https://doi.org/10.5194/hess-16-1047-2012.

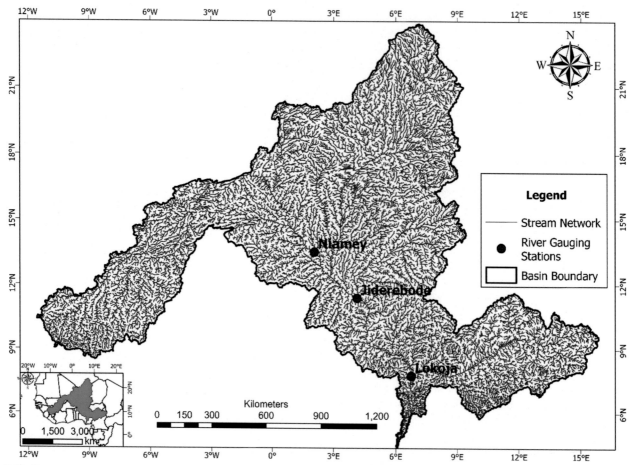

FIGURE 11.1 Niger River Basin hydrographic map and gauging stations (Idolor, 2023). Map lines delineate study areas and do not necessarily depict accepted national boundaries. *Source: From Idolor J.J., Impacts of Climate Change and Land Use/Land Cover Changes on the hydrology of Niger River Basin. (2023).*

11.4 Correlation analysis

The product moment correlation analysis has been adopted in assessing the impact of climate change on hydrological processes especially stream flow discharge or river discharge in many parts of the world including African countries. This was utilized by Sperna Weiland et al. (2012), Arora and Boer (2001), Redelsperger et al. (2006), and Idolor (2023) in assessing the effects of precipitation and temperature characteristics on river discharge of some selected large river basins in Africa. In all these studies, rainfall and temperature parameters selected were correlated with river discharge of some selected large rivers in Africa. For instance, Idolor (2023) employed product moment correlation analysis to determine the relationship between some selected rainfall and temperatures variables as well as river discharge (Q) at four ganging stations within the River Niger Drainage Basin over a period between 1979 and 2020. The results of the study by Idolor (2023) especially in respect of correlation matrix generated for each of the four selected gauging stations in the Niger River Catchment (see Tables 11.3–11.6) displaying the relationship between climate changes variables and river discharge. As evident from the correlation matrix (e.g., see Table 11.3), there was a strong correlation between river discharge (Q), water level (H), and climate variables especially with respect to increasing magnitudes of extreme rainfall and temperatures indices. This further led to increased A-Maximum (AMAX) stream flow and AMAX water level. This further confirmed the findings by Aich et al. (2016). Also, as shown in Tables 11.3–11.6, there was a strong and significant relationship between API and daily stream flow discharge (Q) and daily water level (H). This further revealed the influence of precipitation/rainfall on the hydrological regime of Niger River Basin. The findings of Idolor (2023) in the Niger River catchment further showed that frequency and duration of extreme rainfall and temperature indices such as R95p (very wet days), R99p (extremely wet days), R20mm (very heavy precipitation days), RX5day (maximum consecutive 5-day precipitation), RX1day (maximum 1-day precipitation), SDII (simple daily

TABLE 11.3 Pearson correlation of AMAX streamflow (Guinean) and rainfall indices at Jiderebode Station for the period 1979–2020.

	PR	CDD	CWD	SDII	R10mm	R20mm	R95p	R99p	R95pTOT	R99pTOT	PRCPTOT	RX1day	RX5day	Q
PR	1.0***													
CDD	−0.37*	1.0***												
CWD	0.73***	−0.27	1.0***											
SDII	0.62***	−0.12	0.43**	1.0***										
R10mm	0.97***	−0.26	0.76***	0.61***	1.0***									
R20mm	0.79***	−0.21	0.38*	0.8***	0.77***	1.0***								
R95p	0.79***	−0.24	0.45*	0.9***	0.76***	0.96***	1.0***							
R99p	0.51***	−0.16	0.16	0.83***	0.42**	0.84***	0.86***	1.0***						
R95pTOT	0.47**	−0.23	0.29	0.93***	0.42**	0.71***	0.83***	0.86***	1.0***					
R99pTOT	0.33*	−0.18	0.16	0.8**	0.24	0.59***	0.67***	0.86***	0.91***	1.0***				
PRCPTOT	1.0***	−0.35*	0.74***	0.63***	0.98***	0.79***	0.79***	0.51***	0.47***	0.32*	1.0***			
RX1day	0.56***	−0.25	0.24	0.85***	0.49***	0.76***	0.86***	0.89***	0.86***	0.82***	0.56***	1.0***		
RX5day	0.63***	−0.23	0.39*	0.89***	0.58***	0.75***	0.88***	0.82***	0.88***	0.78***	0.63***	0.95***	1.0***	
Q	0.01	0.15	−0.1	0.36*	−0.02	0.33*	0.29	0.48**	0.4**	0.48**	0.01	0.35*	0.34*	1.0***

*Correlation at 5% significant level. **Correlation at 1% significant level. ***Correlation at 0.1% significant level.
PR, annual total rainfall; CDD, number consecutive dry days; CWD, number of consecutive wet day; SDII, simple daily intensity index. R10mm, number of heavy rainy days; R20mm, number of very heavy rainy days; R95p, number of very wet days; R99p, number of extremely wet days; R95pTOT, rainfall fraction due to very wet days; R99pTOT, total annual wet-day precipitation; PRCPTOT, rainfall fraction due to extremely wet days; RX1day, maximum 1-day precipitation; RX5day, maximum consecutive 5-day precipitation; Q, river discharge (streamflow).

TABLE 11.4 Pearson correlation of AMAX streamflow (Guinean) and rainfall indices at Niamey Station for the period 1979–2020.

	Rainfall	CDD	CWD	SDII	R10mm	R20mm	R95p	R99p	R95pTOT	R99pTOT	PRCPTOT	RX1day	RX5day	Q
Rainfall	1.0***													
CDD	−0.29	1.0***												
CWD	0.71***	−0.24	1.0***											
SDII	0.52***	0.27	0.22	1.0***										
R10mm	0.97***	−0.26	0.75***	0.48**	1.0***									
R20mm	0.74***	−0.01	0.29	0.8***	0.7***	1.0***								
R95p	0.71***	0.04	0.29	0.91***	0.65***	0.96***	1.0***							
R99p	0.36*	0.23	−0.07	0.82***	0.24	0.79***	0.83***	1.0***						
R95pTOT	0.23	0.3	−0.07	0.89***	0.14	0.63***	0.77***	0.86***	1.0***					
R99pTOT	0.12	0.33*	−0.16	0.75***	0.01	0.53***	0.63***	0.87***	0.92***	1.0***				
PRCPTOT	1.0***	−0.28	0.71***	0.53***	0.97***	0.74***	0.71***	0.36*	0.23	0.12	1.0***			
RX1day	0.5***	0.12	0.19	0.79***	0.42**	0.63***	0.75***	0.66***	0.8***	0.75***	0.5***	1.0***		
RX5day	0.52***	0.07	0.24	0.78***	0.46**	0.6***	0.74***	0.59***	0.75***	0.66***	0.52***	0.97***	1.0***	
Q	−0.23	0.24	−0.37*	0.23	−0.29	0.15	0.16	0.41**	0.41**	0.5***	−0.23	0.3*	0.25	1.0***

*Correlation at 5% significant level. **Correlation at 1% significant level. ***Correlation at 0.1% significant level.
PR, annual total rainfall; CDD, number consecutive dry days; CWD, number of consecutive wet day; SDII, simple daily intensity index. R10mm, number of heavy rainy days; R20mm, number of very heavy rainy days; R95p, number of very wet days; R99p, number of extremely wet days; R95pTOT, rainfall fraction due to very wet days; R99pTOT, rainfall fraction due to extremely wet days; PRCPTOT, total annual wet-day precipitation; RX1day, maximum 1-day precipitation; RX5day, maximum consecutive 5-day precipitation; Q, river discharge (streamflow).

TABLE 11.5 Pearson Correlation of AMAX streamflow (Sahelian) and rainfall indices at Niamey Station for the period 1979–2020.

	Rainfall	CDD	CWD	SDII	R10mm	R20mm	R95p	R99p	R95pTOT	R99pTOT	PRCPTOT	RX1day	RX5day	Q
Rainfall	1.0***													
CDD	−0.29	1.0***												
CWD	0.71***	−0.24	1.0***											
SDII	0.52***	0.27	0.22	1.0***										
R10mm	0.97***	−0.26	0.75***	0.48**	1.0***									
R20mm	0.74***	−0.01	0.29	0.8***	0.7***	1.0***								
R95p	0.71***	0.04	0.29	0.91***	0.65***	0.96***	1.0***							
R99p	0.36*	0.23	−0.07	0.82***	0.24	0.79***	0.83***	1.0***						
R95pTOT	0.23	0.3	−0.07	0.89***	0.14	0.63***	0.77***	0.86***	1.0***					
R99pTOT	0.12	0.33*	−0.16	0.75***	0.01	0.53***	0.63***	0.87***	0.92***	1.0***				
PRCPTOT	1.0***	−0.28	0.71***	0.53***	0.97***	0.74***	0.71***	0.36*	0.23	0.12	1.0***			
RX1day	0.5***	0.12	0.19	0.79***	0.42**	0.63***	0.75***	0.66***	0.8***	0.75***	0.5***	1.0***		
RX5day	0.52***	0.07	0.24	0.78***	0.46**	0.6***	0.74***	0.59***	0.75***	0.66***	0.52***	0.97***	1.0***	
Q	−0.02	0.05	−0.08	0.41**	−0.1	0.21	0.33*	0.47***	0.51***	0.5***	−0.02	0.35*	0.38*	1.0***

*Correlation at 5% significant level. **Correlation at 1% significant level. ***Correlation at 0.1% significant level.
PR, annual total rainfall; CDD, number consecutive dry days; CWD, number of consecutive wet day; SDII, simple daily intensity index. R10mm, number of heavy rainy days; R20mm, number of very heavy rainy days; R95p, number of very wet days; R99p, number of extremely wet days; R95pTOT, rainfall fraction due to very wet days; R99pTOT, rainfall fraction due to extremely wet days; PRCPTOT, total annual wet-day precipitation; RX1day, maximum 1-day precipitation; RX5day, maximum consecutive 5-day precipitation; Q, river discharge (streamflow).

TABLE 11.6 Pearson Correlation of AMAX streamflow (Benue) and rainfall indices at Lokoja Station for the period 1979–2020.

	Rainfall	CDD	CWD	SDII	R10mm	R20mm	R95p	R99p	R95p TOT	R99p TOT	PRCP TOT	RX1day	RX5day	Q
Rainfall	1.0***													
CDD	−0.42**	1.0***												
CWD	0.79***	−0.3	1.0***											
SDII	0.6***	−0.11	0.38*	1.0***										
R10mm	0.98***	−0.34*	0.8***	0.6***	1.0***									
R20mm	0.8***	−0.22	0.46**	0.85***	0.79***	1.0***								
R95p	0.76***	−0.25	0.46**	0.92***	0.75***	0.98***	1.0***							
R99p	0.42**	−0.08	0.09	0.85***	0.35*	0.8***	0.85***	1.0***						
R95pTOT	0.39*	−0.19	0.19	0.92***	0.35*	0.7***	0.8***	0.87***	1.0***					
R99pTOT	0.22	−0.1	0.01	0.77***	0.15	0.54***	0.63***	0.85***	0.92***	1.0***				
PRCPTOT	1.0***	−0.4**	0.79***	0.61***	0.99***	0.81***	0.77***	0.42**	0.39*	0.22	1.0***			
RX1day	0.51***	−0.28	0.23	0.87***	0.45**	0.77***	0.85***	0.89***	0.92***	0.86***	0.51***	1.0***		
RX5day	0.58***	−0.28	0.4**	0.89***	0.55***	0.77***	0.86***	0.8***	0.91***	0.78***	0.58***	0.95***	1.0***	
Q	0.36*	−0.18	0.09	0.62***	0.31*	0.62***	0.67***	0.76***	0.63***	0.6***	0.36*	0.72***	0.64***	1.0***

*Correlation at 5% significant level. **Correlation at 1% significant level. ***Correlation at 0.1% significant level.
PR, annual total rainfall; CDD, number consecutive dry days; CWD, number of consecutive wet day; SDII, simple daily intensity index. R10mm, number of heavy rainy days; R20mm, number of very heavy rainy days; R95p, number of very wet days; R99p, number of extremely wet days; R95pTOT, rainfall fraction due to very wet days; R99pTOT, rainfall fraction due to extremely wet days; PRCPTOT, total annual wet-day precipitation; RX1day, maximum 1-day precipitation; RX5day, maximum consecutive 5-day precipitation; Q, river discharge (streamflow).

intensity index), SU25, TN10p, TN90p, and WSDI in addition to daily rainfall amount and API were found to be responsible for extreme hydrological events such as floods and droughts in the Niger Drainage Basin.

In a similar vein, it was also shown in the Lower Niger River Basin at the Lokoja subcatchment in Nigeria by Idolor (2023) that there was a strong relationship between rainfall indices and river discharge as well as water level. It was observed that the peak flow at the Lokoja subcatchment increased significantly due to synergistic effect of combination of rainfall extreme indices (SDII, total annual wet-day precipitation, R99p, RX1day, rainfall fraction due to very wet days, R20mm, and RX5day) and temperature extreme indices (SU25, TN10p, TN90p, and TR20).

Idolor (2023) assessed the interrelationship between rainfall extremes and temperature extremes for four major flood events (Benue, Guinean, Sahelian, and Sudan) within the selected river gauging stations (Jiderebode, Lokoja, and Niamey) of the Niger River Basin. According to Idolor (2023), the Benue flood event occurs at the Lokoja subcatchment of the Lower Niger Basin and the Sahelian flood events occurs at the Niamey subcatchment of the Middle Niger River Basin (Aich et al., 2016; Idolor, 2023). The Sudan flood event occurs at the Jiderebode subcatchment of the Lower Niger River Basin (Idolor, 2023), while the Guinean flood event occurs at the both the Jiderebode subcatchment of the Lower Niger River Basin and the Niamey subcatchment of the Middle Niger River Basin (Idolor, 2023).

In addition, there was also a very strong correlation between extreme rainfall and temperature events. This further account for the increased frequency of recurrent severe flood events at the Lokoja subcatchment of the Niger River Basin. Also, at the Middle Niger River Basin at the Niamey subcatchment, a very strong relationship exists between stream flow discharge and rainfall temperature indices (see Table 11.5).

11.5 Multiple regression analysis

Multiple regression algorithm has also been used in investigating the effects of many precipitation and temperature variables or parameters on the output such as evapotranspiration and river discharge (Q) through the drainage basin outlet of some selected drainage basins in Africa. In these multiple regressions analysis, regressions of a model outputs taken as independent variables, e.g., rainfall amount, maximum (MAX) rainfall intensity, API, MAX temperature, and minimum daily temperature (cool night) on some model outputs such as evapotranspiration (ET) and river discharge. Therefore, linear cause effect links can be considered to be the link discovered through the regression analysis, which can thus be interpreted as the effect or impact of selected precipitation and temperature characteristics on the basin river discharge. In this regard, for each multiple linear regression, the percentage of variance was explained for each predictor/each independent variable and level of significance at $\beta = 0$ (see results in Table 11.7 obtained by Tristian and Polcher (2008). This was also demonstrated in the recent study by Idolor (2023) in the Niger Drainage Basin where MAX rainfall intensity and API were found to be important predictors of stream flow discharge of River Niger. The result on the two main characteristics such as annual precipitation (P) and seasonal length (L) by Tristian and Polcher (2008) in Niame (NIN), Bamboi (BaB), Aniassu (Anc), Koulikoro (KON), and Garoua in the Niger River catchment is shown in Table 11.7.

For instance, according to Tristian and Polcher (2008), at KON station, every rainfall characteristic in the regression is substantial at 1% level of significance ($\alpha = 0.01$), while negative β_p at Eds (Edea) as evident from Table 11.7 means that the same annual rainfall, clustered into multiple events when above 10 mm per day and distributed into less multiple events when below 10 mm per day or into higher intensity events, will result in diminished ET over the basin. Both anomalies (multiple events and a higher intensity of events above the threshold of 10 mm per day for the same annual precipitation) tallies with a rainfall distribution with higher standard deviation but unchanged mean and ultimately to elevated intraseasonal rainfall variability (see Tristian and Polcher, 2008).

TABLE 11.7 Summary of regressions of annual evapotranspiration (E) (in mm J^{-1}) on independent variables P (annual precipitation in mm J^{-1}) and seasonal length (L) (in days).

Regression coefficient	NIN	BaB	KON	AnC	GaB	Eds (Edea)
R^2	0.89	0.49	0.41	0.53	0.12	0.34
β_p (X10^{-1})	3.6**	1.8**	0.60	2.5**	0.60	0.60
β_L (X10^{-3})	2.4**	2.5**	2.8**	1.6	1.9	2.90**

where R^2 is the goodness of fit and β is the regression coefficient: *$\beta \neq 0$ at 5% level and **$\beta \neq 0$ at 1% level of significance.
NIN, Niame; *BaB*, Bamboi; *Anc*, Aniassu; *KON*, Koulikoro; *GaB*, Garoua.

11.6 Models

11.6.1 Global Climate Models

Global Climate Models (GCMs) are sophisticated models that are capable of solving numerical equations to simulate the physical processes in the atmosphere, cryosphere, oceans (hydrosphere), and land surface as well as the responses of the global climate system (Ipcc et al., 2013). According to Laddimath and Patil (2019), the IPCC models are able to describe changes in climate characteristics and patterns as well as provide projections for past and future climate as outputs and reliable weather and climate forecasting, which are made possible by utilizing or employing GCMs. Also, various hydrological parameters/variables can be computed or estimated using these models (see Wilby et al., 2009).

It must be noted that dynamic downscaling (DD) and statistical downscaling (SD) are the main downscaling techniques utilized for climate change impact assessment studies (Laddimath & Patil, 2019). DD can be described as a technique for embedding high-resolution regional climate model into GCM (see Wilby et al., 1998). It is also a physical-based model with very sophisticated design as well as computationally expensive to operate. On the other hand, SD involves deriving the local climate from the large-scale atmospheric characteristics and patterns represented in the GCM output through utilizing the statistical relationship between the large-scale climate output variables called predictors of the various climate change scenarios as well as the local-scale climate variables. The statistical downscaling techniques is increasingly being adopted by the science research because it is easy to implement and computationally cheaper. However, understanding the concept of the climate system and the evolution and development of the drainage basin systems are very important for the development of numerical climate models and hydrological models for simulating hydrological processes within the drainage basin (Eslamian et al., 2017).

Also, since there are close connections between water and land in the hydrologic cycle, drainage basin-scale modeling is fast being adopted as the standard for monitoring and modeling the hydrological processes such as streamflow discharge, modeling nutrient transport, and determining pollutant loads in river systems.

Specifically, the detailed evaluation of global hydrological (i.e., PCR-GLOBWB) model was performed by Van Beek et al. (2011). The PCR-GLOBWB model cell contains two vertical layers and one underlying by groundwater reservoir water infiltrates the cell as precipitation and could also be stored as canopy interception. Also, in the model, evapotranspiration (E) is estimated from potential evapotranspiration and soil moisture conditions. Runoff (runs-off) is estimated as aggregate of noninfiltrating water as well as throughfall (streamflow) and saturated excess surface.

11.6.2 Artificial intelligence models

Artificial intelligence (AI) models such as ANN consist of networks of several processing units of interconnected nodes or neurons, emulating the brain (Haykin, 2009). AI models are capable of determining relationships between input variables such as climate data in forecasting streamflow discharge by means of mathematical computation through gradient descent (Hsu et al., 1997). In recent years, various standard ANN architectures have been utilized in science research for forecasting climate and streamflow discharge such as CNNs, GRU, long- and short-term memory (Duan et al. 2019), and recurrent neural networks. ANNs have no restraints and do not require prior assumptions when compared with conventional hydrological models. AI hydrological models are flexible in simulating hydroclimatic processes. AI models were used to simulate streamflow discharge in River Niger Drainage Basin and forecast the effects of climate change in Niger River Basin (see Idolor, 2023).

11.7 Conclusion

A review of some of the existing studies on the impacts of change in climate (climate change) on river discharge (streamflow) of some large rivers in Africa such as Niger, Orange, Congo, and Zambezi was undertaken in this chapter. The methods of analysis, especially models (e.g., GCMs and correlation and multiple regression algorithms/models) adopted in processing of data (re-analyzed data) using AI, were also discussed in this study and some of its findings related to effects of precipitation/rainfall and temperature parameters/variables were briefly discussed. Most of the rainfall indices and temperature parameters such as MAX rainfall, R > 20mm, API, SDII, and SU25 and TN10p and TN90p, respectively, were found to be positively and significantly related to streamflow/river discharge and water level (H) within the River Niger Drainage Basin in West Africa Subregion (see Idolor, 2023). This chapter, therefore, recommends further study on the impacts/effects of land use/land cover on the hydrological processes (such as streamflow discharge and water level) of African rivers especially those in the Tropical African Regions of the world. This will help in documenting the effects of various land use types on river discharge of African rivers.

References

Aich, V., Koné, B., Hattermann, F. F., & Paton, E. N. (2016). Time series analysis of floods across the Niger River Basin. *Water (Switzerland)*, *8*(4). Available from https://doi.org/10.3390/w8040165, http://www.mdpi.com/2073-4441/8/4/165/pdf.

Arora, V. K., & Boer, G. J. (2001). Effects of simulated climate change on the hydrology of major river basins. *Journal of Geophysical Research Atmospheres*, *106*(4), 3335–3348. Available from https://doi.org/10.1029/2000JD900620, http://onlinelibrary.wiley.com/journal/10.1002/(ISSN) 2169-8996.

Beek., Stich, S., & Vischel, T. (2011). Assessing water balance in the Sahel: Impact of small-scale rainfall variability on runoff. Part II: Idealized modelling of runoff sensitivity. *J. Hydrology*, *33*, 340–355.

Duan, Q., Pappenberger, F., Wood, A., Cloke, H. L., & Schaake, J. C. (2019). *Handbook of hydrometeorological ensemble forecasting*. Berlin, Heidelberg: Springer. Available from https://doi.org/10.1007/978-3-642-40457-3.

Eslamian, S., Safavi, H. R., Gohari, A., Sajjadi, M., Raghibi, V., & Zareian, M. J. (2017). *Climate change impacts on some hydrological variables in the Zayandeh-Rud River Basin. Iran reviving the dying giant: Integrated water resource management in the Zayandeh Rud catchment* (pp. 201–217). Iran: Springer International Publishing. Available from http://www.springer.com/in/book/9783319549200, https://doi.org/10.1007/ 978-3-319-54922-4_13.

Haykin, S. S. (2009). *Neural networks and learning machines*. Prentice Hall/Pearson, 2009.

Hsu, K. L., Gao, X., Sorooshian, S., & Gupta, H. V. (1997). Precipitation estimation from remotely sensed information using artificial neural networks. *Journal of Applied Meteorology*, *36*(9), 1176–1190. Available from http://www.ametsoc.org/AMS/pubs/newlinks3_jrnlpges.html, https://doi.org/ 10.1175/1520-0450(1997)036%3C1176:PEFRSI%3E2.0.CO;2.

Idolor J. J. (2023). Impacts of Climate Change and Land Use/Land Cover Changes on the hydrology of Niger River Basin. 2023.

IPCC. (2007). *Contribution of Working Group 1 to the Fourth Assessment Report of the Intergovernmental Panel on Climate Change. The physical science basis* (pp. 9–96). Cambridge University Press.

Ipcc, T. F., Stocker., Qin, D., Plattner, G.-K., Tignor, M., Allen, S. K., Boschung, J., Nauels, A., & Xia, Y. (2013). *Contribution of working group I to the fifth assessment report of the intergovernmental panel on climate change*. Cambridge: Cambridge University Press, 2013.

Khan, M. S., & Coulibaly, P. (2006). Application of support vector machine in lake water level prediction. *Journal of Hydrologic Engineering*, *11*(3), 199–205.

Laddimath, R. S., & Patil, N. S. (2019). Artificial neural network technique for statistical downscaling of global climate model. *Mapan - Journal of Metrology Society of India*, *34*(1), 121–127. Available from https://doi.org/10.1007/s12647-018-00299-0, http://www.metrologyindia.org/.

Li, Z., Li, Y., Shi, X., & Li, J. (2017). The characteristics of wet and dry spells for the diverse climate in China. *Global and Planetary Change*, *149*, 14–19. Available from https://doi.org/10.1016/j.gloplacha.2016.12.015, http://www.sciencedirect.com/science/journal/09218181.

Mechl, D. S., Nagner, C. R., Rehmel, M. S., Oberg, K. A. & Rainville, F. (2009). Measuring Discharge with Acoustic Doppler Current Profilers from a moving boat, p. 72, Virginia (USA).

Merino, A., Fernández-González, S., García-Ortega, E., Sánchez, J. L., López, L., & Gascón, E. (2018). Temporal continuity of extreme precipitation events using sub-daily precipitation: Application to floods in the Ebro basin, northeastern Spain. *International Journal of Climatology*, *38*(4), 1877–1892. Available from https://doi.org/10.1002/joc.5302.

Mishra, A. K., Singh, V. P., & Jain, S. K. (2010). Impact of global warming and climate change on social development. *Journal of Comparative Social Welfare*, *26*, 239–260.

Murphy, P., Orellana-Alvear, J., Williams, P. & Celleric, R. (2018). Flash Flood Forecasting in an Andean Mountain Catchment: Development of a Step-wise Methodology Based on Water Cycle, pp. 1–24.

Mujere, N., & Eslamian, S. (2014). Climate Change Impacts on Hydrology and Water Resources, in Handbook of Engineering Hydrology. In S. Eslamian (Ed.), *Modeling, Climate Changes and Variability* (Ch. 7, 2, pp. 113–126). USA: Taylor and Francis, CRC Group.

Qaisrani, Z. N., Nuthammachot, N., Techato, K., & Asadullah. (2021). Drought monitoring based on standardized precipitation index and standardized precipitation evapotranspiration index in the arid zone of Balochistan province, Pakistan. *Arabian Journal of Geosciences*, *14*(11), 1–13.

Redelsperger, J. L., Thorncroft, C. D., Diedhiou, A., Lebel, T., Parker, D. J., & Polcher, J. (2006). African monsoon multidisciplinary analysis: An international research project and field campaign. *Bulletin of the American Meteorological Society*, *87*(12), 1739–1746. Available from https:// doi.org/10.1175/BAMS-87-12-1739.

Singh, D., Tsiang, M., Rajaratnam, B., & Diffenbaugh, N. S. (2014). Observed changes in extreme wet and dry spells during the south Asian summer monsoon season. *Nature Climate Change*, *4*(6), 456–461. Available from https://doi.org/10.1038/nclimate2208, http://www.nature.com/nclimate/ index.html.

Sperna Weiland, F. C., van Beek, L. P. H., Kwadijk, J. C. J., & Bierkens, M. F. P. (2012). Global patterns of change in discharge regimes for 2100. *Hydrology and Earth System Sciences*, *16*(4), 1047–1062. Available from https://doi.org/10.5194/hess-16-1047-2012.

Tristian, O., & Polcher, J. (2008). Impacts of precipitation events and land use changes on West African river discharges during the years 1951–2000. *Clim. Dyn*, *31*, 249–262.

Viviroli, D., Archer, D. R., Buytaert, W., Fowler, H. J., Greenwood, G. B., Hamlet, A. F., Huang, Y., Koboltschnig, G., Litaor, M. I., López-Moreno, J. I., Lorentz, S., Schädler, B., Schreier, H., Schwaiger, K., Vuille, M., & Woods, R. (2011). Climate change and mountain water resources: Overview and recommendations for research, management and policy. *Hydrology and Earth System Sciences*, *15*(2), 471–504. Available from https://doi.org/10.5194/hess-15-471-2011.

Chapter 12

Stream morphology changes under climate changes

Bhawana Nigam

Department of Geography, Anugrah Narayan College, Patliputra University, Patna, Bihar, India

12.1 Introduction

In the contemporary era, climate change has become a paramount environmental concern, exerting profound global impacts on natural systems. Among the diverse impacts of climate change, alterations in stream morphology have gained increasing attention due to their significant implications for water resources, ecosystem dynamics, and human communities. Streams, with their intricate network of channels, serve as vital corridors for water flow, sediment transport, and aquatic habitat. However, the changing climate poses numerous challenges to the physical features and characteristics of streams, giving rise to a need for in-depth analysis and understanding (Yousefi et al., 2016).

12.1.1 Study limitations

1. **Data availability and quality:** The study's effectiveness may be constrained by limitations in data availability, particularly regarding historical stream morphology and climate data, which could impact the robustness of the analyses and models used.
2. **Scope of interdisciplinary research:** While advocating for interdisciplinary research, the study's own interdisciplinary approach might encounter challenges in achieving depth across all relevant disciplines due to resource constraints or expertise limitations.
3. **Generalizability:** Findings and strategies developed within the study context may not be universally applicable to all stream systems, as the impacts of climate change on streams can vary significantly based on geographic location, hydrological regime, and local anthropogenic influences.
4. **Temporal and spatial scale:** The study's duration and spatial coverage may not capture longer-term trends or regional variations in stream morphology response to climate change, potentially limiting the comprehensive understanding of these complex interactions.
5. **Stakeholder engagement and implementation:** The effectiveness of sustainable management practices relies heavily on stakeholder engagement and commitment, which can be challenging to achieve fully within the study's scope and timeline.
6. **Unknown future conditions:** Climate change impacts are subject to uncertainty and evolving conditions, and the study's predictions and recommendations may need continual refinement to remain relevant and effective over time.
7. **Financial and institutional constraints:** The feasibility of implementing recommended strategies may be hindered by financial limitations or institutional barriers, impacting the practical application of study outcomes in real-world settings.
8. **Long-term monitoring and evaluation:** The study's ability to assess the long-term effectiveness of proposed strategies may be limited by the availability of resources and ongoing support for monitoring and evaluation efforts beyond the study period.

Acknowledging these limitations is crucial for interpreting the study's outcomes accurately and for informing future research and practical interventions aimed at addressing the challenges posed by climate change on stream morphology and ecosystem health.

Hydrosystem Restoration Handbook. DOI: https://doi.org/10.1016/B978-0-443-29802-8.00012-1

The focus of this chapter is to offer valuable perspectives on how climate changes influence stream morphology in diverse geographical regions. By examining diverse examples from various regions, one can gain an inclusive recognition of the complex interplay between changes in climate and stream morphology. Moreover, exploring the spatial dimensions of these changes allows one to identify common patterns, as well as region-specific responses, contributing to effective management strategies and adaptation plans.

12.2   Methodology

Within this context, the following sections will delve into the key issues and challenges associated with stream morphology under climate change. Through a spatial analysis lens, it will explore the hydrological alterations, erosion and sedimentation dynamics, channel adjustments, thermal regime changes, riparian zone dynamics, and the influence of glacier and snowmelt contributions. By examining these aspects across different geographical spaces, it can elucidate the underlying mechanisms, identify common trends, and address the specific challenges posed by changes in climate on stream morphology.

12.3   Objective

This analysis will help develop a holistic understanding of the impacts of changes in climate on stream morphology, aiding policymakers, researchers, and practitioners in formulating effective strategies to mitigate these effects and foster the resilience of stream ecosystems. By recognizing the challenges ahead, one can work toward sustainable management practices that safeguard the integrity and functionality of streams, ensuring their continued provision of essential ecosystem services and maintaining the well-being of both natural systems and human societies.

12.4   Stream morphology

Stream morphology pertains to the physical attributes and distinctive features exhibited by a stream or river system. It encompasses the shape, size, dimensions, and arrangement of the stream channel, as well as the distribution and composition of sediments within the channel and along its banks. Understanding stream morphology is crucial for comprehending the behavior, function, and dynamics of streams and their ecosystems (Juracek & Fitzpatrick, 2022).

12.5   Key components of stream morphology

12.5.1   Channel shape and pattern

The channel shape refers to the cross-sectional profile of the stream, which can vary from V-shaped to U-shaped, depending on factors such as sediment composition, flow velocity, and stream gradient. Channel pattern refers to the plan view arrangement of the stream, including meandering (sinuous) channels, braided channels, or straight channels.

12.5.2   Streamflow

Streamflow is a critical aspect of stream morphology, encompassing both the quantity (discharge) and quality (water chemistry) of water flowing through the channel. Streamflow dynamics, including base flow (continuous flow between precipitation events) and peak flows (high flows during storm events), influence sediment transport, channel erosion, and the overall shape of the stream.

12.5.3   Sediment dynamics

Streams transport and deposit sediments, shaping the channel and influencing the stream morphology. The size, composition, and distribution of sediments within the stream channel are crucial components of stream morphology. Sediments can range from fine particles like silt and clay to coarser materials like sand, gravel, and even boulders. The stability of the stream banks, as well as the vegetation and landforms within the riparian zone (the area adjacent to the stream), plays a significant role in stream morphology. Vegetation within the riparian zone helps stabilize banks, mitigates erosion, and influences sediment input into the stream.

12.5.4 Channel adjustments

Streams inherently possess the natural inclination to adapt their channel morphology based on variations in sediment supply, water flow, and various environmental factors. Channel adjustments can occur through processes such as erosion, deposition, widening, narrowing, meander migration, and avulsion (sudden channel shifting). These adjustments influence stream morphology and the overall shape and dynamics of the stream.

12.6 Importance of understanding stream morphology

Understanding stream morphology is crucial for the following reasons.

12.6.1 Ecosystem functioning

Stream morphology influences habitat availability for aquatic organisms, including fish, insects, and plants. It affects water quality, nutrient cycling, and the availability of food and shelter within the stream ecosystem.

12.6.2 Water resource management

Stream morphology influences water storage, flow patterns, and flood risk. Knowledge of stream morphology is essential for sustainable water resource management, including water supply planning, floodplain management, and maintaining water quality.

12.6.3 Land use planning and engineering

Understanding stream morphology is vital for designing infrastructure, such as bridges, culverts, and levees, that interact with stream systems. Furthermore, it plays a vital role in informing land use planning decisions by considering the probable effects of human activities on river morphology.

12.6.4 Conservation and restoration

Stream morphology assessments provide valuable information for conservation initiatives and restoration projects aimed at enhancing ecological integrity, protecting sensitive habitats, and improving water quality.

A comprehensive understanding of stream morphology allows for a more accurate assessment of how changes in climate, anthropogenic activities, and natural processes impact streams. This knowledge enables effective management strategies, conservation efforts, and restoration actions to preserve and sustain the health and functionality of stream ecosystems.

12.7 Understanding the effects of changes in climate

The effects of changes in climate on stream morphology can be substantial, resulting in alterations to the overall structure and functioning of river systems (Sangam et al., 2020). Here, we present some of the perspectives on stream morphology changes under the climate change.

12.7.1 Hydrological alterations

The climate change influences precipitation patterns, causing shifts in the timing, concentration, and duration of rainfall and melting of snow. These alterations in the hydrological regime can result in changes in streamflow patterns, such as increased peak flows, altered base flows, and shifts in the timing of high- and low-flow events. These hydrological changes can have direct impacts on stream morphology.

12.7.2 Erosion and sedimentation

Changes in streamflow patterns can influence erosion and sediment transport processes within streams. Increased peak flows and higher sediment loads can lead to enhanced erosion and sedimentation, altering channel morphology. Streams may experience increased bank erosion, widening or narrowing of channels, and changes in sediment deposition patterns. This can affect habitat availability for aquatic species and may increase the risk of flooding.

12.7.3   Channel adjustments

Streams have a natural tendency to adjust their channels to accommodate changes in flow and sediment dynamics. Amidst climate change, the frequency and magnitude of extreme events like floods and droughts could see an escalation. These events can trigger channel adjustments, including channel widening, deepening, or avulsion (sudden shift of the main channel to a new course). These adjustments can alter the overall morphology of the stream and impact adjacent ecosystems and human infrastructure.

12.7.4   Thermal regimes

Climate changes can influence water temperatures in streams. Rising air temperatures can lead to increased water temperatures, particularly during periods of low flow. Higher water temperatures can impact aquatic organisms' physiology, behavior, and distribution. Changes in stream morphology, such as channel narrowing or reduced shading from vegetation, can further exacerbate thermal stress in streams.

12.7.5   Riparian zone dynamics

Climate change can also impact the vegetation and dynamics of riparian zones, which are the areas adjacent to streams. Changes in temperature and precipitation patterns can affect riparian vegetation growth and composition. Alterations in vegetation can influence bank stability, sediment input, and shading of the stream, ultimately affecting stream morphology.

12.7.6   Glacier and snowmelt contributions

In regions where glaciers and snowpack exert a substantial influence on streamflow generation, changes in climate-induced reductions in glacier volume and changes in snow accumulation and melt patterns can have profound effects on stream morphology. Glacier retreat can reduce sediment supply to downstream reaches, altering channel morphology and sediment dynamics.

12.8   Hydrological alterations in streams caused by climate change

a. The research on "climate change and water resources in the Columbia Basin, British Columbia, Canada, and Washington, USA" (Stewart et al., 2000) examines changes in hydrological regimes, including streamflow, snowpack, and water availability, due to projected change in the climate. "The study utilized a basin hydrology model to estimate changes in natural streamflow under various scenarios. These scenarios indicated potential shifts towards earlier seasonal peaks, along with possible reductions in total annual flow and lower minimum flows. The results from both exercises demonstrated similar outcomes, indicating a trend towards reduced reliability in meeting objectives for power production, fisheries, and agriculture" (Stewart et al., 2000). However, the impact on flood control objectives varied across scenarios, with some showing no significant changes while others displaying reduced reliability.

This analysis shows that although "the existing high level of development and management in the Columbia Basin persists, vulnerabilities persist, and potential impacts from natural streamflow changes due to global climate change cannot be ignored" (Stewart et al., 2000).

The lessons underscore the complexity of the interactions between the Columbia Basin and its changing hydrological conditions. It serves as a reminder that despite significant development and management efforts, vulnerabilities remain, and proactive measures are necessary to deal with the potential consequences of climate change on water resources within the basin.

1. In the study "Hydrologic understanding of Indian sub-continental river basins to climate change" (Mishra & Lilhare, 2016), the "Soil Water Assessment Tool" (Mishra & Lilhare, 2016) was utilized to assess the hydrologic sensitivity of Indian subcontinental river basins. The findings indicate that most of the Indian subcontinental river basins are projected to encounter a shift toward a warmer and wetter climate in the future. It was observed that alterations in

precipitation and temperature have a greater influence on surface runoff than evapotranspiration. These results suggest that the hydrologic cycle in these basins is liable to strengthen under the predicted future climate conditions.

2. Another study estimated the "potential alterations to the flow regime due to variation in climate and reservoir actions for most of the tributaries of the San Joaquin River Basin, California, United States" (Muskey L. et al., 2022). The study specifically examined the sensitivity of freshwater aquatic ecosystems to alterations in the flow routine caused by human activities, including river control and the effects of climate change caused by atmospheric warming. The findings demonstrated that both changes in climate and reservoir operations play a substantial role in alterations in the flow routine of rivers worldwide. The modification of natural flow patterns, manipulated by these drivers, can have profound implications for freshwater aquatic ecosystems.

These studies provide perspectives into the hydrological alterations resulting from changes in climate in different regions. They utilize various methodologies, including hydrological modeling and climate projections, to assess the impacts on water resources. These studies can serve as valuable references for understanding the potential hydrological changes caused by climate change and their effects on water management and adaptation strategies.

12.9 Erosion and sedimentation in streams caused by climate change

The study "Climate change impacts on sediment yield and soil erosion in a watershed is a study on the Gaoping River, in Taiwan" (Chen et al., 2020) aimed to assess "the effect of climate change on sediment yield difference, sediment transport, and erosion deposition distribution at the water divide scale" (Chen et al., 2020). To supervise the assessment, the researchers utilized the "Physiographic Soil Erosion Deposition (PSED) model" (Cheng et al., 2020a, 2020b). The findings of the study revealed a significant increase in soil erosion and deposition volume under the A1B-S change in the climate circumstances. Moreover, the condition was found to worsen with increasing return periods. Specifically, the total erosion amount and sediment yield in the dividing line displayed increases in the range from 4% to 25% and 8% to 65%, respectively. Additionally, deposition volumes showed an increase in the range from 2% to 23%. These outcomes highlight the significant influence of climate change unevenness on the watershed, leading to raised sediment yields. This exacerbates the potential impacts of natural disasters on the region.

1. "Climate change as a significant driver of soil erosion and sediment delivery to water bodies: An analysis of the Elbe Basin in Germany" has been reviewed by the researchers (Uber et al., 2022). A spatially distributed erosion rate and sediment delivery simulation using the "WaTEM/SEDEM model" was conducted using data from 193 locations within the Elbe River Basin (Uber et al., 2022). It was found that the greatest erosion rates were observed in hilly arable lands located in the middle and northeast of the basin. Despite variations among climate models, emission scenarios, and erosion model suspicions, there is a high probability of soil erosion and sediment delivery in the future. Study results show that soil erosion and sediment delivery could increase up to 14% using the median values of climate models and behavioral erosion models (Uber et al., 2022). These results highlight the magnitude of considering climate change in soil erosion and sediment delivery assessments.

2. Another study meant to explore "the effects of climate change on sediment production and transfer processes in hillslopes and channels, particularly in temperature-sensitive Alpine environments" (Hirschberg et al., 2020). The researchers employed a combination of a "stochastic weather generator model" and the latest climate change projections to feed a "hillslope-channel sediment cascade model," focusing on the Ill graben, a significant "debris-flow system in the Swiss Alps" (Hirschberg et al., 2020). By utilizing this approach, they were able to calculate the climate change impacts and associated doubts on sediment yield and the episode of debris flows at an "hourly temporal resolution." The study concluded that the estimated changes in rainfall and air temperature would lead to a decrease in both sediment yield (by 48%) and the occurrence of debris flows (by 23%). This decrease can be attributed to a decline in sediment amount from hillslopes, primarily driven by "frost-weathering processes." In order to emphasize how sensitive expected changes in sediment output and debris flow hazard are to basin elevation, the researchers also ran a model experiment, highlighting the significance of evaluating natural risks and hazards in mountainous situations.

12.10 Channel adjustments in streams caused by climate change

a. The paper inspects "the channel dynamics of the Basento River in Italy" (de Musso et al., 2020) over the past 150 years, during which human actions such as "hydraulic efforts, gravel mining, and changes in climates have had a major impact on the river." The study quantifies the changes in channel width and depth through historical maps, aerial images, and geomorphological measurements. The findings show that the channel significantly narrowed throughout the 20th century, especially from the 1950s to the 1990s. Natural causes, such as changes in flood frequency, intensity, and duration, have greatly influenced channel adjustment, especially since the late 1990s. This study highlights the importance of quantitative approaches in understanding the complex interactions between human activities, changes in climates, and river channel dynamics, which can have significant implications for river management and restoration.

b. Another study investigates how hydrology and morphology in Myanmar's Chindwin River Basin have changed as a result of climate changes (Sangam et al., 2020). The study makes forecasts about an increase in yearly runoff and more catastrophic flooding using climate models. Considerable alterations in morphology on the left embankment of upstream regions are anticipated, posing a danger of flooding and impacting riverine populations. The study highlights the importance of mathematical modeling to understand climate impacts, allowing policymakers to develop strategies for adaptation and mitigation to safeguard lives and livelihoods in the region.

c. The research centers on the Lower Mississippi River (Cheng et al., 2020b) and investigates historical channel adjustments in direct response to changes in climate and flood events. By studying the river's past behavior, the study delves into how changing climatic conditions and flood occurrences have influenced the river's channel configuration over time. Through this approach, the study clarifies the intricate interactions between climate-related variables and the river's morphological transformations, offering priceless insights into the changing characteristics of river networks and their reaction to environmental changes in this particular region.

12.11   Thermal regimes in streams caused by climate change

In a warming world, "the effect of warming temperatures on surface water oxygen saturation is a major concern" (Matthews & Zimmerman, 1990). The impact of climate change on dissolved oxygen (DO) saturation levels has received less attention in studies than the rise in "river water temperatures (RWTs)" (Matthews & Zimmerman, 1990), which has received more attention. To predict RWTs in seven significantly polluted river catchments in India on a monthly scale, this study created a hybrid deep learning model employing "long- and short-term memory and k-nearest neighbor bootstrap resampling" (Matthews & Zimmerman, 1990). According to the findings, under the Representative Concentration Pathway 8.5 scenario, summer RWTs in these basins might rise by 3.1°C to 7.8°C by the years 2071−2100. The capacity for DO saturation can be reduced by 2%−12% as a result of such changes, with every 1°C increase in RWT resulting in a 2.3% fall in DO levels of saturation over Indian streams and rivers. According to an assessment of NASA Earth Exchange Global Daily Downscaled Estimates of air temperature with a Representative Concentration Pathway 8.5 scenario, the Ganga, Sabarmati, Tungabhadra, Musi, and Narmada basins are expected to be 3.1°C, 3.8°C, 5.8°C, 7.3°C, and 7.8°C, respectively, for the years 2071−2100.

1. In water-limited mountainous regions like the Sierra Nevada (Ficklin et al., 2013), climate-driven changes in hydrology can significantly impact water quality and ecosystem health. In a study on this topic, the impact of increased release of "greenhouse gases on dissolved oxygen (DO) concentrations, river temperature, and sediment transport was evaluated using forecasts from climate models" (Ficklin et al., 2013). According to the findings, summer and spring river temperatures might rise by 1°C to 5.5°C by 2100, with subbasins of the southern Sierra Nevada at low height experiencing the largest increases. This temperature rise, coupled with changes in streamflow patterns, leads to a decrease in DO levels (around 10%) and sediment concentrations (around 50%). Consequently, several native indicator species may face challenges in survival due to these changes. The study emphasizes the importance of understanding these impacts at a detailed level to develop effective adaptive management strategies for water quality in water-limited mountainous systems.

2. Climate change is expected to worsen thermal extremes in river systems (Cheng et al., 2020b), impacting fish mortality and thermoelectric power plants. The study focused on the southeastern United States and evaluated thermal events based on duration, intensity, and (Akhtar et al., 2008; Chen et al., 2020; Cheng et al., 2020a, 2020b; de Musso et al., 2020; Dwire et al., 2018; Escanilla-Minchel et al., 2020; Ficklin et al., 2013; Hirschberg et al., 2020; Jeelani et al., 2012; Juracek & Fitzpatrick, 2022; Matthews & Zimmerman, 1990; Mishra & Lilhare, 2016; Muskey L. et al., 2022; Nakamura, 2022; Sangam et al., 2020; Stewart et al., 2000; Uber et al., 2022) severity. According to the results, climate change will make temperature extremes worse in both controlled and uncontrolled river networks, while the regulated system will see considerably less of an impact. Reservoir management has a limited ability to reduce some thermal effects (12.2% to 26.0% enhancement). The number of river segments suffering high temperature and low flow occurrences (hydrologic hot-dry events) will, nevertheless, rise by 21.4% by the 2080s

under the RCP8.5 scenarios despite stronger reservoir stratification. The expected duration of these events is 10.3 days annually, which is a significant increase over earlier times.

12.12 Riparian zone dynamics in streams caused by climate change

Riparian groundwater-dependent ecosystems, wetlands, and habitats can be found at various elevations in Blue Mountains of Oregon, USA (Dwire et al., 2018). Although these habitats only make up a small fraction of the environment, they are extremely valuable for conservation because they offer vital habitats for a variety of flora and fauna. However, climate change poses a considerable threat to these special environments, primarily due to predicted alterations in snowpack and hydrologic patterns shortly. According to the study, livestock grazing, and water alterations have made many riparian habitats less effective. These factors make it harder for them to handle any additional pressure that a hotter environment can bring. Short-term changes are expected to occur in places near springs and small streams, and some riparian zones and wetlands might gradually get smaller over time. More resistant to drought conifers and shrubs may replace riparian hardwood species as a result of increasing temperatures and dropping soil moisture. The frequency and size of wildfires originating in highland forests may also affect the distribution of riparian plants.

1. Japan's riparian woodlands serve as essential green and blue infrastructure, providing a range of ecological services such as water transportation, protection from the sun, debris from leaves supply, insect input, monitoring of water quality, and animal corridors (Nakamura & Nakamura, 2022). Unfortunately, many woods have already been harmed by stressors brought on by people, such as logging, development of agriculture, river supervision, and dam building. Forecasts on the climate for Japan in the 21st century show prospective changes, such as an increase in mean precipitation of nearly 10%, an increase in the frequency of high-magnitude floods, and a decrease in the run-off of snowmelt floods. Further effects on riparian forests and related ecosystems may result from these changes.
2. Climate change will likely cause changes in the distributions of riparian species as temperature and rainfall patterns alter stream habitats (Matthews & Zimmerman, 1990). To support wildlife, management interventions need spatial forecasts of suitable habitats under projected climatic conditions. However, this poses challenges due to the diverse responses of riparian species to changing conditions and the need to assess various streamflow and temperature characteristics that affect habitat suitability.

To address these challenges, the study on six watersheds in southern California selects species representatives from clusters of species based on environmental attributes and assesses modifications to suitability for habitat using temporary, ecologically specific metrics that characterize temperature and streamflow regimes (Matthews & Zimmerman, 1990). To determine habitat choices, stream-specific environmental indicators created from species observations in specific reaches and long-term trends are employed rather than just climatic factors.

The study forecasts changes in habitat appropriateness by the end of the century by integrating species distribution data from neighborhood surveys with modeled stream flow and temperature projections. The relevance of hydrological and river temperature fluctuates by cluster, with the ranges of low-elevation, warm-water species extending, and those of high-altitude, cold-water species decreasing.

12.13 Glacier and snowmelt contributions in streams caused by climate change

The melting of snow and glacier/ice melt were examined as major variables affecting the seasonal flow of Himalayan rivers in a study (Jeelani et al., 2012). They used a "water budget model" to recreate the hydrological patterns in the Liddar watershed, which is situated in the "western Himalaya, India, for the 20th century (1901−2010) and under future IPCC A1B climate change scenarios, to examine the long-term contributions of these processes to river hydrology" (Jeelani et al., 2012). Long-term precipitation and temperature data analysis in the area showed an upward trend in temperatures ($0.08°C$ per year) and a rise in precipitation (0.28 mm per year) throughout the 20th century, with discernible seasonal changes. In particular, the winter months showed the greatest warming along with a decline in the precipitation process, whereas the spring months had an increase in precipitation. These modifications have sped-up snowmelt, hastened departure, and considerable seasonal fluctuations in water supply.

The observed climatic trends from 1901 to 2010 and their anticipated effects (https://www.ipcc.ch/pdf/special-reports/spm/sres-en.pdf) point to a worrying outlook for the region's water supplies. The quick runoff and changes in snow distribution have drastically changed the timing and volume of water flow in the Himalayas. The effects of these changes are especially worrying, since they may result in a considerable rise in spring runoff and a subsequent decrease in water availability during the summer. These changes in water flow may make it difficult to manage the area's water supplies, which might hurt irrigated agriculture throughout the summer.

According to Akhtar et al. (2008), the Hindu Kush Himalaya mountains are crucial in supplying water to almost 1.9 billion people in Asia. However, as temperatures rise and snow and glaciers melt more quickly in this area, the effects of climate change are becoming more and more obvious. The security of water, food, energy, and livelihoods for populations throughout Asia is seriously threatened by these changes. The effects of climate change on the cryosphere and mountain habitats are discussed in this chapter, along with how they affect river systems and societal dynamics in the mountains, hills, and plains. The effects of climate change are already being felt in high mountains and hills, where large losses and damages are occurring as a result of notable changes in agropastoral systems and a rise in the severity of floods and droughts.

It gets harder to separate the climate change indication from various other ecological and management elements as one proceeds downstream. However, the chapter describes how numerous areas in the hills and plains, such as hydropower, irrigation, towns, businesses, and the ecosystem as a whole, can be impacted by climate change in the mountains. Additionally, it talks about the possible effects of climate change, such as a rise in catastrophes and emigration from impacted areas.

Chile holds 3.8% of the world's non-Antarctic and non-Greenland glacier area (Escanilla-Minchel et al., 2020). A severe megadrought in the nation's south-central area has had a substantial impact, increasing the country's reliance on water supplies from snow and glacier melting during dry years. Rising temperatures, less solid-state precipitation, and a faster rate of glacier and snow storage loss in the Chilean Andes are all results of recent climatic change. Research on a glacial-naval watershed in the central Andes was undertaken using hydrological modeling to better manage water resources in glacier-fed inhabited regions. It looked at hydrological systems and evaluated how various climate change scenarios will affect water output. Depending on the emission scenarios, the results predict a drop in the mean annual discharge of 18.1% to 43.3% by 2050 and 31.4% to 54.2% by 2100. Reduced precipitation lessens the contribution of snowmelt and rainfall to discharge, whereas glacier melt helps counteract the drying trend. A long-term danger to the region's water supply exists beyond the peak water, which will happen around 2040 and will be caused by glacier thinning.

12.14   Conclusion

In conclusion, climate change is a significant environmental challenge that profoundly impacts stream morphology across various geographical spaces. The alterations in precipitation patterns and hydrological regimes lead to changes in streamflow dynamics, erosion, and sedimentation processes. These changes, in turn, influence channel adjustments, thermal regimes, and riparian zone dynamics.

The studies highlighted in this analysis demonstrate that climate change can result in increased soil erosion, sediment yield, and sediment transport rates, posing risks to water resources, ecosystems, and communities. Additionally, shifts in streamflow patterns can lead to channel adjustments, affecting the overall shape and functioning of streams. Moreover, rising temperatures can have an adverse influence on the thermal regimes of streams, influencing aquatic habitat and species distribution. Sustainable water resource management, planning for land use, and conservation activities depend on understanding the effects of climatic change on stream morphology. These results must be considered by decision-makers, researchers, and practitioners as they develop adaptable policies that promote the adaptive abilities of stream ecosystems and protect the availability and quality of water for human populations. Furthermore, addressing the challenges posed by climate change on stream morphology requires collaboration between stakeholders, interdisciplinary research, and the implementation of sustainable management practices. By recognizing the complexities of the interactions between climate change and stream systems, one can develop robust strategies to mitigate impacts, protect ecosystems, and ensure that critical ecosystem services continue to be provided. Overall, the perspectives gained from these studies underscore the importance of proactive planning and adaptation in the face of climate change. By integrating scientific information with conservation and restoration efforts, one can work toward sustaining the health and functionality of streams, supporting both natural systems and human well-being in the face of this pressing environmental challenge.

References

Akhtar, M., Ahmad, N., & Booij, M. J. (2008). The impact of climate change on the water resources of the Hindukush-Karakoram-Himalaya region under different glacier coverage scenarios. *Journal of Hydrology, 355*, 148–163.

Chen, C. N., Samkele, T. S., & Tsai, H. C. (2020). Climate change impacts on soil erosion and sediment yield in a watershed. *Water, 12*(8), 2247. Available from https://doi.org/10.3390/w12082247.

Cheng, Y., Voisin, N., Yearsley, R. J., & Nijssen, B. (2020a). Thermal extremes in regulated river systems under climate change: an application to the southeastern U.S. rivers. *Environmental Research Letters*, *15*(9), 1−10. Available from https://doi.org/10.1088/1748-9326/ab8f5f.

Cheng, Y., Voisin, N., Yearsley, J. R., & Nijssen, B. (2020b). Reservoirs modify river thermal regime sensitivity to climate change: A case study in the southeastern United States. *Water Resources Research*, *56*(6), e2019WR025784. Available from https://doi.org/10.1029/2019WR025784.

Dwire, K. A., Mellman-Brown, S., & Gurrieri, J. T. (2018). Potential effects of climate change on riparian areas, wetlands, and groundwater-dependent ecosystems in the Blue Mountains, Oregon, USA. *Climate Services*, *10*, 44−52.

Escanilla-Minchel, R., Alcayaga, H., Soto-Alvarez, M., Kinnard, C., & Urrutia, R. (2020). Evaluation of the impact of climate change on runoff generation in an Andean glacier watershed. *Water*, *12*(12), 3547.

Ficklin, D. L., Stewart, I. T., & Maurer, E. P. (2013). Effects of climate change on stream temperature, dissolved oxygen, and sediment concentration in the Sierra Nevada in California. *Water Resources Research*, *49*, 2765−2782. Available from https://doi.org/10.1002/wrcr.20248.

Hirschberg, J., Fatichi, S., Bennett, G., McArdell, B., Lane, S., & Molnar, P. (2020). Climate change impacts on sediment yield and debris-flow activity in an Alpine catchment. *Journal of Geophysical Research: Earth Surface*, *126*(1). Available from https://doi.org/10.1029/2020JF005739.

Jeelani, G., Feddema, J. J., Veen der van, J. C., & Stearns, L. (2012). Role of snow and glacier melt in controlling river hydrology in Liddar watershed (western Himalaya) under current and future climate. *Water Resources Research*, *48*(12). Available from https://doi.org/10.1029/2011WR011590.

Juracek, K. E., & Fitzpatrick, F. A. (2022). Geomorphic responses of fluvial systems to climate change: A habitat perspective. *River Research and Applications*, *38*(4), 757−777. Available from https://doi.org/10.1002/rra.3938.

Matthews, W. J., & Zimmerman, E. G. (1990). Potential effects of global warming on native fishes of the southern Great Plains and the Southwest. *Fisheries*, *15*(6), 26−32.

Mishra, V., & Lilhare, R. (2016). Hydrologic sensitivity of Indian sub-continental river basins to climate change. *Global and Planetary Change*, *139*, 78−96.

de Musso, N. M., Capolongo, D., Caldara, M., Surian, N., & Pennetta, L. (2020). *Water 2020*, *12*(1), 307.

Muskey L, M, Facincani, F. D., Rallings, A. M., Rheinheimer, D. M., Medellin-Azurara, J., & Viers, J. H. (2022). Assessing Hydrological Alteration Caused by Climate Change and Reservoir Operations in the San Joaquin River Basin, California. *Frontiers in Environmental Science*, *10*, In this issue. Available from https://doi.org/10.3389/fenvs.2022.765426.

Nakamura, F. (2022). Riparian forests and climate change: Interactive zone of green and blue infrastructure. In F. Nakamura (Ed.), *Green infrastructure and climate change adaptation. ecological research monographs*. Singapore: Springer.

Sangam, S., Naditha, I., Thanapon, P., Somchai, C., Sarawut, N., & Muhammad, B. (2020). A multi-modeling approach to the assessment of climate change impacts on hydrology and river morphology in the Chindwin River Basin, Myanmar. *Catena*, *188*, 104464.

Stewart, I. T., Dyck, H. J., Sandford, R. W., et al. (2000). Climate change and Water Resources in the Columbia Basin, British Columbia, Canada, and Washington, USA: Impacts and adaptive strategies. *Water International*, *25*(2), 253−272.

Uber, M., Rossler, O., Astor, B., Hoffmann, T., Oost, V., & Hillerbrand G, K. (2022). Climate change impacts on soil erosion and sediment delivery to German federal waterways: A case study of the Elbe Basin. *Atmosphere*, *13*(11), 1752. Available from https://doi.org/10.3390/atmos13111752.

Yousefi, N., Khodashenas, S. R., Eslamian, S., & Askari, Z. (2016). Estimating the width of the stable channels using multivariable mathematical models. *Arab Journal of Geosciences*, *9*, 321. Available from https://doi.org/10.1007/s12517-016-2322-0.

Part V

Case studies

Chapter 13

Stream flow restoration case studies in the United States and Canada

Saeid Eslamian and Mousa Maleki

Department of Water Sciences and Engineering, College of Agriculture, Isfahan University of Technology, Isfahan, Iran

13.1 Introduction

Streamflow recharge (SFR) refers to the practice of intentionally recharging surface water into aquifers or storing water in soils to augment river and streamflows. As water scarcity increases across North America due to population growth and climate change, SFR is emerging as a promising multibenefit approach to enhance water supplies, mitigate floods and droughts, restore aquatic ecosystems, and promote more resilient water management.

Historically, many indigenous communities recognized the dynamic interaction between surface and subsurface waters, intentionally spreading excess river flows into riparian zones, floodplains and gravel beds. In recent decades, resource managers have increasingly utilized SFR techniques to supplement natural recharge lost to urbanization, land use changes, and infrastructure like dams. At the same time, new aquatic reconnection projects are helping to reverse hydrologic disconnection in overdrained landscapes (Gaaloul et al., 2021).

This chapter examines SFR case studies across the Canada and United States to identify evolving best practices. Case studies are organized by region to highlight place-based innovations and cross-border learning opportunities. The overarching objectives are to illustrate how SFR strategies can be tailored to diverse hydroclimatic settings, integrated into holistic watershed governance, and optimized through social and technological progress. By exploring both historical and emerging applications of SFR, we aim to guide more widespread development and adaptive management of this important solution.

In North America's hydrological mosaic, surface–groundwater interactions are dynamic phenomena connecting terrestrial and aquatic realms through complex feedback loops inhererent to Earth's hydrosphere. Traditionally, indigenous societies intuitively understood these interdependencies, strategically managing flows via natural infrastructure to support mutually reinforced ecological and cultural objectives.

Contemporary circumstances demand innovative solutions to escalating water supply risks exacerbated by a changing climate. As populations burgeon and development fragments former floodplains, the need to replenish diminished aquatic reserves becomes increasingly critical. SFR presents a holistic approach to restoring hydrologic balance at both local and watershed scales. Rather than relying solely on built infrastructure to regulate natural variability, SFR leverages the hydrologic restorative abilities of soils, sediments, and ecohydrologic corridors.

The case studies explored herein offer valuable insights into place-based tailoring of techniques to geographic circumstances. Regional climatic, geological, and socioeconomic particularities shape the most impactful SFR permutations. By investigating diverse applications across diverse geohydrologic settings, we can discern guiding design principles adaptable to new contexts. Operational and regulatory challenges inherent to transboundary waters further necessitate cooperative solutions and knowledge-sharing between riparian partners.

An integrative lens considering not just engineered but also ecological dimensions of the hydrosphere allows for more prudent and resilient long-term management. As climatic uncertainties intensify, a portfolio balancing gray, green, and natural infrastructure best equips communities and landscapes to roll with hydrologic fluctuations. The following case analyses thereby aim to illuminate pathways toward integrated watershed governance optimizing the hydrologic functions and societal values of SFR.

Hydrosystem Restoration Handbook. DOI: https://doi.org/10.1016/B978-0-443-29802-8.00013-3

13.2   Pacific Northwest projects

The Pacific Northwest region of North America encompasses diverse hydrological landscapes, from coastal estuaries and floodplains to mountainous headwaters and inland valleys. Several case studies highlighted in this section illustrate how researchers in this region have applied emerging best practices in SFR science to gain valuable insights into its complex coastal and inland watershed systems. In particular, collaborative studies involving federal, state, and local agencies have worked to quantify recharge rates across different terrain, better understand interactions between surface and groundwater, and identify opportunities to enhance both habitat protection and water resource sustainability under a changing climate.

13.2.1   Fraser Valley Storage Assessment (British Columbia)

The Fraser Valley Storage Assessment in British Columbia is a significant project aimed at assessing the storage needs and potential solutions for water resources in the Fraser Valley region. The Fraser Valley region is known for its fertile agricultural lands and diverse ecosystems, but it also faces challenges related to water management, especially during dry periods. This assessment project seeks to address these challenges by evaluating the current storage capacity, identifying potential storage sites, and recommending strategies for improving water storage in the region (Gasser et al., 2016). The overview of Fraser Valley in British Columbia is shown in Fig. 13.1.

The assessment involves a comprehensive analysis of the region's water resources, including surface water and groundwater availability. It takes into account factors such as precipitation patterns, hydrological conditions, and water demand from various sectors, including agriculture, industry, and municipalities. By understanding the water supply and demand dynamics, the project aims to identify gaps in water storage infrastructure and propose viable solutions (Reinesch et al., 2022).

One of the key objectives of the Fraser Valley Storage Assessment is to explore options for increasing water storage capacity. This may involve the construction of new reservoirs, expansion of existing storage facilities, or implementing

FIGURE 13.1   Fraser Valley in British Columbia (author).

innovative storage methods such as ASR. The project team conducts detailed studies to evaluate the feasibility, environmental impacts, and economic viability of different storage options. Stakeholder engagement is an essential component of the assessment, ensuring that the perspectives and interests of local communities, indigenous groups, and other relevant parties are considered (Curry & Zwiers, 2018).

The outcomes of the Fraser Valley Storage Assessment will provide valuable insights to policymakers, water managers, and stakeholders involved in water resource planning and management. The project's recommendations will help inform decision-making processes related to water infrastructure investments, land use planning, and water allocation strategies. By enhancing the region's water storage capacity, Fraser Valley can better withstand periods of water scarcity, support sustainable agricultural practices, and protect the health of its ecosystems (Middleton & Allen, 2014).

13.2.2 Willamette River Basin Program (Oregon)

The Willamette River Basin Program in Oregon is a comprehensive initiative focused on managing and protecting the water resources of the Willamette River Basin. The Willamette River Basin is a vital water source for the region, providing water for drinking, agriculture, industry, and supporting diverse ecosystems. The program aims to address various water-related challenges, including water quality degradation, habitat loss, flood management, and water supply reliability (Tague & Grant, 2004).

One of the primary goals of the Willamette River Basin Program is to improve water quality within the basin. To achieve this, the program implements measures to reduce pollution from various sources, including urban runoff, agricultural activities, and industrial discharges. It also emphasizes the restoration and protection of riparian areas and wetlands, which act as natural filters and provide habitat for local wildlife. Through collaborative efforts with stakeholders, the program promotes sustainable land management practices and encourages the adoption of best management practices to minimize water pollution (Herrera et al., 2014).

Another important aspect of the program is flood management. The Willamette River Basin is prone to flooding, which poses risks to human settlements, infrastructure, and the environment. The program employs strategies such as floodplain mapping, levee system improvements, and flood forecasting to enhance the region's resilience to floods. It also emphasizes the importance of maintaining natural floodplain areas and promoting nature-based solutions for flood mitigation. By reducing flood risks and enhancing floodplain functionality, the program aims to protect lives, property, and critical ecosystems.

Furthermore, the Willamette River Basin Program focuses on ensuring water supply reliability in the face of population growth and climate change. The program evaluates water availability and demand, identifies potential water supply gaps, and explores options for increasing water storage and efficiency. It also promotes water conservation practices and encourages water users to adopt sustainable water management strategies (Herrera et al., 2014).

13.2.3 Puget Sound Low-Flow Pilot (Washington)

The Puget Sound Low-Flow Pilot project in Washington aims to address the challenges posed by low-flow conditions in the Puget Sound region. Low-flow conditions occur when water levels in rivers and streams drop significantly, often during dry periods or due to water abstractions. These conditions can have adverse effects on aquatic ecosystems, water quality, and water-dependent industries such as fisheries and recreation (Reidy, 2004).

The pilot project focuses on developing and implementing strategies to mitigate the impacts of low-flow conditions. It involves monitoring and modeling low-flow dynamics, assessing the ecological and socioeconomic consequences, and testing innovative approaches to maintain adequate flow levels in rivers and streams.

One of the key components of the Puget Sound Low-Flow Pilot is the exploration of water conservation and efficiency measures. By promoting water conservation practices and implementing efficiency measures, such as improved irrigation techniques and water-efficient appliances, the project aims to reduce water demand and alleviate stress on water resources during low-flow periods (Stanley et al., 2010).

Additionally, the project investigates the potential for water storage and augmentation strategies. This may involve exploring options for increasing water storage capacity, such as constructing small reservoirs or implementing managed aquifer recharge (MAR) systems. By strategically storing water during periods of high flow, it becomes possible to release it during low-flow periods to maintain adequate stream flows and support ecosystem health.

The Puget Sound Low-Flow Pilot also emphasizes the importance of stakeholder engagement and collaboration. It involves working closely with local communities and indigenous groups to ensure their perspectives and knowledge are incorporated into the project's design and implementation. By involving stakeholders in the decision-making process, the project aims to foster a sense of ownership and shared responsibility for the protection and sustainable management of water resources in the Puget Sound region (Bahnick, 2018).

The outcomes of the Puget Sound Low-Flow Pilot project will provide valuable insights and recommendations for managing low-flow conditions in the Puget Sound region. The project's findings can inform policy development, water resource planning, and management strategies to better protect aquatic ecosystems, sustain water-dependent industries, and ensure the long-term resilience of the region's water resources.

In conclusion, the Pacific Northwest projects discussed in this section demonstrate the region's commitment to addressing water resource challenges and promoting sustainable water management practices. The Fraser Valley Storage Assessment in British Columbia aims to improve water storage capacity in the region, thereby enhancing water availability during dry periods. The Willamette River Basin Program in Oregon focuses on water quality improvement, flood management, and ensuring water supply reliability. The Puget Sound Low-Flow Pilot in Washington seeks to mitigate the impacts of low-flow conditions through water conservation, storage, and stakeholder collaboration. These projects exemplify the importance of integrated water resource management approaches that consider ecological, social, and economic factors to ensure the sustainable use and protection of water resources in the Pacific Northwest (Stanley et al., 2010).

13.3 Prairie Plains

The expansive Prairie Plains region spans inland watersheds across several western US states and Canadian prairie provinces. SFR research conducted here confronts unique hydrological conditions in this transitional zone between mountain headwaters and interior basins. Studies highlighted provide insights into recharge mechanisms operating through complex plains and valley aquifer systems. Notable projects leverage remote sensing and Geographic Information System (GIS) techniques to map recharge potential across varied land uses, while others involve stakeholder partnerships to reconcile groundwater usage with ecological flows. Highlights include collaborative efforts examining impacts of irrigation and agriculture on aquifer replenishment. Together, these case studies advance understanding of recharge dynamics specific to plains hydroscapes and inform sustainable management of their integral surface and subsurface water resources.

13.3.1 Assiniboine River Managed Aquifer Recharge project (Manitoba)

The Assiniboine River MAR project in Manitoba is a significant initiative aimed at enhancing water availability and sustainability in the region. The Assiniboine River Basin is an important water source for various sectors, including agriculture, industry, and municipalities. However, the basin faces challenges related to water scarcity and variability, particularly during dry periods. The MAR project seeks to address these challenges by implementing MAR techniques to replenish groundwater resources (Unduche et al., 2018).

MAR involves diverting excess surface water during periods of high flow and directing it into underground aquifers for storage. This stored water can later be extracted during dry periods to supplement water supplies. The Assiniboine River MAR project involves the construction of infrastructure, such as diversion channels, infiltration basins, and recharge wells, to facilitate the recharge process.

The project also involves comprehensive monitoring and modeling efforts to assess the impacts of MAR on groundwater levels, water quality, and ecosystem health. By understanding the hydrological dynamics and potential environmental implications, the project aims to ensure the sustainability and effectiveness of the recharge activities (Unduche et al., 2018).

The Assiniboine River MAR project is a collaborative effort involving various stakeholders, including government agencies, local communities, and water users. It emphasizes the importance of stakeholder engagement and cooperation to ensure the project's success. By involving stakeholders in the decision-making process, the project aims to foster a sense of ownership and shared responsibility for the sustainable management of water resources in the Assiniboine River Basin (Pomeroy et al., 2010).

13.3.2 Qu'Appelle Valley Long-Term Program (Saskatchewan)

The Qu'Appelle Valley Long-Term Program in Saskatchewan is a comprehensive and integrated initiative aimed at addressing water management challenges in the Qu'Appelle River Basin. The Qu'Appelle River Basin is a vital water resource for the region, supporting various industries, recreational activities, and ecological habitats. However, the basin faces issues related to water quality degradation, flooding, and water allocation conflicts. The Long-Term Program seeks to mitigate these challenges through a coordinated and long-term approach to water resource management (Jasechko et al., 2017).

One of the key objectives of the Qu'Appelle Valley Long-Term Program is to improve water quality within the basin. The program focuses on reducing nutrient loading, sedimentation, and other pollutants that contribute to water quality degradation. It involves implementing best management practices in agriculture, wastewater treatment, and urban stormwater management to minimize pollution sources. The program also emphasizes the restoration and protection of riparian areas and wetlands, which play a crucial role in improving water quality and providing habitat for wildlife.

Flood management is another important aspect of the Qu'Appelle Valley Long-Term Program. The program aims to reduce the impacts of flooding on human settlements, infrastructure, and natural systems. It involves measures such as floodplain mapping, levee system improvements, and flood forecasting to enhance the region's resilience to floods. The program also recognizes the importance of maintaining natural floodplain areas and promoting nature-based solutions for flood mitigation (Haig, 2018).

Furthermore, the Long-Term Program focuses on water allocation and governance issues within the Qu'Appelle River Basin. It addresses the challenges associated with competing water demands from various sectors, including agriculture, industry, and municipalities. The program promotes collaboration among stakeholders and the development of water allocation frameworks that consider ecological, social, and economic factors. By fostering dialogue and cooperation, the program aims to achieve a balanced and sustainable approach to water allocation in the basin.

13.3.3 Platte River Cooperative Agreement (Nebraska, Colorado, and Wyoming)

The Platte River Cooperative Agreement is a landmark agreement between the states of Nebraska, Colorado, and Wyoming aimed at managing and protecting the water resources of the Platte River Basin. The Platte River Basin is a vital water source for these states, supporting agriculture, municipal water supplies, industry, and important ecological habitats. The Cooperative Agreement represents a collaborative effort to address water-related challenges and promote sustainable water management practices in the basin (Jasechko et al., 2017).

One of the primary objectives of the Platte River Cooperative Agreement is to maintain and enhance habitat for endangered species, particularly the endangered whooping crane and the threatened piping plover. The agreement recognizes the ecological importance of the Platte River Basin and the need to balance water use with the protection of critical habitats. It includes provisions for water releases and flow management to support the ecological needs of these species during their migration and breeding seasons.

The agreement also addresses water allocation issues and promotes water conservation practices. It encourages the development and implementation of water management plans that promote efficient water use and minimize wasteful practices. The states involved in the agreement work together to optimize water allocation and balance the needs of different sectors, including agriculture, municipalities, and industry (Haig, 2018).

Furthermore, the Platte River Cooperative Agreement recognizes the importance of adaptive management and scientific research. It emphasizes the need for ongoing monitoring, assessment, and research to inform decision-making processes and improve water management practices. The agreement promotes collaboration among scientists, water managers, and stakeholders to enhance understanding of the basin's hydrological dynamics and address emerging water resource challenges.

The Platte River Cooperative Agreement also includes provisions for drought management and response. The states collaborate to develop drought contingency plans and implement measures to mitigate the impacts of drought on water supplies and ecosystems. By working together and sharing resources, the states aim to enhance their resilience to drought conditions and ensure the sustainable use of water resources during periods of water scarcity (Haig, 2018).

In addition, the Platte River Cooperative Agreement recognizes the importance of public engagement and involvement. It encourages transparency, public participation, and collaboration with stakeholders in decision-making processes related to water management in the basin. This inclusive approach ensures that diverse perspectives are considered and that the interests of various stakeholders, including agricultural producers, environmental organizations, and local communities, are taken into account.

The Platte River Cooperative Agreement serves as a model for interstate water management and cooperation. By establishing a framework for collaborative decision-making and sustainable water management practices, the agreement promotes the long-term health and sustainability of the Platte River Basin. It demonstrates the importance of interstate cooperation and the shared responsibility for protecting and managing water resources in a complex and interconnected system.

In conclusion, the Prairie Plains projects discussed in this section highlight the efforts of various stakeholders to address water management challenges and promote sustainable practices in the region. The Assiniboine River MAR project in Manitoba focuses on MAR to enhance water availability during dry periods. The Qu'Appelle Valley Long-Term Program in Saskatchewan addresses water quality, flood management, and water allocation issues in the Qu'Appelle River Basin. The Platte River Cooperative Agreement between Nebraska, Colorado, and Wyoming aims to manage and protect the water resources of the Platte River Basin through habitat conservation, water allocation, and drought management. These projects demonstrate the importance of collaboration, stakeholder engagement, and integrated approaches to ensure the sustainable use and protection of water resources in the Prairie Plains region.

13.4 Great Lakes Initiatives

The Great Lakes Initiatives encompass a range of collaborative efforts and programs aimed at addressing various environmental and water management challenges within the Great Lakes Basin. This section highlights three specific initiatives: Credit Valley Conservation Authority (CA) leadership in Ontario, the Milwaukee Metropolitan Sewerage District (MMSD) in Wisconsin, and the Saginaw Bay Watershed Sediment and Nutrient Reduction (SFR) planning in Michigan.

13.4.1 Credit Valley Conservation Authority (Ontario)

The Credit Valley CA, located in Ontario, plays a vital role in the protection and management of the Credit River watershed and its surrounding natural resources. The authority is responsible for implementing various initiatives that focus on watershed planning, flood management, water quality improvement, and natural heritage conservation. Its leadership and collaborative approach have been instrumental in promoting sustainable practices and raising awareness about the importance of protecting the Credit River watershed (Conservation, 2012).

The Credit Valley CA works closely with local municipalities, community organizations, and stakeholders to develop and implement watershed management plans. These plans aim to balance the needs of development and growth with the preservation of natural ecosystems and water resources. The authority also engages in educational programs and outreach activities to promote public awareness and involvement in watershed stewardship.

13.4.2 Milwaukee Metropolitan Sewerage District

The MMSD is an organization responsible for wastewater management and flood control in the Milwaukee metropolitan area. Facing challenges related to combined sewer overflows, urban runoff, and water pollution, the MMSD has implemented innovative strategies to improve water quality and protect the health of Lake Michigan. One notable initiative led by the MMSD is the "GREENSEAMS" program. This program focuses on the restoration and preservation of natural areas, such as wetlands, floodplains, and stream corridors, within the Milwaukee River Basin. By enhancing the natural infrastructure, the program helps to reduce stormwater runoff, improve water quality, and provide habitat for wildlife. The MMSD also invests in green infrastructure projects, including rain gardens, permeable pavements, and green roofs, to manage stormwater at its source. These practices help to reduce the volume and velocity of stormwater runoff, minimize the strain on the sewer system, and improve water quality in local water bodies (Christensen, 2004). Industries and Infrastructure of MMSD are shown in Fig. 13.2.

13.4.3 Saginaw Bay Watershed Sediment and Nutrient Reduction (streamflow recharge) planning (Michigan)

The Saginaw Bay Watershed Sediment and Nutrient Reduction (SFR) planning initiative in Michigan focuses on addressing sediment and nutrient pollution in the Saginaw Bay and its contributing watersheds. Excessive sediment and nutrient inputs can result in harmful algal blooms, degraded water quality, and ecological imbalances in the bay. The SFR planning initiative brings together various stakeholders, including state and federal agencies, local governments, agricultural producers, and environmental organizations, to develop and implement strategies for reducing sediment and nutrient loads in the watershed. These strategies may include implementing best management practices on agricultural lands, improving stormwater management, restoring wetlands, and promoting sustainable land use practices. The initiative also emphasizes the importance of monitoring and research to inform decision-making processes and track the effectiveness of implemented measures. By collecting data and conducting assessments, stakeholders can better understand the sources and impacts of sediment and nutrient pollution and develop targeted strategies to address them (Welty, 2005).

The SFR planning initiative recognizes the interconnectedness of the watershed and the need for a collaborative approach to achieve significant and long-lasting improvements in water quality. It highlights the importance of

FIGURE 13.2 Milwaukee Metropolitan Sewerage District Industries and Infrastructure (author).

engaging stakeholders, promoting voluntary participation, and providing technical and financial support to encourage the adoption of sustainable practices.

In conclusion, the Great Lakes Initiatives discussed in this section demonstrate the collaborative efforts and innovative approaches undertaken to address water management and environmental challenges within the Great Lakes Basin. The Credit Valley CA's leadership in Ontario promotes sustainable practices and raises awareness about the importance of watershed stewardship. The MMSD in Wisconsin implements programs and projects to improve water quality and reduce stormwater runoff. The Saginaw Bay Watershed Sediment and Nutrient Reduction (SFR) planning initiative in Michigan focuses on reducing sediment and nutrient pollution in the Saginaw Bay through collaborative strategies and monitoring efforts. These initiatives highlight the significance of regional cooperation, stakeholder engagement, and science-based approaches in protecting and preserving the Great Lakes ecosystem.

13.5   Mid-Atlantic experiments

The Mid-Atlantic region of the United States is a hub for environmental experimentation and innovation. In this section, we will explore three notable experiments taking place in the region: Chesapeake Bay nutrient trading, Christina Basin aquifer storage, and Passaic River restoration.

13.5.1   Chesapeake Bay nutrient trading (Washington, District of Columbia, Maryland, and Virginia)

The Chesapeake Bay is an iconic and ecologically important estuary, but it faces significant challenges due to nutrient pollution. Excessive amounts of nitrogen and phosphorus from various sources, including agriculture and wastewater treatment plants, have led to harmful algal blooms and degraded water quality in the bay. To address this issue, the states of Washington, DC, Maryland, and Virginia have implemented a nutrient trading program. Nutrient trading offers a market-based approach to reducing nutrient pollution. Regulated entities, such as wastewater treatment plants and farmers, can participate in the program by obtaining nutrient credits. These credits represent a certain amount of nutrient reductions achieved beyond the required targets. Entities that exceed their reduction goals can sell their excess credits to entities that are struggling to meet their targets. This system creates a financial incentive for entities to find cost-effective ways to reduce nutrient pollution (Neff et al., 2000).

The nutrient trading program promotes collaboration among stakeholders and encourages the adoption of innovative practices. For example, farmers can implement best management practices, such as cover cropping or precision nutrient application, to reduce nutrient runoff from their fields. They can then generate nutrient credits that can be sold to wastewater treatment plants or other entities. This approach incentivizes the use of sustainable agricultural practices and fosters a sense of shared responsibility for protecting the Chesapeake Bay ecosystem.

13.5.2 Christina Basin aquifer storage (Delaware)

The Christina Basin in Delaware faces challenges related to water supply availability, particularly during dry periods. To address this issue, the state has implemented an aquifer storage and recovery (ASR) project in the region. ASR involves capturing excess water during periods of high flow and injecting it into underground aquifers for storage. The stored water can then be recovered during periods of low supply or high demand. In the Christina Basin, the ASR project aims to enhance the reliability and sustainability of water supplies, particularly for agricultural and municipal users (Musial, 2015).

The ASR project involves the construction of injection wells and monitoring systems to enable the recharge and recovery of water from the aquifers. During times of abundant water, such as heavy rainfall events, excess surface water or treated wastewater is injected into the aquifers. Later, when water demand exceeds supply, the stored water can be retrieved and used. By utilizing ASR, the Christina Basin can reduce reliance on surface water sources and better manage water resources. The project promotes long-term water supply resilience, reduces the need for costly infrastructure expansions, and contributes to overall water sustainability in the region. It also helps mitigate the impacts of droughts and increases the availability of water for agricultural irrigation, municipal use, and environmental needs.

13.5.3 Passaic River restoration project (New Jersey)

The Passaic River, located in New Jersey, has a rich history but has suffered from industrial pollution and urban development over the years. To restore and revitalize the river, a comprehensive restoration project is underway. The Passaic River restoration project focuses on multiple components, including pollution remediation, habitat restoration, and community engagement. Efforts are being made to clean up contaminated sediments, reduce pollution sources, and improve water quality in the river. This includes the remediation of areas affected by legacy industrial activities and the implementation of stormwater management practices to reduce urban runoff (Plan, 2006). A shot of Passaic River is shown in Fig. 13.3.

In addition to pollution remediation, the restoration project aims to restore and enhance habitat along the river. Wetland restoration, floodplain management, and riparian zone preservation efforts are being undertaken to improve the overall ecological health and resilience of the Passaic River system. These habitats play a crucial role in supporting

FIGURE 13.3 Passaic River in New Jersey (author).

biodiversity, filtering pollutants, and providing natural flood mitigation. Community engagement plays a significant role in the Passaic River restoration project. Local stakeholders, including community groups, environmental organizations, and residents, are actively involved in the decision-making process and contribute to the project's planning and implementation. This inclusive approach ensures that the restoration efforts align with the needs and aspirations of the communities living along the river.

The Passaic River restoration project exemplifies the importance of addressing complex environmental challenges through a multifaceted approach that combines pollution remediation, habitat restoration, and community engagement. By revitalizing the river ecosystem, the project aims to provide long-term benefits for both the environment and the communities that depend on the Passaic River. In conclusion, the Mid-Atlantic region of the United States is witnessing several experiments and initiatives aimed at addressing environmental challenges and promoting sustainability. Chesapeake Bay nutrient trading, Christina Basin aquifer storage, and Passaic River restoration exemplify the innovative approaches being undertaken to tackle nutrient pollution, ensure water resource availability, and restore degraded ecosystems. These experiments demonstrate the commitment of various stakeholders to protect and preserve the natural resources of the Mid-Atlantic region for future generations.

13.6   New England trials

New England, located in the northeastern United States, is known for its rich natural beauty and diverse ecosystems. The region is also a hotbed for environmental trials and experiments aimed at addressing pressing challenges and promoting sustainable practices. In this section, we will explore three notable trials taking place in New England: Ipswich River adaptive management, Presumpscot River Community Project, and Lake Champlain Basin-Wide Strategy.

13.6.1   Ipswich River Adaptive Management (Massachusetts)

The Ipswich River, located in Massachusetts, is a vital waterway that supports a variety of plant and animal species. However, the river faces challenges such as water scarcity, habitat degradation, and the impacts of climate change. To address these issues, the Ipswich River Adaptive Management (IRAM) project was initiated. The IRAM project employs an adaptive management approach to enhance the river's health and resilience. Adaptive management involves continuously monitoring and evaluating the outcomes of management actions and adjusting strategies based on new information and changing conditions.

The IRAM project brings together a diverse group of stakeholders, including government agencies, environmental organizations, scientists, and local communities. Collaboratively, they develop and implement strategies to mitigate water scarcity, restore habitats, and manage competing water demands. The project focuses on balancing the needs of human water users, such as agriculture and municipalities, with the ecological needs of the Ipswich River ecosystem. Through the IRAM project, innovative techniques for water conservation and habitat restoration are being tested. For example, water-efficient irrigation practices are being promoted among farmers, and streamflow augmentation measures are being explored to maintain healthy water levels in the river during dry periods. Additionally, efforts are being made to reduce pollution inputs and improve water quality. The IRAM project serves as a model for integrated water resource management and demonstrates the power of collaboration and adaptive approaches in addressing complex environmental issues (Woods, Horsley, & Mackin, 2002).

13.6.2   Presumpscot River Community Project (Maine)

The Presumpscot River, located in Maine, is a significant waterway that has been impacted by industrial activities in the past. The Presumpscot River Community Project aims to restore and revitalize the river while actively engaging and involving the local community in the restoration efforts. The project is a collaborative initiative between various stakeholders, including local residents, environmental organizations, government agencies, and businesses. By involving the community in decision-making processes and restoration activities, the project seeks to build a sense of ownership and stewardship among the residents.

The Presumpscot River Community Project focuses on multiple aspects of river restoration, including water quality improvement, habitat restoration, and recreational development. Efforts are being made to reduce pollution inputs, restore fish passage, and enhance riparian vegetation along the riverbanks. These measures aim to improve both the ecological health of the river and its usability for recreational purposes. The project also emphasizes the importance of environmental education and community outreach. By organizing workshops, educational programs, and volunteer activities, the project aims to raise awareness about the river's value, promote sustainable practices, and encourage community members to actively participate in the restoration efforts (Craig & Coalition, 2010).

The Presumpscot River Community Project exemplifies the significance of community engagement in environmental restoration and management. By involving local residents as active partners, the project not only enhances the ecological integrity of the river but also fosters a sense of pride and connection to the natural resources among the community.

13.6.3 Lake Champlain Basin-Wide Strategy (Vermont and New York)

Lake Champlain, situated at the border of Vermont and New York (Fig. 13.4), is one of the largest freshwater bodies in the United States. However, the lake faces challenges such as nutrient pollution, algal blooms, and degraded water quality. To address these issues, the Lake Champlain Basin Program has developed a comprehensive, basin-wide strategy. The Lake Champlain Basin-Wide Strategy brings together multiple stakeholders, including state and federal agencies, Native American tribes, municipalities, and nonprofit organizations. The strategy focuses on reducing nutrient pollution, improving water quality, and enhancing the overall ecological health of the lake and its surrounding watershed.

Key components of the strategy include agricultural best management practices, stormwater management, and wastewater treatment upgrades. These measures aim to reduce nutrient inputs into the lake, particularly phosphorus, which is a major driver of algal blooms. The strategy also emphasizes the importance of monitoring and research to better understand the lake's ecosystem dynamics and guide management decisions. In addition to pollution reduction efforts, the strategy promotes habitat restoration, shoreline stabilization, and conservation practices to protect and enhance the lake's natural habitats. It recognizes the importance of preserving the lake's biodiversity and providing habitat for fish, waterfowl, and other wildlife. The Lake Champlain Basin-Wide Strategy is a testament to the commitment of multiple stakeholders to address complex environmental challenges collaboratively. By adopting a basin-wide approach, the strategy recognizes the interconnectedness of the lake and its watershed and seeks to ensure the long-term sustainability and resilience of the Lake Champlain ecosystem (Howland, 2015).

In conclusion, New England is a region characterized by its commitment to environmental conservation and sustainability. The trials and experiments taking place in the region, such as the Ipswich River adaptive management, Presumpscot River Community Project, and Lake Champlain Basin-Wide Strategy, showcase the innovative approaches being undertaken to address water resource management, restore degraded ecosystems, and engage local communities. These trials exemplify the power of collaboration, adaptive management, and community involvement in promoting environmental stewardship and ensuring the long-term health and resilience of New England's natural resources.

FIGURE 13.4 Lake Champlain, situated at the border of Vermont and New York (author).

13.7   Western programs

The Western United States is home to vast and diverse landscapes, including iconic rivers and water bodies that face unique challenges. In this section, we will explore three notable programs taking place in the region: Colorado River Basin salinity control, Truckee River Operating Agreement, and Rio Grande multiparty collaboration.

13.7.1   Colorado River Basin salinity control (Colorado, Utah, and Wyoming)

The Colorado River Basin is a critical water source that supplies water to millions of people and supports ecosystems across several states, including Colorado, Utah, and Wyoming. However, the basin faces challenges related to salinity, which refers to the concentration of dissolved salts in the water. Excessive salinity can have detrimental effects on agricultural productivity, water quality, and ecosystem health. To address this issue, the states within the Colorado River Basin have implemented a salinity control program.

The salinity control program focuses on reducing the amount of salts entering the Colorado River system. This involves implementing best management practices in agriculture, such as improving irrigation efficiency and managing drainage water. By reducing the amount of salt-laden water that enters the river, the program aims to maintain water quality and protect downstream users and ecosystems. The program also promotes research and monitoring to better understand the sources and impacts of salinity in the Colorado River Basin. This information is used to guide management decisions and develop targeted strategies to control salinity effectively (Rumsey et al., 2021).

13.7.2   Truckee River Operating Agreement (California and Nevada)

The Truckee River, located in California and Nevada, is a vital water resource for both states. However, competing water demands and the impacts of climate change have put strain on the river's water availability. To address these challenges and ensure sustainable water management, California and Nevada have developed the Truckee River Operating Agreement (TROA). The agreement brings together various stakeholders, including water agencies, Native American tribes, environmental organizations, and local communities.

The TROA provides a framework for collaborative decision-making and allocation of water resources in the Truckee River Basin. It establishes operating rules and procedures to balance the needs of different water users, including municipal, agricultural, and environmental interests. The agreement also emphasizes the importance of environmental restoration and water conservation. It includes provisions for habitat restoration, fish passage improvements, and the enhancement of riparian zones along the Truckee River. These measures aim to improve ecological conditions and ensure the long-term health of the river ecosystem. Through the Truckee River Operating Agreement, California and Nevada are demonstrating the power of collaborative water management in a region where water resources are highly valued and contested. The agreement promotes cooperation, balances competing interests, and supports the sustainable use of water in the Truckee River Basin (Scott, 2006).

13.7.3   Rio Grande Multiparty Collaboration (Colorado, New Mexico, and Texas)

The Rio Grande is a major river that flows through Colorado, New Mexico, and Texas, providing water for both irrigation and municipal use. However, the river faces challenges related to water scarcity, drought, and interstate water disputes. To address these challenges and foster cooperation among the states, the Rio Grande Multiparty Collaboration was established. This collaborative effort brings together representatives from Colorado, New Mexico, Texas, as well as federal agencies and Native American tribes.

The multiparty collaboration aims to find mutually beneficial solutions to water management and allocation in the Rio Grande Basin. It encourages dialogue, information sharing, and coordinated decision-making among the parties involved. Through this collaborative approach, the states seek to avoid costly legal battles and develop sustainable water management strategies. The collaboration also recognizes the importance of environmental flows and ecosystem health in the Rio Grande Basin. Efforts are being made to protect and restore critical habitats, improve water quality, and promote the recovery of endangered species. The Rio Grande Multiparty Collaboration is an exemplar of interstate cooperation and multistakeholder engagement in water management. By working together, the states are addressing common challenges, finding equitable solutions, and ensuring the long-term viability of the Rio Grande as a vital water resource for both human and ecological needs (Stanger, 2013).

In conclusion, the Western United States is home to innovative programs and collaborations focused on addressing water management challenges and promoting sustainability. The Colorado River Basin salinity control program, Truckee River Operating Agreement, and Rio Grande Multiparty Collaboration demonstrate the power of cooperation, adaptive management, and stakeholder involvement in ensuring the long-term health and resilience of Western water resources. These programs serve as models for other regions facing similar water challenges and emphasize the importance of balancing human needs with the preservation of ecosystems in the face of growing water scarcity and climate change.

13.8   Emerging Best Practices in streamflow recharge studies

SFR plays a crucial role in maintaining the health of aquatic ecosystems and ensuring sustainable water supplies for human needs. SFR studies aim to understand the dynamics of water movement between surface waters and groundwater systems, with a particular focus on quantifying recharge rates, identifying recharge mechanisms, and assessing the impacts of human activities on SFR processes. In recent years, there have been significant advancements and emerging best practices in SFR studies in the United States and Canada (Arnold & Allen, 1999). This scientific text explores these emerging best practices and their implications for water resource management and ecological conservation. Streamflow recharge studies are essential for several reasons. First, they provide valuable insights into the availability and sustainability of water resources. Understanding the recharge rates and mechanisms allows water managers to make informed decisions regarding water allocation and planning for future water needs. Second, SFR studies contribute to the protection and restoration of aquatic ecosystems. Streamflow recharge is crucial for maintaining base flows, supporting habitat connectivity, and sustaining the health of riparian vegetation. By quantifying SFR rates and identifying key recharge areas, scientists can prioritize conservation efforts and implement targeted strategies for ecological restoration.

13.8.1   Integration of hydrological and geophysical techniques

One of the emerging best practices in SFR studies is the integration of hydrological and geophysical techniques. Traditional methods for quantifying SFR rates, such as water balance modeling and tracer studies, have limitations in spatial coverage and accuracy. However, by combining hydrological measurements with geophysical techniques such as electrical resistivity tomography and ground-penetrating radar, researchers can obtain a more comprehensive understanding of subsurface hydrological processes.

Geophysical techniques provide valuable information about subsurface properties, such as soil moisture content, hydraulic conductivity, and the presence of preferential flow paths. This information can be used to refine SFR models, improve recharge estimation, and identify areas of high recharge potential.

13.8.2   Remote sensing and geographic information system applications

Advancements in remote sensing technologies and geographic information system (GIS) have revolutionized SFR studies. Remote sensing data, such as satellite imagery and aerial photographs, can be used to map land cover types, vegetation dynamics, and surface water features. These data help identify areas with high SFR potential, such as floodplains and wetlands, and assess changes in land use and land cover that may impact recharge processes.

GIS-based modeling tools enable the integration of multiple datasets, such as topography, soil properties, and climate data, to estimate recharge rates at various spatial scales. These tools facilitate the identification of recharge hotspots, the evaluation of recharge variability across different landscapes, and the assessment of the impacts of land management practices on SFR processes.

13.8.3   Collaborative and multidisciplinary approaches

Effective SFR studies require collaboration among scientists, water managers, and stakeholders from different disciplines. Collaborative and multidisciplinary approaches help leverage diverse expertise and resources, leading to more robust study designs, data collection methods, and interpretation of results.

Engaging local communities and indigenous knowledge holders in SFR studies is particularly crucial. Indigenous communities often have deep-rooted connections to water resources and possess valuable traditional knowledge about local hydrological processes. Incorporating indigenous perspectives and knowledge systems enhances the accuracy and relevance of SFR studies and fosters a sense of ownership and stewardship among local communities.

13.8.4   Implications for water resource management and ecological conservation

The emerging best practices in SFR studies have significant implications for water resource management and ecological conservation. Accurate estimation of recharge rates and identification of recharge areas can inform water allocation decisions, groundwater management strategies, and the design of sustainable water supply systems. Understanding the impacts of human activities on SFR processes, such as land use changes and water extraction, can guide land management practices and inform policies aimed at minimizing negative impacts on water resources.

Moreover, incorporating SFR studies into ecological conservation efforts can enhance the protection and restoration of aquatic ecosystems. By identifying critical recharge areas and understanding the hydrological connectivity between surface waters and groundwater systems, conservation practitioners can prioritize conservation actions, restore flow regimes, and enhance habitat connectivity.

Emerging best practices in SFR studies in the United States and Canada are advancing our understanding of the complex processes involved in streamflow recharge. Integration of hydrological and geophysical techniques, application of remote sensing and GIS tools, and adoption of collaborative and multidisciplinary approaches are enhancing the accuracy and spatial coverage of SFR studies. These advancements have implications for water resource management and ecological conservation, providing valuable insights for sustainable water allocation, protection of aquatic ecosystems, and the development of effective land management practices. Continued research and implementation of these best practices will contribute to the long-term sustainability of water resources and the preservation of freshwater ecosystems in the United States and Canada.

Table 13.1 summarizes several emerging best practices that have been implemented in recent streamflow recharge studies, as discussed in Section 13.8. The table presents these best practices, including the integration of hydrological and geophysical techniques, applications of remote sensing and GIS modeling, and collaborative multidisciplinary approaches. It then outlines the benefits these approaches provide for both water resource management and ecological conservation objectives. As innovative methods and tools are increasingly employed to study recharge processes, researchers are gaining broader and more accurate insights into freshwater resources and aquatic habitats. The emerging practices highlighted in Table 13.1 enhance SFR quantification and enable more informed decision-making around sustainable water allocation and environmental flow protection.

TABLE 13.1 Emerging best practices in streamflow recharge studies: applications and implications (author).

Emerging best practices	Description	Benefits for water resource management	Benefits for ecological conservation
Integration of hydrological and geophysical techniques (e.g., electrical resistivity tomography, ground-penetrating radar)	Combining field measurements with subsurface imaging techniques provides a more comprehensive understanding of recharge processes and improves recharge estimation	Refined recharge models can inform groundwater management strategies and water allocation decisions	Identifying areas of high recharge potential can guide conservation efforts and restoration of flow regimes
Application of remote sensing (satellite imagery, aerial photos) and GIS modeling	Mapping land use/cover changes and integrating spatial datasets estimates recharge across landscapes	Assessing impacts of land use changes on recharge guides sustainable land management practices	Identifying recharge hotspots prioritizes conservation actions to preserve recharge areas critical for aquatic habitats.
Collaborative, multidisciplinary approaches incorporating local/indigenous knowledge	Leverages diverse expertise and fosters community participation/stewardship	Engaging stakeholders in research enhances relevance of findings and nonexpert perspectives	Indigenous knowledge of hydrology aids accuracy and recognizes traditional ecological knowledge in conservation planning
Continued advancement of these best practices	Ongoing research refines quantification methods and spatial coverage of recharge processes	More robust recharge understanding informs groundwater/surface water policy and management decisions	Hydrological connectivity insights between habitats guide adaptive restoration goals for sustaining freshwater ecosystems

13.9 Conclusion

In conclusion, this chapter has provided a comprehensive overview of SFR case studies in the United States and Canada. Through a detailed examination of various projects and initiatives, we have gained valuable insights into the diverse approaches employed to enhance streamflow recharge in different geographical and hydrological settings.

The case studies presented here demonstrate the importance of understanding local hydrological conditions, geological formations, and land use patterns when designing and implementing SFR projects. From the MAR efforts in California's Central Valley to the innovative stormwater management strategies in urban areas of Canada, each case study highlights the need for tailored solutions that address specific regional challenges. Furthermore, the successes and lessons learned from these case studies underscore the potential for SFR to play a crucial role in sustainable water resource management. By replenishing groundwater reserves and stabilizing streamflow, SFR projects contribute not only to the resilience of ecosystems but also to the reliability of water supplies for both human and environmental needs.

It is evident that collaboration between stakeholders, including government agencies, researchers, communities, and private sector entities, is essential for the successful implementation of SFR projects. The integration of interdisciplinary expertise and the engagement of local communities are key factors in achieving long-term sustainability. As we move forward, it is imperative to continue advancing research and technology in the field of streamflow recharge. Additionally, policy and regulatory frameworks should be developed or refined to support and incentivize SFR initiatives. This will ensure that SFR remains a viable and effective tool in the broader context of water resource management. Ultimately, the case studies presented in this chapter serve as valuable references for practitioners, policymakers, and researchers interested in the application of streamflow recharge techniques. By building upon the knowledge and experiences shared here, we can work toward a more resilient and sustainable water future for the United States, Canada, and beyond.

References

Arnold, J. G., & Allen, P. M. (1999). Automated methods for estimating baseflow and ground water recharge from streamflow records 1. *Journal of the American Water Resources Association*, *35*(2), 411−424. Available from https://doi.org/10.1111/j.1752-1688.1999.tb03599.x.

Bahnick, M., (2018). Evaluation of the Stream Function Assessment Methodology (SFAM) in watersheds of the Puget Sound lowlands.

Christensen, E.R., (2004). Milwaukee Metropolitan Sewerage District (MMSD) Stormwater Monitoring Program 2000−2004.

Craig, M., Coalition, P.R.W., (2010). Presumpscot Watershed Initiative, Implementing Water Quality Improvements to Support Vital Fisheries, Final Report.

C.V. Conservation. (2012). Credit Valley Conservation.

Curry, C. L., & Zwiers, F. W. (2018). Examining controls on peak annual streamflow and floods in the Fraser River Basin of British Columbia. *Hydrology and Earth System Sciences*, *22*(4), 2285−2309. Available from https://doi.org/10.5194/hess-22-2285-2018, http://www.hydrol-earth-syst-sci.net/volumes_and_issues.html.

Gaaloul, N., Eslamian, S., Katlane, R., (2021). Tunisian experiences of traditional water harvesting, conservation, and recharge. In Eslamian, S., Eslamian, F. (Eds.). Handbook of Water Harvesting and Conservation. Vol. 2: Case studies and application examples (Ch. 12, pp. 171−198). Wiley, New Jersey, USA.

Gasser, P.-Y., Smith, C. A. S., Brierley, J. A., Schut, P. H., Neilsen, D., Kenney, E. A., & Yang. (2016). The use of the land suitability rating system to assess climate change impacts on corn production in the lower Fraser Valley of British Columbia. *Canadian Journal of Soil Science*, *96*(2), 256−269. Available from https://doi.org/10.1139/cjss-2015-0108.

Haig, H. A. (2018). *Analysis of lakewater isotopes in the northern Great Plains: Insights from long-term monitoring and spatial surveys* (2018). Canada: The University of Regina.

Herrera, N. B., Burns, E. R., & Conlon, T. D. (2014). Simulation of groundwater flow and the interaction of groundwater and surface water in the Willamette Basin and Central Willamette Subbasin. Oregon. US Department of the Interior, US Geological Survey.

Howland, W. G. (2015). The Lake Champlain Basin Program: Its history and role. *Vermont Journal of Environmental Law*, *17*, 2015.

Jasechko, S., Wassenaar, L. I., & Mayer, B. (2017). Isotopic evidence for widespread cold-season-biased groundwater recharge and young streamflow across central Canada. *Hydrological Processes*, *31*(12), 2196−2209. Available from https://doi.org/10.1002/hyp.11175, http://onlinelibrary.wiley.com/journal/10.1002/(ISSN)1099-1085.

Middleton, M. A., & Allen, D. M. (2014). Vulnerability assessment for groundwater dependent streams. *Final report*, 2014.

Musial, C. T. (2015). Dynamic Surface Water-groundwater Exchange in Tidal Freshwater Zones: Insights from the Christina River Basin.

Neff, R., Chang, H., Knight, C. G., Najjar, R. G., Yarnal, B., & Walker, H. A. (2000). Impact of climate variation and change on Mid-Atlantic Region hydrology and water resources. *Climate Research*, *14*(3), 207−218. Available from https://doi.org/10.3354/cr014207, http://www.int-res.com/journals/cr/cr-home/.

Plan, M.W., (2006). Lower Passaic River Restoration Project.

Pomeroy, J., Fang, X., Westbrook, C., Minke, A., Guo, X., & Brown, T. (2010). *Prairie hydrological model study final report*. Saskatoon, Saskatchewan: Centre for Hydrology, University Saskatchewan.

Reidy, C. A. (2004). Variability of hyporheic zones in Puget Sound lowland streams (Doctoral dissertation).

Reinesch, A., Fausak, L., Joseph, A., Kylstra, S., & Lavkulich, L. (2022). An integrated framework for regional assessment of water, energy, and nutrients from food loss of selected crops in the Lower Fraser Valley, Canada. *Agricultural Sciences*, *13*(5), 633–657. Available from https://doi.org/10.4236/as.2022.135042.

Rumsey, C. A., Miller, O., Hirsch, R. M., Marston, T. M., & Susong, D. D. (2021). Substantial declines in salinity observed across the Upper Colorado River Basin during the 20th century, 1929–2019. *Water Resources Research*, *57*(5), p.e2020WR028581.

Scott T.R. (2006). Review of Truckee River operating agreement negotiations and related modeling efforts. In Operating Reservoirs in Changing Conditions, Proceedings of the Operations Management 2006 Conference (pp. 312–323). Available from https://doi.org/10.1061/40875(212)31.

Stanger, W. F. (2013). The Colorado River Delta and Minute 319: A Transboundary Water Law Analysis. *Environs: Envtl. L. & Pol'y J*, *37*, 2013.

Stanley, S., Grigsby, S., & Hruby, T. (2010). Olson, Puget Sound Watershed Characterization Project: Description of Methods, Models and Analysis, 2010.

Tague, C., & Grant, G. E. (2004). A geological framework for interpreting the low-flow regimes of Cascade streams, Willamette River Basin, Oregon. *Water Resources Research*, *40*(4), W043031–W043039. Available from https://doi.org/10.1029/2003WR002629, http://onlinelibrary.wiley.com/journal/10.1002/(ISSN)1944-7973.

Unduche, F., Tolossa, H., Senbeta, D., & Zhu, E. (2018). Evaluation of four hydrological models for operational flood forecasting in a Canadian Prairie watershed. *Hydrological Sciences Journal*, *63*(8), 1133–1149.

Welty, N. R. (2005). Exploring relationships between land use and echohydrology using multivariate statistics and process-based models.

Woods, S., Horsley, S., & Mackin, K. (2002). Balancing the water budget in the Ipswich River Watershed. *Proceedings of the Water Environment Federation*, *2002*(2), 1098–1109. Available from https://doi.org/10.2175/193864702785664932.

Chapter 14

PATRICOVA viewer for teaching the risk of flooding: a resource to improve resilience to natural hazards

Álvaro-Francisco Morote[1], Jorge Olcina[2] and Saeid Eslamian[3]

[1]Department of Experimental and Social Sciences Education, Faculty of Teaching Training, University of Valencia, Valencia, Spain, [2]Department of Regional Geographical Analysis and Physical Geography, University of Alicante, Alicante, Spain, [3]Department of Water Sciences and Engineering, College of Agriculture, Isfahan University of Technology, Isfahan, Iran

14.1   Introduction

In the European region, hazards of an atmospheric nature in the last years have been generating growing human and economic losses (Pérez-Morales et al., 2021). On a global scale, according to the Centre for Research on the Epidemiology of Disasters (Centre for Research on the Epidemiology of Disasters [CRED], 2019), these phenomena are considered the most hazardous (43% of the total) and those that affect the most people. Besides, worldwide it is estimated that the annual victims of floods may rise to 300,000 in 2050 and reach 390,000 from the 1980s of the current 21st century (Intergovernmental Panel on Climate Change [IPCC], 2014). The European Environmental Agency (European Environment Agency [EEA], 2017) has found that between the period 2000 and 2014, there were some 2000 fatalities in the European continent due to floods, affecting about 8.7 million people. In Spain, as Olcina (2018) has shown, 2 million people live in floodables areas, registering 526 deaths due to this phenomenon during the period 1995−2015. If the study area of the work presented here (Valencian territory) is taken into account, according to the PATRICOVA, it has been found that 600,000 inhabitants (12% of the people) live in flood areas. Of this population, 30,000 have their residence in areas with a high risk. Regarding the educational field, a fact that highlights the interest of taking these risks into account is that, officially, in 327 educational centers, some element or part of the center is located in a flood zone (Plan de Acción Territorial sobre Prevención del Riesgo de Inundación en la Comunitat Valenciana [PATRICOVA], 2015).

At international scale, different works, recently have highlighted the interest of studying flood risk issues at all educational levels (Ahmad & Numan, 2015; Lechowicz & Nowacki, 2014; McWhirter & Shealy, 2018; Morote et al., 2022). In Spain, most of the investigations on this subject (from didactics) have been carried out in the field of natural sciences (Díez-Herrero, 2015; Garzón et al., 2009). However, from the *Didactics of Geography*, it is not usual to find this type of publication (Cuello, 2018; Cuello & García, 2019; Ollero, 1997). In the Valencian Community, some work has been carried out on the social representations of future teachers in primary education (Morote & Hernández, 2020, 2021; Morote & Souto, 2020) as well as teaching proposals on field trips (Morote, 2017; Morote & Pérez, 2019).

The interest in dealing with the risk of flooding in the school environment is due to different causes: (1) it is a geographical subject (that should be taught in geography in the second year of baccalaureate), according to the current national curriculum (Royal Decree 1105/2014, of December 26) and the Valencian Autonomous Community (Decree 87/2015, of June 5); (2) the Valencian region (object of study) has become a region at risk (Calvo, 2001), both due to its atmospheric conditions in the context of current global warming and due to the increase in risk (urbanization, waterproofing and encroachment of flood zones); (3) the erroneous perception of society about its supposed domain of nature, which would allow reaching "0 risk" in the face of natural hazards, a totally unlikely issue (Olcina, 2017); and 4) the effects of global warming predict that these hazards (heavy rains) will be more intense and frequent in the future (Intergovernmental Panel on Climate Change [IPCC], 2022).

Hydrosystem Restoration Handbook. DOI: https://doi.org/10.1016/B978-0-443-29802-8.00014-5

The teaching of flood risks in the school stage is a content that must be addressed, especially in the subject of social sciences and/or geography (Morote & Souto, 2020). However, despite the importance of the topic, in the scientific literature there are few works that have been dedicated to the subject. However, in the international context, there are different studies about the teaching of floods in schools (Jacobi, 2005; Lee et al., 2019; Lozina & Pagliaricci, 2015; Mudavanhu, 2015; Shah et al., 2020; Tsai et al., 2020; Valdanha & Jacobi, 2021; Zhong et al., 2021).

In the European context, the study of Lechowicz and Nowacki (2014), or Williams et al. (2017) can be referenced. In relation to the territorial area under study (Spain), works have been published in the past years: (1) proposals and experiences of field trips (Morote, 2017; Morote & Pérez-Morales, 2019); (2) social representations of school students (Hernández-Ruiz et al., 2020) and teachers in training (Morote & Hernández, 2020; Morote et al., 2021; Morote & Souto, 2020); and (3) or from the press (Cuello, 2018).

The objective of this research is to present some didactic activities to work about floods from the viewer offered by the PATRICOVA. These activities are aimed at the second year of baccalaureate (17−18 years old) (stage in which geography is studied as an optional subject) and are intended to promote among the students capacities to interpret the territory, especially its immediate territory in the learning process. With this, the objective is to achieve greater knowledge of floods by society, which will allow for more resilient territories and societies in the face of extreme episodes, as a measure of adaptation to the current process of climate change.

14.2   Sources and methodology

To develop the proposed aims, first, the current national curriculum for compulsory secondary education and baccalaureate (Royal Decree 1105/2014 of December 26) has been consulted. In Spain, it should be noted that the contents are transferred to the autonomous communities that are liable for specifying them in their own curriculum. Although, in this work, a specific area of study of the Valencian territory (the Vega Baja region) is proposed, what is intended is that it be a model and guide to be developed in other areas. Therefore, for the design of the activities, national rules have been taken into account. It is a general proposal that can even be taught in ESO (12−16 years) or even primary education (6−11 years), especially for the third cycle (fifth and sixth years). It is, therefore, a matter for each teacher to adapt it to their course and region under study. Likewise, these proposals can serve as a model for other international territories that have similar climatic and urban characteristics.

In geography (second year of baccalaureate), the contents related to the risk of flooding are inserted in Block 3 "Climate diversity and vegetation" and Block 4 "Hydrography." It is worth noting the interest that the curriculum gives when dealing with these phenomena, since it can be interpreted that not only the climatic factor (hazard factor) but also the human component (vulnerability factor) is considered, as stated in Block 5 "The natural landscapes and nature−society interrelationships." In this way, the geography proposed in the second year of baccalaureate has as its fundamental purpose to provide a global and interrelated interpretation of each geographical phenomenon and to offer the mechanisms that serve to facilitate answers and explanations to the problems posed by the Spanish territory. Regarding the resource that is proposed to be used in this proposal (PATRICOVA viewer), it should also be noted that, in the Royal Decree, in Block 1 "Geography and the study of geographic space," it is proposed to work on different skills and competencies related to cartographic interpretation.

To complete the sessions established here, the teacher must prepare and provide an explanatory text with the contents under study. As an example, different information is presented below that will help students understand and interpret the territory of analysis (the region of Vega Baja del Segura, Alicante).

14.3   Study area: the region of the Vega Baja of the Segura (Alicante, Spain) as a laboratory of the territory

The study area (*Vega Baja of the Segura*) forms part of the southernmost territory of Alicante (Valencian Community), in what some regional divisions have called, in a broad sense, the *Bajo Segura* (Fig. 14.1). It is integrated into the climatic area of the southeast, being one of the regions with the lowest rainfall in the Iberian Peninsula and Europe (Gil & Olcina, 2017). There is a lack of rainfall, with intense and prolonged droughts, without prejudice to irregular floods of high, and even exceptional, hourly intensity (Gil & Olcina, 2017). An example of this has been the episode of September 12−13, 2019 with the cold drop or Isolated Depression at High Levels (DANA), reaching very high intensities and precipitations, characteristic of the torrential rains of this region, exceeding the average values of annual

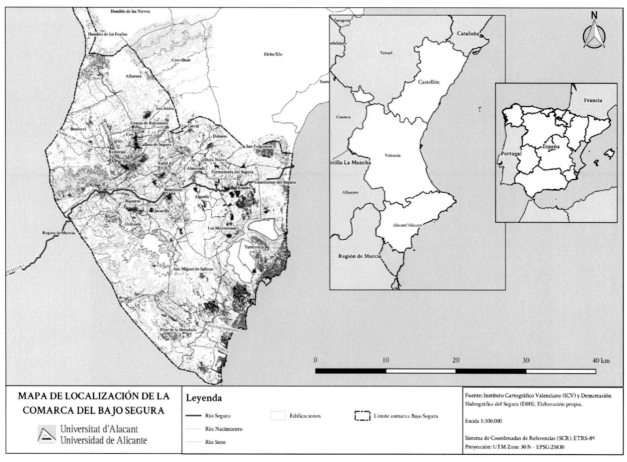

FIGURE 14.1 Study area (Vega Baja region, Alicante) (own elaboration). Map lines delineate study areas and do not necessarily depict accepted national boundaries.

precipitation. For example, at the Orihuela observatory, 500 mm were recorded in the entire rainy episode (September 12−13) with hourly intensities of 200 mm in one hour (September 12) (Fig. 14.2).

The driest period is the summer interval that extends to some months of spring and autumn, in such a way that it has a long duration (5−6 months) and intensity. The average annual precipitations reaches values close to and below 300 mm (Orihuela, 317 mm; Guardamar del Segura, 287 mm and Torrevieja, 271 mm). The most favorable months for the rains are those of autumn and spring, with marked peaks (Gil & Olcina, 2017). The number of rainy days per year is scarce and ranges between 30 and 50. As for temperatures, the coldest monthly average does not fall below 8°C, presenting a thermal amplitude greater than 18°C and an annual average lower than 16°C (Gil & Olcina, 2017). The average annual temperatures are around 18°C, especially due to the mild winter temperatures (Olcina & Moltó, 2019). In summary, the *Vega Baja of the Segura* region is characterized by mild temperatures, without major thermal changes except when exceptional temperature events occur (cold and heat waves), with very low average annual rainfall (200−375 mm). Also, with a pronounced interannual and intra-annual irregularity, as well as a very prominent temporal (hourly) concentration, with the genesis of frequent episodes of torrential rain, with very dry summers and the development of drought events (Gil & Olcina, 2017).

Regarding human occupation, the *Vega Baja of the Segura* is a risky land, occupied since historical times, but profoundly modified since the second half of the last 20th century with the implementation of land uses in areas that present extreme natural behavior. In it converge: (1) a traditional agriculture around the Segura riverbed, implanted since Muslim times; (2) a modern, technified agriculture, which has spread since the 1970s of the 20th century thanks to the arrival of flows from the Tajo-Segura Aqueduct (ATS); (3) a very widespread tourist activity in the coastal strip; and (4) the construction of numerous internal and external connection road infrastructures (national highways and highways). All this has resulted in a high value of vulnerability to the danger, also very high, of floods that is registered in this territory.

Regarding the Segura river (Fig. 14.3), in its middle and lower reaches, is a typically Mediterranean river with a low average flow and very irregular interannual behavior and significant floods due to torrential rains. This same behavior, although with a much lower flow and even zero for months or years, corresponds to the ravines and ravines of the lower section of the Segura river in the province of Alicante (Abanilla ravine, Derramador ravine, as the most

FIGURE 14.2 Aerial image of Orihuela (September 12–13, 2019). *Source: From Valencia plaza (2021).*

FIGURE 14.3 Segura river as it passes through the city of Orihuela. *Source: Photograph of the authors (2020).*

outstanding courses), which experience flash floods with high drag capacity in case of heavy rainfall. La *Vega Baja*, from the hydrological point of view, is flooded either by the floods of the Segura river itself or by floods coming from the Guadalentín river, which coincide with the floods of the Segura river itself or by the more localized participation of the floods of the Abanilla or Derramador ravines, which generally provide complementary flows to the Segura flood itself. The overflow of the greatest socioeconomic and territorial consequences in contemporary history took place on October 14–15, 1879, the well-known "Santa Teresa flood," which caused more than 1000 fatalities (300 in Orihuela) and the destruction of 5762 homes. Since the end of the Civil War, the region has known very notable floods that have given rise to various proposals, until the development of the flood of November 1987, which led to the urgent approval of the Plan for the Defense of River Basin Avenues.

14.4   Didactic proposals to work on the risk of flooding: the case of the Vega Baja of the Segura

The proposals presented here are integrated into an activity that includes two sessions with a duration of 55 min as established by Royal Decree 1105/2014 (geography subject of the second year of the baccalaureate). Regarding the geographical framework, the Alicante region of the Vega Baja has been proposed. The objective of this proposal is for the students to make a diagnosis of this region based on the physical–ecological conditions and territorial and demographic planning as well as knowing the risk areas with the consultation of the PATRICOVA viewer (Table 14.1) (Table 14.2).

In the first session, three exercises are proposed (see Table 14.1). In the first, based on the text and the reference cartography provided (national topographic map available in the viewer of the Valencian Cartographic Institute, ICV) and the explanation provided by the teacher, the students must recognize and explain the climate and functioning of the main fluvial course that articulates this space (Segura river) and its main tributaries and characteristics (regime, fluvial, water uses, climate, etc.). Exercise No. 2 is intended for students to analyze the consequences in the territory of the main historical floods that occurred in Orihuela. In the third exercise, the student, from the PATRICOVA viewer, must recognize the villages that are affected by the risk of flooding. For this proposal, special attention will be paid to the areas affected by the Segura river (Figs. 14.4 and 14.5). In addition, in this third exercise, the learners must prepare a table (sheet facilitated by the teacher) and register all the affected villages and indicate the types of risk levels that affect them (from "Very low" to "Very high").

In the second session, and with the municipalities file provided in the first session, the students must search the National Institute of Statistics (https://www.ine.es/), according to the municipalities identified as having a risk of flooding (Segura river): (1) the total number of registered inhabitants and differentiate between Spanish and foreigners (number and percentage) (exercise 4) and (2) total dwellings and differentiate between typologies (main, secondary, and empty) (exercise No. 5). Regarding the demographic aspect, it is interesting to recognize the percentage of the foreign population with respect to the total to see if the student describes any opinion or analysis about the possible impact on

TABLE 14.1 Curriculum proposal to treat the risks of flooding in the geography of second year of baccalaureate. Case study of the Vega Baja region (Alicante) (own elaboration).

Contents	Evaluation criteria	Learnings standards
– Territorial characteristics of the study area according to climatic conditions, river courses, relief, etc. – Analysis of the historical episodes of heavy hourly intensity rains and their consequences. – Characteristics of territorial planning and urban uses. – Population and housing. – Use of the PATRICOVA cartographic viewer and analysis of risk areas.	– Diagnose the main territorial and physical–ecological characteristics (climatic conditions, river courses, relief). – Analyze the historical episodes of heavy hourly intensity rains that caused serious flooding problems. – Recognize the territorial configuration according to the planning process and urban uses. – Use the PATRICOVA cartographic viewer correctly and recognize the main areas affected by the risk of flooding.	– Diagnose the main territorial and physical–ecological characteristics (climatic conditions, river courses, relief). – Recognize the main historical episodes of strong hourly intensity rains that caused serious flooding problems. – Understand how the territorial configuration has been produced, taking into account the process of planning and urban uses. – Manage the PATRICOVA cartographic viewer and recognizes the main areas affected by the risk of flooding.

TABLE 14.2 Proposed exercises for sessions 1 and 2. "Territorial diagnosis of the Vega Baja. A region where the risk of flooding constitutes a structural feature" (own elaboration).

Session 1

— **Exercise No. 1**. Identify and describe the climate and the main river courses of the study area.
— **Exercise No. 2**. Analyze and describe the consequences that occurred in the study area coinciding with the main historical floods (20th to 21st centuries).
— **Exercise No. 3**. Use the PATRICOVA GIS viewer, which municipalities are affected by the risk of flooding? Based on the file provided by the teacher, register all the municipalities in the region affected by the risk of flooding the Segura river according to the different levels of risk assigned by PATRICOVA.

Session 2

— **Exercise No. 4**. Check the National Statistics Institute (https://www.ine.es/). From the municipalities identified in exercise No. 3, record the number of inhabitants registered in these localities. Next, disaggregate them between Spaniards and foreigners. Do you think that this differentiation may have any impact on the risk?
— **Exercise No. 5**. Check the National Statistics Institute (https://www.ine.es/). It registers the number of registered households (last census of 2021) and disaggregates the dwellings by type (main, secondary, and empty) of the identified municipalities. Which municipalities have a greater number of second homes? What are they related to?

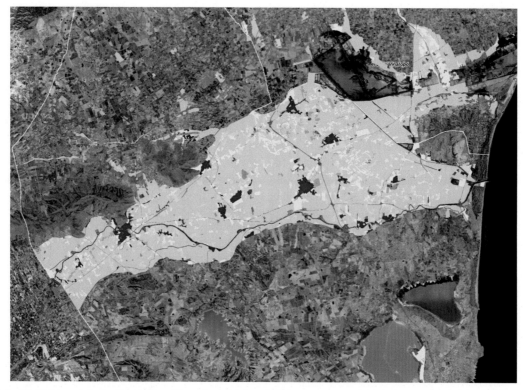

FIGURE 14.4   Flood risk areas in the Vega Baja region (Alicante). Note: Risk levels: red (very high), orange (high), yellow (medium), green (low), and blue (very low). The red colors coincide with the main population centers (own elaboration). *Source: From (PATRICOVA, 2022).*

the risk of this population cohort due to the lack of knowledge of the territory by this contingent that resides in this part of the south of the province of Alicante. As for the dwellings, it will be assessed if the students know how to interpret and relate the secondary dwellings with urbanizations and with the foreign population. Also, from this analysis, the descriptions and opinions on whether the Spanish population knows the territory where they reside will be taken into account. Therefore it is about analyzing whether the students identify the prejudices about the foreign population in comparison with the Spanish population.

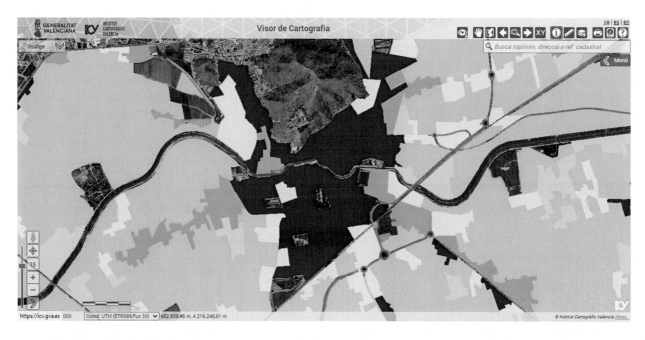

FIGURE 14.5 Flood risk areas in the city of Orihuela. Note: Risk levels: red (very high), orange (high), yellow (medium), green (low), and blue (very low). The red colors coincide with the main population centers (own elaboration). *Source: From (PATRICOVA, 2022).*

14.5 Conclusions

This chapter aims to provide examples, by way of case studies, that can be implemented in the classroom for a better interpretation of the risk of flooding, in the context of the geography subject of the second year of baccalaureate. For this, a specific area of study has been chosen and the consultation of a viewer of the GIS (free software) of the Valencian administration. In addition, it is intended that students know and become familiar with this type of resources (development of digital skills) and learn to locate and understand their closest environment. It is a proposal that can be adapted to other Spanish regions as long as there is basic information for it (flood risk mapping). In fact, the possibility of consulting and using the viewer of the National Flood Zone Mapping System opens the possibility of adapting this proposal to other Spanish territories, even if they do not have a specific plan on a regional scale.

The activities designed here should also serve to expand the training of future teachers who teach geography (the second year of baccalaureate). In this sense, it has been found that most classes use a technical methodology based on lectures and reproduction and memorization activities (Morote, 2020; Parra & Morote, 2020). The exercises that are proposed have the sense of problematizing and practicing a critical geography and breaking with the stereotypes and little scientific rigor that, on occasions, feed the main resources used in the classroom (school manual) (Morote & Olcina, 2020). The PATRICOVA viewer is an effective and very simple tool to implement in classrooms and even adapt these sessions to other lower levels (primary and secondary education). In addition, these activities are complemented by the use of other resources and sources.

The proposal to work on the risk of flooding in the classroom from the information contained in the web viewer of a public administration has the guarantee of the rigor that is understood about the official information on a subject as sensitive as the risk in said portal. It should be remembered that the official flood maps are the risk legal accreditation document for use in territorial planning processes. Also, other limitation is that the teaching staff is not familiar with the use of these cartographic viewers, or that there are difficulties in accessing the network, or the computer rooms do not exist for their development. However, it can be prepared, by the teaching staff, as a practical class with explanation through PowerPoint presentations or digital whiteboard, where appropriate, and active participation of the students. It is also interesting that the purpose of the activity is the participation of the students, which can materialize at the baccalaureate educational level, with a sharing in the classroom of the solutions that are proposed for the trouble of flooding in the area of residence.

As a proposal for the future, there is the need to promote these activities, both in active geography teachers and in training (masters in secondary and baccalaureate teacher training), as well as the adaptation of these proposals to the rest of educational levels. There is an interesting challenge that should allow, through education in territorial and environmental values, to create more resilient societies in the face of the dangers of nature, whose future effect is estimated to be more harmful in the framework of global warming.

References

Ahmad, S., & Numan, S. M. (2015). Potentiality of disaster management education through open and distance learning system in Bangladesh open university. *Turkish Online Journal of Distance Education, 16*(1), 249−260.

Calvo, F. (2001). *Sociedades y territorios en riesgo*. Barcelona: Ediciones del Serbal.

Centre for Research on the Epidemiology of Disasters (CRED). (2019). Economic Losses, Poverty & Disasters (1998−2017). Retrieved from: https://www.emdat.be/.

Cuello, A. (2018). Las Inundaciones del invierno 2009-2010 en la prensa, un recurso educativo para las ciencias sociales. *Revista de Investigación en Didáctica de las Ciencias Sociales, 2*, 70−87.

Cuello, A., & García, F. F. (2019). ¿Ayudan los libros de texto a comprender la red fluvial de la ciudad? *Revista de Humanidades, 37*, 209−234.

Díez-Herrero, A. (2015). Buscando riadas en los árboles: Dendrogeomorfología. *Enseñanza de las ciencias de la tierra: Revista de la Asociación Española para la Enseñanza de las Ciencias de la Tierra, 23*(25), 272−285.

European Environment Agency (EEA). (2017). Climate change, impacts and vulnerability in Europe 2016. An indicator-based report, Luxemburgo, Retrieved from: https://www.eea.europa.eu/publications/climate-change-impacts-and-vulnerability-2016.

Garzón, G., Ortega, J. A., & Garrote, J. (2009). Las avenidas torrenciales en cauces efímeros: ramblas y abanicos aluviales. *Enseñanza de las Ciencias de la Tierra, 17*(3), 264−276.

Gil, A., & Olcina, J. (2017). *Tratado de climatología*. Alicante: Publicaciones de la Universidad de Alicante.

Hernández-Ruiz, M., Miguel García-Pozuelo, M., Díez-Herrero, A., & Carrera, C. (2020). Mejora de la percepción y conocimiento infantil sobre el riesgo de inundaciones: Programa 'Venero Claro-Agua' (Ávila). In I. López Ortiz, J. Melgarejo, & P. Fernández (Eds.), *Riesgo de inundación en España: Análisis y soluciones para la generación de territorios resilientes* (pp. 1201−1210). Spain: University of Alicante.

Intergovernmental Panel on Climate Change (IPCC). (2014). Climate Change 2013 and Climate Change 2014 (3 vols.) Retrieved from: http://www.ipcc.ch/.

Intergovernmental Panel on Climate Change (IPCC). (2022). Climate Change 2022: Impacts, Adaptation and Vulnerability. Retrieved from: https://www.ipcc.ch/report/ar6/wg2/

Jacobi, P. R. (2005). Impactos socio-ambientales urbanos del riesgo de la búsqueda de la sustentabilidad: el caso de la Región Metropolitana de São Paulo. *Ciudad y territorio: Estudios territoriales, 145-146*, 671−682.

Lechowicz, M., & Nowacki, T. (2014). School education as an element of natural disaster risk reduction. *Prace i Studia Geograficzne, 55*, 85−95.

Lee, Y., Kothuis, B.B., Sebastian, A., Brody, S. (2019). Design of transformative education and authentic learning projects: Experiences and lessons learned from an international multidisciplinary research and education program on flood risk reduction. In *ASEE Annual Conference and Exposition*.

Lozina, A. A., & Pagliaricci, F. (2015). La escuela desde el barrio. *Extensión en red, 6*, 28−35.

McWhirter, N., & Shealy, T. (2018). Case-based flipped classroom approach to teach sustainable infrastructure and decision-making. *International Journal of Construction Education and Research*, 1−21.

Morote, A. F. (2017). El Parque Inundable La Marjal de Alicante (España) como propuesta didáctica para la interpretación de los espacios de riesgo de inundación. *Didáctica Geográfica*, 18, 211−230.

Morote, A. F. (2020). Recuerdos y experiencias de la Geografía Escolar. Caso de estudio de la asignatura de Geografía de las Regiones del Mundo (University of Alicante). In R. Roig-Vila (Ed.), *Investigación e innovación en la Enseñanza Superior. Nuevas aportacions desde la investigación e innovación educativas* (pp. 334−343). Barcelona, Spain: Ediciones Octaedro.

Morote, A. F., & Hernández, M. (2020). Social representations of flooding of future teachers of primary education (social sciences): A geographical approach in the Spanish Mediterranean Region. *Sustainability*, 12(15), 1−14. Available from https://doi.org/10.3390/su12156065.

Morote, A. F., Hernández, M., Eslamian, S. (2022), Flood risk instruction measures: Adaptation from the school. In Eslamian, S., Eslamian, F. (Eds.). Handbook of disaster risk reduction for resilience: Disaster and social aspects (Ch. 13, pp. 313−328). Springer Nature. Switzerland AG.

Morote, A. F., & Hernández, M. (2021). Water and Flood Adaptation Education: From Theory to Practice. *Water Productivity Journal*, 1(3), 31−40. Available from https://doi.org/10.22034/wpj.2021.264887.1025.

Morote, A. F., Hernández, M., & Olcina, J. (2021). Are future school teachers qualified to teach flood risk? An approach from the geography discipline in the context of climate change. *Sustainability*, 13(15), 8560. Available from https://doi.org/10.3390/su13158560, 1-22.

Morote, A. F., & Olcina, J. (2020). El estudio del cambio climático en la Educación Primaria: una exploración a partir de los manuales escolares de Ciencias Sociales de la Comunidad Valenciana. *Cuadernos Geográficos*, 59(3), 158−177. Available from https://doi.org/10.30827/cuadgeo.v59i3.11792.

Morote, A. F., & Pérez, A. (2019). La comprensión del riesgo de inundación a través del trabajo de campo: Una experiencia didáctica en San Vicente del Raspeig (Alicante). *Vegueta. Anuario de la Facultad de Geografía e Historia*, 19, 609−631.

Morote, A. F., & Pérez-Morales, A. (2019). La comprensión del riesgo de inundación a través del trabajo de campo: Una experiencia didáctica en San Vicente del Raspeig (Alicante). *Vegueta. Anuario de la Facultad de Geografía e Historia*, 19, 609−631.

Morote, A. F., & Souto, X. M. (2020). Educar para convivir con el riesgo de inundación. *Estudios Geográficos*, 81(288), 1−14. Available from https://doi.org/10.3989/estgeogr.202051.031.

Mudavanhu, C. (2015). The impact of flood disasters on child education in Muzarabani District, Zimbabwe. *Jamba: Journal of Disaster Risk Studies*, 6(1), 138.

Olcina, J. (2017). La enseñanza del tiempo atmosférico y del clima en los niveles educativos no universitarios: propuestas didácticas. In R. Sebastiá, & E. Tonda (Eds.), *Enseñanza y aprendizaje de la Geografía para el siglo XXI* (pp. 119−148). Alicante, Spain: Servicio de Publicaciones de la Universidad de Alicante.

Olcina, J. (2018). Verdades y mentiras sobre el riesgo de inundaciones en el litoral mediterráneo: balance de medio siglo. *Jornada Sobre Fenómenos Meteorológicos Extremos en el Mediterráneo*, 11 de diciembre de 2018. Valencia, AEMET.

Olcina, J., & Moltó, E. A. (2019). *Climas y tiempos del País Valenciano*. Publicaciones de la Universidad de Alicante.

Ollero, A. (1997). Crecidas e inundaciones como riesgo hidrológico. Un planteamiento didáctico. *Lurr@lde*, 20, 261-283.

Parra, D., & Morote, A. F. (2020). Memoria escolar y conocimientos didáctico-disciplinares en la representación de la educación geográfica e histórica del profesorado en formación. *Revista Interuniversitaria de Formación del Profesorado*, 95(34.3), 11−32. Available from https://doi.org/10.47553/rifop.v34i3.82028.

Pérez-Morales, A., Gil, S., & Quesada, A. (2021). Do we all stand equally towards the flood? Analysis of social vulnerability in the Spanish Mediterranean coast. *Boletín de la Asociación de Geógrafos Españoles*, 88, 1−39. Available from https://doi.org/10.21138/bage.2970.

Plan de Acción Territorial sobre Prevención del Riesgo de Inundación en la Comunitat Valenciana (PATRICOVA). (2015). Memoria. Retrieved from: https://politicaterritorial.gva.es/es/web/planificacion-territorial-e-infraestructura-verde/patricova-docs.

Plan de Acción Territorial sobre Prevención del Riesgo de Inundación en la Comunitat Valenciana (PATRICOVA). (2022). Visor PATRICOVA. Retrieved from: https://visor.gva.es/visor/.

Shah, A. A., Gong, Z., Ali, M., Sun, R., Naqvi, S. A. A., & Arif, M. (2020). Looking through the Lens of schools: Children perception, knowledge, and preparedness of flood disaster risk management in Pakistan. *International Journal of Disaster Risk Reduction*, 50, 101907. Available from https://doi.org/10.1016/j.ijdrr.2020.101907.

Tsai, M. H., Chang, Y. L., Shiau, J. S., & Wang, S. M. (2020). Exploring the effects of a serious game-based learning package for disaster prevention education: The case of Battle of Flooding Protection. *International Journal of Disaster Risk Reduction*, 43, 101393. Available from https://doi.org/10.1016/j.ijdrr.2019.101393.

Valdanha, D., & Jacobi, P. R. (2021). Etnoconservación y educación ambiental en Brasil: Resistencias y aprendizaje en una comunidad tradicional. *Praxis & Saber*, 12(28), 11443.

Williams, S., McEwen, L. J., & Quinn, N. (2017). As the climate changes: Intergenerational action-based learning in relation to flood education. *The Journal of Environmental Education*, 48(3), 154−171.

Zhong, S., Cheng, Q., Zhang, S., Huang, C., & Wang, Z. (2021). An impact assessment of disaster education on children's flood risk perceptions in China: Policy implications for adaptation to climate extremes. *Science of the Total Environment*, 757, 143761. Available from https://doi.org/10.1016/j.scitotenv.2020.143761.

Chapter 15

Lake rehabilitation case studies in China

Jihui Fan[1], Majid Galoie[2], Artemis Motamedi[3], Ning Niu[1,4] and Saeid Eslamian[5]

[1]Institute of Mountain Hazards and Environment, Chinese Academy of Sciences, Chengdu, Sichuan, P.R. China, [2]Civil Engineering Department, Imam Khomeini International University, Qazvin, Iran, [3]Civil Engineering Department, Buein Zahra Technical University, Qazvin, Iran, [4]University of Chinese Academy of Sciences, Beijing, P.R. China, [5]Department of Water Sciences and Engineering, College of Agriculture, Isfahan University of Technology, Isfahan, Iran

15.1 Introduction

More than 70% of the Earth is covered by oceans, and a small part of the Earth (about 150 million km^2) is occupied by lakes and rivers (O'Sullivan, and Reynolds, 2005). A lake is "a standing body of water occupying a basin and usually having no connection with the sea" (Forel, 1901, the founder of limnology, pp. 2−3) and filled with either saline or fresh water (Servos et al., 1989). The formation of lakes is sometimes due to catastrophes such as volcanic eruptions, floods, landslides, and avalanches, human activities, and sometimes over a long period of time. Also ponds, river, glacial, or karst lakes may dry up over time and turn into swamps. Although no one can count the exact number of bodies of water on Earth (O'Sullivan, and Reynolds, 2005), the total number of lakes in the world in 1990 was estimated to be about 1.25 million, with a total area of 2.6 million km^2. There are 7.2 million water bodies reported with an area of $0.01−0.1$ km^2 (Meybeck et al., 1995).

At present, the number of natural lakes on the globe is probably at least 2 million. The number of lakes in China (excluding Tibet region) is estimated to be about 2300, in Austria 6000, in Finland and Sweden about 100,000, in Russia (including Siberia) about 500,000, and in Canada nearly 1 million or more (O'Sullivan, and Reynolds, 2005).

The Caspian Sea (Kaspijsko more) is the largest lake in the world (Wikipedia), located at 41°40′0″N, 50°40′0″E, between Europe and Asia in the north of the mountainous Iranian plateau. It has an area of 3,71,000 km^2, which is as same as the area of Japan, but it is saline.

The next largest lake, Lake Superior in the United States, has an area of 82,100 km^2, and the others in order of size are as follows: Aralskoye More, Lake Victoria, Lake Huron, Lake Michigan, Lac Tanganyika, Ozero Baykal in Russia (which is the deepest lake in the word with the maximum depth around 1741 m and represents over 10% of the total fresh water in the world's lakes), Great Bear Lake, Great Slave Lake, Lake Erie, Lake Winnipeg, Lake Malawi, Ozero Balkhash, Lake Ontario, Ladozhskoye Ozero, Lac Tchad, Tonle Sap, and Lac Bangweolo with the total area of around 5,41,900 km^2 (Herdendorf, 1982).

The biology of lakes is undoubtedly influenced by geography, climate, rivers, sources of incoming water and its pollution, humans, and hydrology and hydrography of impounded water (O'Sullivan, and Reynolds, 2005).

15.1.1 The use of lake and reservoir

There are two types of lakes, natural and artificial. Artificial lakes are mostly created for irrigation purposes, like the ancient "Moeris" lake in Egypt dates back to 6000 years. Sri Lanka has also been known for its dams for more than 2000 years and has more than 9000 artificial lakes. Many artificial lakes were created by mining materials such as gravel, coal, and so on. Lakes and reservoirs play a crucial role in the economy of many regions, and the most common and widespread uses are as follows: drinking water, mills, agricultural ponds, bathing, recreation (not bathing), water sports (not bathing such as boating, surfing, etc.), fish ponds, landscaping, irrigation, industrial processes, transport, energy (generation and cooling), flood management control, fire ponds, purification basins, view sight, tourism, commercial processes, human well-being, and human satisfaction.

Hydrosystem Restoration Handbook. DOI: https://doi.org/10.1016/B978-0-443-29802-8.00015-7

15.1.2 The global water budget

Climate changes, the "El Nino" in the Pacific, and recent global warming have led to increased desertification and the shrinking of many lakes. The ocean and dryland water balance has been summarized in Table 15.1, and question mark explained in detailed are still the subject of conflicting opinions (O'Sullivan, and Reynolds, 2005).

The sources (inputs) of water in lakes are (i) precipitation, (ii) inflowing streams and runoff, and (iii) inflowing groundwater. The outflows are (iv) evaporation, (v) intake, and (vi) seepage to groundwater. Watershed morphology and climatic conditions are closely related to these inputs and outputs. The most dynamic component of this balance is precipitation, which varies in time and space (O'Sullivan, and Reynolds, 2005; Winter, 1995) (Fig. 15.1).

15.1.3 Determining water budgets over a lake

i. Precipitation: Measuring precipitation is a simple task. The catch efficiency of these measuring gauges is high when the precipitation falls vertically and there is less wind. In the windy areas, using the Alter shields would overcome the problem. It should be noted that averaging, Theissen polygons and isohyets are three common averaging methods for calculating the amount of precipitation over a region.

ii. Evaporation: Although the interaction among temperature, vapor pressure, and radiation in "energy budget method" is the way to predict evaporation, the most common instrument for measuring evaporation is evaporation pan. In order to use pan data for lakes, pan-to-lake coefficients need to be determined. This value is 0.7 for annual estimates of evaporation (Kohler et al., 1955) and vary from 0.5 to 2 for monthly estimation (Hounam, 1973).

iii. Inflow and outflow streams: Estimating inflow and outflow is possible by installing the gauges upstream and downstream of the lake. In this case, the value of the groundwater input between the gauge and the edge of the lake should be considered.

TABLE 15.1 World water balance (10^3 km^3 per year) (International Lake Environment Committee and United Nations Environment Program).

Water balance parameter	10^3 km^3 per year
Precipitation on ocean surfaces	347
Evaporation from ocean surfaces	383
Precipitation on land surfaces	99
Evaporation from land surfaces	63
Annual runoff (rivers)	28
Annual runoff (surface and subsurface groundwater)	8?

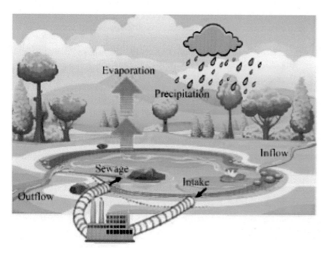

FIGURE 15.1 Schematic diagram of the water budget parameters of the lake.

iv. Groundwater: Lakes may have seepage outflow, inflow, or both (Born et al., 1974; Winter, 1976). Due to the elevation of the lake and surrounded area, groundwater recharge or receive is caused. There are many researches near the lakes related to preferential flow, nonlinear seepage, focused recharge, and transpiration-induced seepage from lakes, and the interaction of the geology of lake beds and climate conditions has been studied.

Any imbalance between inputs and outputs causes a change in the volume and the water level of the lake. Gauges are usually installed to measure the water level of the lake accurately.

15.1.4   Determining the residence time of lakes

Residence time is a parameter that is usually calculated by dividing lake volume by the total losses of water (or the rate of outflow/inflow), and it is important in eutrophication researches (Vollenweider, 1976). Water residence time is usually short for mountainous, riverine, or glacial lakes because normally large amount of loss occurs during water flushing or outflow. The residence times in these types of lakes measured in years. In contrast, for lakes with the small loss of evaporation, the residence time is much longer and measured in decades.

15.2   Lake rehabilitation

In 1970, lake rehabilitation was in a "youthful" stage of development, and numerous significant researches have been done in the proceeding decades (Peterson & Martin-Robichaud, 1988). Nowadays, due to the lack of water in urban areas, lake rehabilitation has been become a global interest. By increasing the awareness of environmental protection, researching on lake rehabilitation and lake restoration has been focused with more details in many countries (González del Tánago et al., 2012). Due to complicated ecosystems of the lakes, lake renewal and their improvement is a complicated phenomenon. The relationships between physical, chemical, and biological characteristics of each lake force us to do research on a lake-by-lake basis (Peterson & Martin-Robichaud, 1988). Usually, the lake condition is improved by two different methods: direct and indirect via treating the lake or watershed or both (Singh, 1982; Zamparas & Zacharias, 2014). Actually, the suggested management practices of lake improvement are to prevent the pollution rather than correction. It means that the major point sources should be detected first and the water quality should be monitored well before entering the lake body. Usually, the most visible pollution in the lake and the prim cause of the depletion of dissolved oxygen (DO) is about high nutrient inputs through agricultural, industrial, and untreated domestic sewage discharges (Zaragüeta & Acebes, 2017); therefore they should be controlled before entering the lake.

Most pollutants in the lakes comes from the industrial outfalls and soil fertilizers; however, pollutants transport through the atmosphere and wind should not be omitted. Heavy metals and pollutants from the production of cement, steel, and radioactive and fossil fuel are the common examples of pollution that move into lakes.

15.2.1   Lake rehabilitation techniques

Controlling the nutrient is an important task to achieve long-term improvements in the lake; however, if a lake suffers from nutrient influx, there are some techniques to help in lake rehabilitation. These are discussed in the following subsections.

15.2.1.1   Dilution/flushing

Dilution/flushing of lakes is an act to replace nutrient-rich water with nutrient-poor waters. This dilution is done by two different methods:

a. diverting a stream or adding quantities of nutrient-poor surface waters into the lake and
b. pumping out the water from the lake and diluting the nutrient-rich water with more nutrient-poor groundwater recharge.

15.2.1.2   Aeration

In situations where nutrient control is not feasible and there is no other opportunity to dilute the lake, aeration can provide good relief. It means that the problem will not be solved completely, but the results can be improved. Aeration is usually used in biological treatment systems. When air rises, the level of DO in water also rises, this is suitable for the growth of aerobic microorganisms. Aeration would control the unpleasant taste, color, and smell of water and reduce gases that cause pH changes. It is also a suitable method to maintain thermal stratification.

This study characterizes the quality and hydrological properties of the Xinglong Lake as a case study. The authors believe that they have been able to present an accurate reflection of the knowledge on this lake and its function on the ecosystems and economy of the new area.

15.3 Sichuan Tianfu New Area

With the rapid development of social economy, the process of urbanization has also increased. The population and land use in old cities have reached saturation capacity, and so new areas have come up (Ouyang et al., 2018; Zhang, 2022). Compared with European and American countries, industrial development in China started in the 1980s. Therefore, in recent years, the Chinese government has begun to pay more attention to the "urgency of building eco-parks." In October 1989, China held the first academic symposium on landscape ecology, which focused on the need for ecological lake construction and rehabilitation. Ecology can be divided into two areas: restoration and ecological knowledge. Its main purpose is to repair the damage caused by humans. Restoration ecologists define restoration as a time-consuming task, for example, restoring biodiversity, species, and so on, which takes time to repair. Human interventions and natural disasters usually cause ecological damage (Fan et al., 2017; Hou et al., 2022).

In 1992, the first national-level new area, Shanghai Pudong New Area, was established. To date, there are a total of 19 new areas in China. In these new areas, wetlands have various functions such as water conservancy, water purification, climate regulation, flood and pollution control, and biodiversity conservation. The new Sichuan Tianfu area is the 11th new area approved in October 2014. It is an important support for China's development of the "Yangtze River Economic Belt."

Xinglong Lake is located in the Tianfu New District. It serves as a water balance regulation hub and was built in November 2013 (Zhang, 2022).

15.4 Ecological construction of Xinglong Lake

Xinglong Lake (Fig. 15.2) is a small and artificial shallow lake in the new Tianfu area of Chengdu, Sichuan City, China. The catchment area is now mostly urban, but it has changed from a hilly ecosystem to an urban area.

In this area, the annual average temperature is 18°C (presented in Fig. 15.3), with the highest temperature in August (27. 5°C) and the lowest temperature in January (10°C), the average annual precipitation is 1100 mm, and the annual average wind speed is 2.8 m s^{-1}.

FIGURE 15.2 Xinglong Lake location in Tianfu new area, Chengdu, China. Map lines delineate study areas and do not necessarily depict accepted national boundaries.

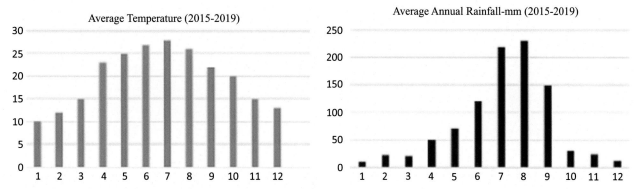

FIGURE 15.3 Xinglong district meteorological records: (left) average monthly temperature and (right) average annual rainfall. *From (Zhang, J. (2002). Exploration of ecological construction of urban parks. Master thesis. Sichuan Agricultural University. p. 71 [In Chinese])*

FIGURE 15.4 (left) Xinglong Lake in 2012 (10 years ago, as mentioned on the photograph) and (right) in 2023 *The photograph has been taken by the authors.*

It is an artificial lake with an area of about 4.33 km^2, making it the largest body of water in the central districts of Chengdu.

The water of Xinglong Lake mainly comes from Luxi River, a tributary of Minjiang River, and eventually flows into Jinjiang River via Huanglong River. This river belongs to the left branch of the Fuhe River in the river system of the Dujiangyan Irrigation Project and is also the second largest river flowing through the Tianfu New Area. The area where Xinglong Lake is located was originally a flood retention basin in the Luxi River Basin, and the planning team took full advantage of the characteristics of the low-lying terrain to form a lake. As a comprehensive aquatic ecology improvement project integrating flood control, ecology, landscape, among others, it is the "ecological green lung" in the Tianfu New Area and an important node for ecological conservation and water conservancy regulation of the Luxi River (Hou et al., 2022; Wang, 2022).

Xinglong Lake is divided into four parts, with the northern shore being the recreation area, the southern shore focusing on nature and culture, the eastern shore focusing on the harbor and scientific innovation culture, and the western shore serving as the gateway to the new area.

It is interesting to note that this lake is entirely man-made. As shown in Fig. 15.4, Xinglong Lake was a farmland in 2012.

Looking at the Google Earth photographs from 2010 to 2023, the transformation of the rural area and farmland into the lake is very clear. In Fig. 15.5, the steps can be traced.

The construction of the Xinglong Lake started with a change in the structure and pathway of the Luxi River in "Tianfu New District," Chengdu. To do this, a dyke was constructed upstream of the entrance to the Xinglong Lake. Fig. 15.6 shows the detour of the Luxi River and the construction of two tributaries to Xinglong Lake. On the Goggle Earth images, the authors have drawn a yellow arc to show the entrance river during the last 10 years. This makes it easy to find the old and new location of the lake. In all the images, it is important to focus on the restoration of the northern area of the lake, which has been completely reconstructed. The river path (the light blue path on the first picture) was completely changed and diverted into a huge artificial channel after it was fed into the lake. It can be seen in the picture labeled 2015. This constructed path can be followed in Fig. 15.6.

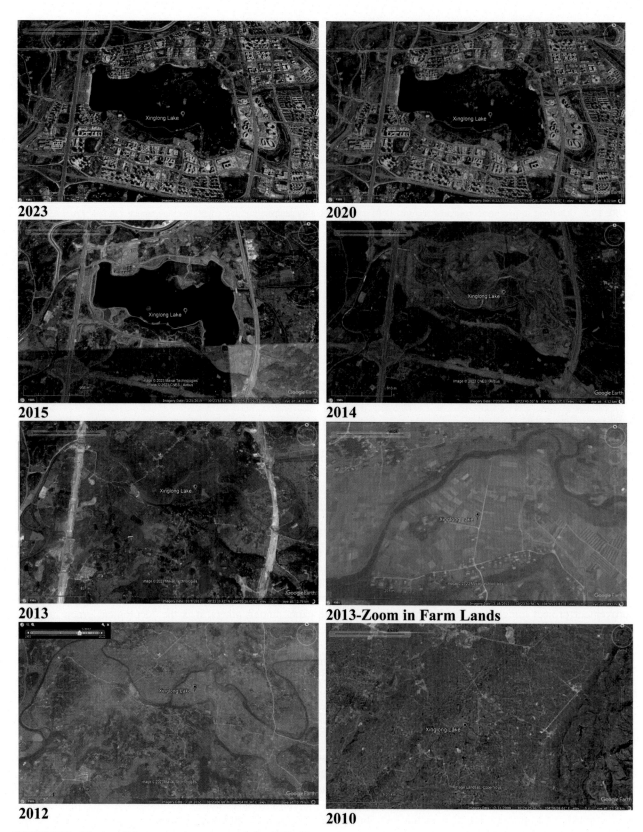

FIGURE 15.5 Xinglong Lake construction processes during 2010−23.

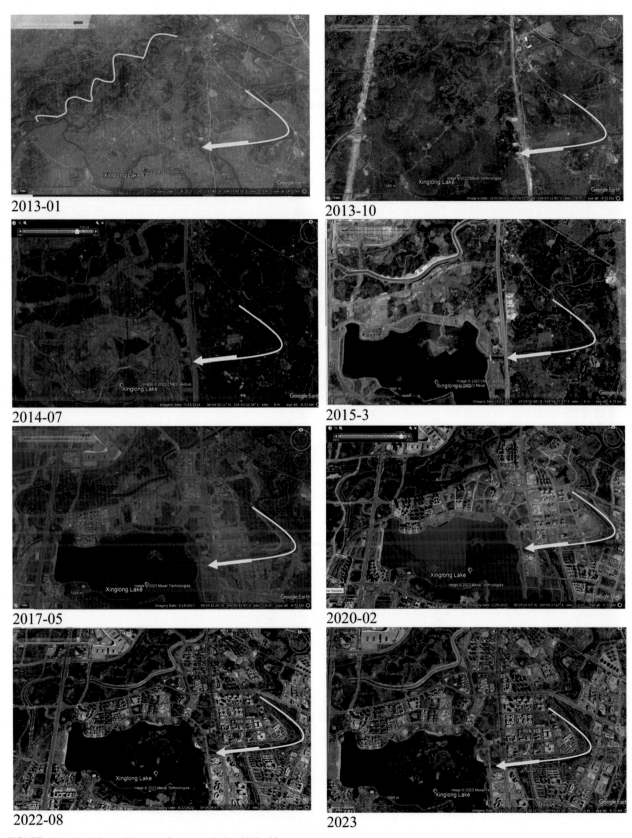

2013-01

2013-10

2014-07

2015-3

2017-05

2020-02

2022-08

2023

FIGURE 15.6 Luxi River Diversion Processes during 2013—23.

15.4.1 The suitable location to construct Xinglong Lake

Before choosing the location, it is necessary to know exactly the nature of the surrounding terrain and nature. The location is extremely important. It should be located near the source of the water system and in a low-lying area. According to hydrological modeling, the flood water should be protected. Soil infiltration in the area should be relatively weak. An area with high rainfall and low evaporation should be selected to save more water. As for the geographical location, the area should have a convenient transportation system. Sightseeing can performed on bicycles and public transportation such as buses, subways, and so on. In addition, parking lots and multiple routes at peak times should be considered. Xinglong Lake has all these facilities and the selected place had all the advantages (Hou et al., 2022).

15.4.2 Sponge city and rainfall harvesting

According to the concept of "sponge city" around Xinglong Lake, there are many suitable points to build a "rainwater garden." Permeable pavement in the Tianfu new development area would capture rain and control runoff and water accumulation on the road (Fan et al., 2021; Galoie et al., 2022).

15.5 Purpose of Xinglong Lake

Adjusting the urban climate, creating a "lung of the city" in the Tianfu New Area, optimizing the living environment, and improving the regional ecology, economy, and people's welfare are the main goals of the redevelopment of Xinglong Lake (Hou et al., 2022).

15.5.1 Flood controlling in Xinglong Lake

Xinglong Lake was created in a valley depression of a tributary of the Luxi River. During heavy rain in summer, this site was used as a large flood retention area. To achieve the goals of the new district, the water of the Luxi River was diverted to the southwest to build the present-day Xinglong Lake.

In order to control the runoff and pollution of Xinglong Lake, many measures were taken to prevent domestic and municipal sewage from flowing directly into the Luxi River and the lake. To control the discharge, the flood diversion project was also considered, which is summarized in Fig. 15.7.

In terms of protecting urban safety, the safety function of flood control and storage during the flood season is emphasized, as well as the ability to control floods. Flood and sediment control and improving urban safety prevention is one of the most important keys to Xinglong Lake rehabilitation.

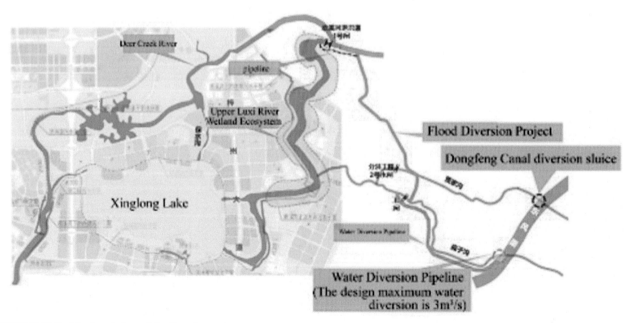

FIGURE 15.7 Xinglong Lake and the diversion systems.

15.5.2 Social and ecotourism service around Xinglong Lake

The construction of Xinglong Lake began in November 2013. In recent years, ecotourism has gradually won the favor of the public. It has become one of the main targets and an important basis for the ecological planning and economic benefits of Tianfu New Area. The government has provided adequate protection. From the perspective of the lake, its biological and ecological function is of great significance.

In recent years, various innovative projects have been carried out in the lake and the Luxi River area, such as Chengdu Scientific Research Center of Chinese Academy of Sciences and Chengdu Expo City, which is becoming an important innovation center in western China (Fu et al., 2023a, 2023b; Ouyang et al., 2018; Tan et al., 2020; Tan et al., 2021).

Public services are listed as follows:

- the modified of blue professional running track with high quality of material and color (Figure 15.8 (1)),
- gray walking track (Figure 15.8 (1)),
- public service building construction,
- children art center,
- entertainments,
- cultural scene creation,
- restaurants,
- malls,
- sand beach for children (Figure 15.8 (2)),
- sports parks,
- basketball court (Figure 15.8 (3)),
- skating and bike court (Figure 15.8 (4)),
- museum,
- Zhongxin underwater bookstore (Figure 15.8 (5)), and
- ferry race.

Xinglong Lake will not only become a "Green Heart of Tianfu," combining multiple functions such as flood control, ecology, and landscape but also bring new impetus, new life, and new economy to the new region. Xinglong Lake brings many ecological benefits: it provides a suitable environment for breeding various animals and plants and brings natural vitality to the city. Evaporation leads to increase in humidity and the reflection of sunlight leads to lowering of temperature (Deji et al., 2024).

In 2020, more projects were started around the lake, which are now completed and are a very beautiful sight. Around the lake, there is a beautiful and harmonious pattern of water and land. Many flowers have been placed on the grass, and the trees are shaped in an organic, ecological sequence.

Since 2018, the much-acclaimed underwater bookstore at Xinglong Lake in Chengdu was officially opened to the public on October 30, 2021 after three years of planning and construction. The area is around 500 m^2 and divided into reading, exhibition, and meditation salons (Fig. 15.8). The Xinglong area is presented in Fig. 15.9.

15.5.3 Economic benefits of Xinglong Lake

The excellent natural ecological environment around the lake attracts local and foreign tourists for recreation and entertainment. This leads investors from the surrounding area to plan and build the area.

This allows the economy in this region to develop rapidly. Many well-equipped offices, stores, apartments, shopping malls, and public facilities are settling in Chengdu Tianfu New Area, creating a high-quality living and working environment around the lake.

15.6 Special optimization plan

15.6.1 Special optimization plan for water ecological reconstruction

In 2017–20, many researches were conducted to improve the quality of the Luxi River Basin. Some researches were related to the eutrophication of the water and are still being carried out. During these years, water quality was monitored at many stations upstream of the lake (Fig. 15.9 in Shi et al., 2022) and at one station at the outlet. During 2017–20, the water quality was reported to be IV class with variations in total nitrogen.

(1)

(2)

FIGURE 15.8   Xinglong Lake Area (Zhao and Lu, 2023). *All have been taken by authors and No. 5 is available in Zhao., Z., Lu, Y. (2023). Chengdu Xinglong Lake Zhongxin underwater bookstore [In Chinese]. Journal of Idea and Criticism. pp 105–107).*

(3)

FIGURE 15.8 (Continued).

(4)

(5)

To protect the ecology, underwater plants, fish, and shellfish have been successively planted and released into the lake. An efficient ecological composite purification system has also been built. A scene of harmonious coexistence between man and nature and man and animals has developed in this lake.

In 2017, three years after the construction, sustainable water ecosystem was gradually improved, and the water quality reached the surface water category IV, and the transparency reached 1.5 m. In 2020, six years after the construction of the project, the main index of water quality was improved and reached II class. Through daily and monthly inspections, the quality is now class I.

In this lake, controlling the amount of submerged plant, which was artificially constructed, is very important because controlling the submerged plant in the early stage causes self-diversification in the later stages in future. Also, controlling the population of phytoplankton is important because their high population decreased the light sharply and limits the growth of submerged plants (Tan et al., 2020; Tan et al., 2021).

15.6.2 Acidification control in Xinglong Lake

After Gjessing's research in 1994, when the lake was divided into two parts by a plastic curtain and one part was acidified, the HS content (humic substances), TOC (total organic carbon), and color were measured. The results showed that

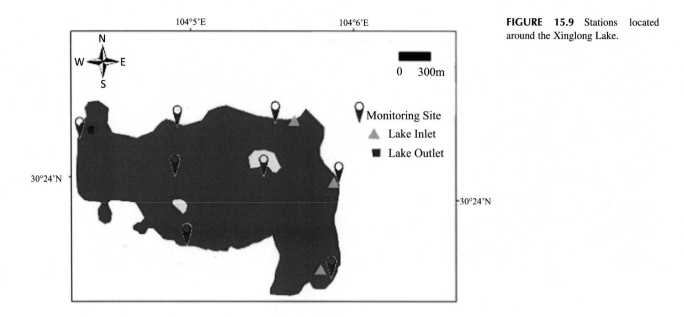

FIGURE 15.9 Stations located around the Xinglong Lake.

low pH increases the transparency of the lake. Changes in pH cause Fe and especially aluminum to hydrolyze and precipitate along with colored organic matter. In this situation, dissolved organic matter breaks down into lighter substances (Schindler et al., 1997). Because of the importance of pH in lakes, this parameter is measured daily in Xinglong Lake.

15.6.3 Controlling algae populations in Xinglong Lake

To support and maintain phytoplankton populations in the lake, it is necessary to control algae growth and conduct some regular experiments in the lake. Algae require temperature, light, and nutrients for their growth. Photosynthetic production, P, and heterotrophic respiration, R, are included in Eq. (15.1). Algal populations decrease DO and cause an imbalance in the aerobic biological treatment of the lake. A sudden increase in phosphorus input can cause algal growth, but nitrogen appears to be the limiting parameter. The results of measuring these parameters and calculating the N/P ratio show the shifts in the species composition of the algal communities. Low ratios lead to growth of blue-green algae, while high ratios lead to lower growth. Therefore phosphorus control and treatment of wastewater discharges to the lake could limit algal growth more. In Xinglong Lake, there are many stations for measuring nitrogen and phosphorus. In general, phosphorus entering lakes in a variety of ways (e.g., washed-in agricultural fertilizers, untreated wastewater, adsorbed to clays, and debris flows) should be carefully monitored (O'Sullivan, and Reynolds, 2005).

$$106\,CO_2 + 16\,NO3^- + HPO4^{-2} + 122H2O + 18H^+$$
$$P\downarrow\uparrow R$$
$$C_{106}H_{263}O_{110}N_{16}P_1 + 138O_2$$

(15.1)

Algal growth and ecosystem mainly depend on water depth. Fig. 15.10 shows the water depth in Xinglong Lake. The diversity of different fish species should also be studied carefully.

It should be noted that attention must be paid to the safety of the lake. The standard depth of the artificial lake near the shore should not exceed 0.7 m. The depth is referred to in Fig. 15.9.

The ecological restoration project can effectively reduce the species of cyanobacteria (especially the species with algal toxins) in the water body of Xinglong Lake and increase the species of diatoms. The ecological restoration technology can regulate and improve the ecological composition of eutrophic water bodies and improve the population structure of phytoplankton.

15.6.4 Biodiversity and maintenance of the lake ecological restoration

The relationship between fish, shrimp, and snail population density should be fully controlled in Xinglong Lake. The water quality and the regulation of the structure and quantity of fish in the water according to the quantity and species of fish is the most important goal to maintain the ecological balance (Tian, 2020).

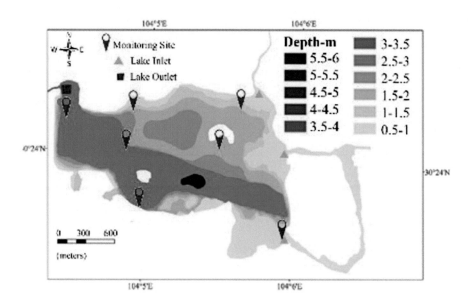

FIGURE 15.10 Water depth in Xinglong Lake.

15.7 Strengths, weaknesses, opportunities, and threats analysis

Strengths, Weaknesses, Opportunities, and Threats (SWOT) analysis is a strategic planning technique used to help an organization identify SWOT related to project planning. In Xinglong Lake the results are as follows:

1. **Strength:** Xinglong Lake is currently the largest artificial lake in Chengdu, which is rich in natural vegetation. It provides a good foundation for the subsequent resource development.
2. **Weakness:** The landscape belt around the lake is relatively long; in some parts, the traffic performance is poor.
3. **Opportunity:** It is located in an important geographical location in Tianfu New District, Chengdu, a key urban development area.
4. **Challenge:** A high monitoring system is required to control water pollution in future.

15.8 Smart water management

Intelligent water management is also carried out for Xinglong Lake. Sometimes, a series of water quality problems occur (Galoie et al., 2023). In response to the temporal and spatial monitoring of the ecosystem of Xinglong Lake and to analyze and simulate the dynamic changes of nitrogen, phosphorus, and planktonic organisms, a high-precision water quality model was developed for Xinglong Lake. This model was named PCLake. The PCLake model was developed by the team of Janse, a Dutch limnological modeler. It is a comprehensive ecological model suitable for shallow water lake, like Xinglong (Shi et al., 2022).

Through this intelligent water model and comprehensive water quality analysis, it would be easy to promote pollution load control at the intake. The PCLake model could also be used to closely monitor research on quantitative simulation and climate impacts in the future.

The results in Table 15.2 were confirmed by the "Ecological Environment and Urban Management Bureau of Tianfu New Area in Sichuan." The station is located at longitude E104.08220 and latitude N30.40030.

Monthly monitoring data include water temperature, pH, DO, electrical conductivity, transparency, permanganate index, Chemical Oxygen Demand (COD), Biological Oxygen Demand over 5 days (BOD5), ammonia nitrogen, P, TN (Total Nitrogen Bound), copper, zinc, fluoride, selenium, arsenic, mercury, cadmium, chromium, lead, cyanide, phenol, petroleum, anionic surfactant, sulfide, coliform, and chlorophyll. All measuring methods are based on the handbook of "Lake Eutrophication Survey Specification (Second version)" (Table 15.3).

Using digital management to strengthen the intelligent management of urban rivers and lakes can effectively ensure the long-term healthy operation of urban rivers and lakes and the high ecological value.

The intelligent management system mainly uses technologies such as digital twin networking, virtual simulation, and digital graphs to control platforms and typical business applications. This intelligent management of hydrology,

TABLE 15.2 Monitoring results in January, April, summer, and autumn (2022).

Monitoring items	Unit	Monitoring results January 2022	Monitoring results April 2022	Monitoring results July 2022	Monitoring results October 2022	Standard limits
		Winter	Spring	Summer	Autumn	
Water temperature	C	9.8	18.2	27.3	25.1	/
pH	–	8.21	8.21	8.27	8.27	6–9
DO	mg lit^{-1}	9.83	7.92	9.06	7.63	7.5
Electrical conductivity	ms cm^{-1}	217	192	175	195	/
Transparency	Cm	400	182	229	266	/
Permanganate index	mg L^{-1}	2.6	2.5	2.9	2.9	4
COD	mg L^{-1}	10	9	19	7	15
BOD5	mg L^{-1}	1	2.3	1.4		3
Ammonia	mg L^{-1}	Not detected	0.086	0.069	0.038	0.15
Total phosphorus	mg L^{-1}	0.02	0.01	0.05	0.02	0.025
Total nitrogen	mg L^{-1}	0.28	0.5	0.7	0.52	0.5
Copper	mg L^{-1}	Not detected	Not detected	Not detected	Not detected	0.01
Zinc	mg L^{-1}	Not detected	Not detected	Not detected	Not detected	0.05
Fluoride	mg L^{-1}	0.35	0.342	0.369	0.375	1
Selenium	mg L^{-1}	Not detected	Not detected	Not detected	Not detected	0.01
Arsenic	mg L^{-1}	Not detected	Not detected	Not detected	Not detected	0.05
Mercury	mg L^{-1}	Not detected	Not detected	Not detected	Not detected	0.00005
Cadmium	mg L^{-1}	Not detected	Not detected	Not detected	Not detected	0.001
Chromium	mg L^{-1}	Not detected	Not detected	Not detected	Not detected	0.01
Lead	mg L^{-1}	Not detected	Not detected	Not detected	Not detected	0.01
Cyanide	mg L^{-1}	Not detected	Not detected	Not detected	Not detected	0.005
Phenol	mg L^{-1}	Not detected	Not detected	Not detected	Not detected	0.002
Petroleum	mg L^{-1}	Not detected	Not detected	Not detected	Not detected	0.05
Anionic surface active	mg L^{-1}	Not detected	Not detected	Not detected	Not detected	0.2
Sulfide	mg L^{-1}	Not detected	Not detected	Not detected	Not detected	0.05
Coliform	unit L^{-1}	86	1.6×10^2	1.6×10^3	3.8×10^2	200
Chlorophyll	mg L^{-1}	0.004	0.002	Not detected	Not detected	/

environment, and ecology of the water system could enable dynamic real-time integration of all elements and realize flood control and disaster reduction (Zhang, Wang, et al., 2022 in Fig. 15.11).

The simulation and prediction results of the main water quality factors in Xinglong Lake based on PCLake are shown in Figs. 15.12–15.14 (Shi et al., 2022). The results show that nitrogen concentrations decrease significantly in spring and summer. This is due to the decomposition of aquatic plants in autumn and winter, and the nitrogen in the

TABLE 15.3 Water quality indicators during ecological restoration (2020–22).

Stage	Total phosphorus mg L^{-1}	Ammonia mg L^{-1}	COD mg L^{-1}	Total nitrogen mg L^{-1}	Permanganate index mg L^{-1}	Transparency (cm)
Before ecological restoration 2020/09	0.03	0.19	13.33	1.34	3.05	40.5
Initial restoration 2021/09	0.05	0.28	15.81	0.83	2.75	243.0
One year after restoration 2022/09	0.02	0.11	19.00	0.84	2.90	267.0

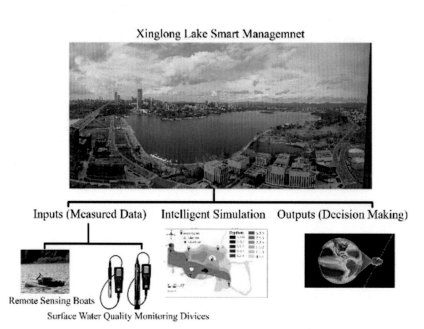

FIGURE 15.11 Xinglong Lake digital twin system.

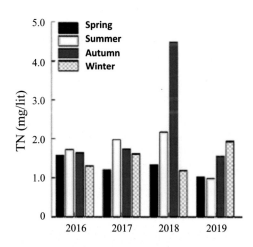

FIGURE 15.12 Seasonal variation of water quality before upgrading Xinglong lake water ecology project. *From Shi, T., Chen, Y., Liu, Z., Zhang, H., Wang, H., Fan, H., Ding, Y. (2022). Numerical simulation and comprehensive evaluation of key water quality factors in urban shallow lakes: Xinglong Lake case study. In* Proceedings of the China Water Conservancy Society. *Yellow River Water Conservancy Press: China Water Conservancy Society [In Chinese].*

plants is released back to the water body, while the aquatic plants in spring and summer grow gradually and can absorb some of the nitrogen and phosphorus in the water body. If measures are taken to control the lake, the concentrations of TN will decrease; otherwise, if no measures are taken to control the lake, the concentrations of TN in Xinglong Lake will increase in the next four years. The transparency would also decrease significantly (Shi et al., 2022).

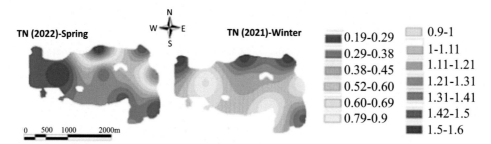

FIGURE 15.13 Variation of water quality after improving Xinglong Lake water project (2021−.22). *From Shi, T., Chen, Y., Liu, Z., Zhang, H., Wang, H., Fan, H., Ding, Y. (2022). Numerical simulation and comprehensive evaluation of key water quality factors in urban shallow lakes: Xinglong Lake case study. In* Proceedings of the China Water Conservancy Society. *Yellow River Water Conservancy Press: China Water Conservancy Society [In Chinese].*

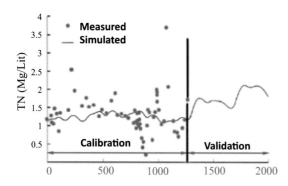

FIGURE 15.14 Xinglong Lake water quality factor simulation and prediction results. *From Shi, T., Chen, Y., Liu, Z., Zhang, H., Wang, H., Fan, H., Ding, Y. (2022). Numerical simulation and comprehensive evaluation of key water quality factors in urban shallow lakes: Xinglong Lake case study. In* Proceedings of the China Water Conservancy Society. *Yellow River Water Conservancy Press: China Water Conservancy Society [In Chinese].*

15.9 Achievements

1. The water quality is stable, and the self-purification ability of the lake is improved.

    Since the completion of the Xinglong Lake Comprehensive Water Quality Improvement Project, the water quality of the lake area has reached class II water standards, and some indicators have reached class I water standards. The deepest transparency can reach 4 m, which has exceeded the expected governance goals. Through the construction of a clear water ecosystem, submerged plants have covered 70% of the water area of the lake area. Xinglong Lake has successfully changed from the original "algae-type" muddy water lake to a "grass-type" clear water lake.

2. The integrity of aquatic organisms and biodiversity have increased.

    By enriching the topography of the lake bottom, there are 11 different water depths in Xinglong Lake. It has the same topography of natural lakes. Then, different aquatic organisms are connected to each other to save the Xinglong Lake ecosystem. In addition, the perfect water ecosystem of Xinglong Lake has also attracted more than 40 species of waterbirds, including ruddy shelducks, mallard ducks, bone-topped chickens, red-billed gulls, and lesser grebes.

3. Green belt has been constructed around the lake.

    To protect of water ecosystem, green belt has been constructed around Xinglong Lake and construction activity areas have been arranged. This belt creates tour paths with different interests and experiences. It is equipped with a complete public service supporting system, which has attracted a lot of popularity and many tourists to Tianfu New District.

15.10 Conclusions

Since the industrial revolution, the economy has developed rapidly. People usually see only the benefits of the industrial revolution, but the damage to the environment is usually ignored. The change of the environment makes it difficult for animals and plants to interact with each other. It is important to protect the environment because only then can people and nature coordinate and cooperate in their common development.

Sichuan Tianfu New District was planned to develop in 2010, with an estimated population of about 6 million people. The planned area is 1578 km^2. Decision-makers derived the concept of "protecting the ecological environment" and put forward the idea of creating a lake in the area. In October 2014, Sichuan Tianfu New District was officially recognized as a national new district.

Xinglong Lake is an artificial lake created in 2013 and located in Sichuan Tianfu New Area. It was formed by damming the Luxi River and forming a tributary on the left bank of the Fuhe River in the Dujiangyan water system. It is the largest lake in Chengdu with the area of about 4.33 km^2 and an 8.8 km long green circuit. The area of the park and green areas around the lake is 13 km^2, and the lake can hold 10 million m^3 of water. Xinglong Lake is divided into four parts, and the northern part is very new and has been almost completely rebuilt. The southern part has been focused on nature. The eastern part is known as the city of scientific innovation, and the western part is the gateway to the new area of the future. The green road around the lake is a great path for jogging and running. It is rich in ecological services and ecotourism. The primary purpose of building the lake is to protect biodiversity, preserve landscape resources, improve the ecosystem, enhance the economy, promote tourism, and ultimately restore the natural ecosystem.

The water quality was poor during 2016 and 2020; however, after the completion of the second water quality improvement in 2021, the overall water quality of the lake reached class III. Currently, total phosphorus and total nitrogen concentrations are good, but control measures need to be taken or nitrogen and phosphorus concentrations will increase. Before 2021, Xinglong Lake was eutrophic, and since 2022, there has been no further eutrophication (Hou et al., 2022).

Around the lake, there are many playgrounds, recreational facilities, art halls, restaurants, and shopping centers that improve human welfare. The water quality has reached the European standard of "Blue" water and is known as the "urban green heart" of Tianfu New District.

15.11 Suggestions

To enrich the theory of ecological construction of urban parks, the following suggestions and opinions were given for the future construction of ecological parks:

1. Site visits to the ecological parks of Xinglong need to be conducted through photographs, questionnaires, and so on to find out the problems in the planning and construction of the parks.
2. The relevant literature on urban ecological parks in China and abroad needs to be researched to collect more useful information and further knowledge.
3. Daily and high-tech pollution field measurements need to be conducted to model and optimize better ecological design.
4. Further exploration of Xinglong Lake and River should be performed using a digital twin platform to enable intelligent simulation and more accurate decision-making.
5. The long-term water quality and long-term maintenance of the lake are required.
6. The possible damage caused by changes in the water quality of Xinglong Lake in the future should be considered. It is recommended to control external pollution, maintain front wetlands, and reduce nutrient load. Aquatic plants should be planted to build a good water ecosystem.

Acknowledgments

The authors would like to express special thanks for the support provided by CAS President's International Fellowship Initiative; (Grant Nos. 2022VCC0004 and 2022VCC0005) and the National Natural Science Foundation of China (Grant Nos. U2240216-1 and 41871072) and Supported by Sichuan Science and Technology Program (2024YFHZ0246).

References

Born, S. M., Smith, S. A., & Stephenson, D. A. (1974). *The hydrogeologic regime of glacial-terrain lakes, with management and planning applications. Inland lake renewal and management demonstration project* (p. 73) Madison: University of Wisconsin Press.

Deji, Y., Fan, J., Galoie, M., & Motamedi, A. (2024). Impact of large dam reservoirs on slight local season shifting (case study: Three Gorges Dam). *International Journal of Hydrology Science and Technology, 18*(2), 186–200. Available from https://doi.org/10.1504/IJHST.2024.140313.

Fan, J., M. Galoie, A. Motamedi, Y. Liao, S. Eslamian (2021). Rainwater conservation practices in China. In Handbook of Water Harvesting and Conservation: Case Studies and Application Examples (pp. 261-281). Wiley.

Fan, X., Zhang, X., & Fan, X. (2017). Research on the construction of an artificial ecological lake water quality monitoring system. *Automation Technology and Application, 36*(5), 111–115..

Forel, F.A. (1901). Handbuch der Seenkunde: Allgemeine Limnologie (249). J. Engelhorn, Stuttgart

Fu, K., Chai, X., Wan, L., Zhao, B., Xue, Y., & Tang, T. (2023a). Evolution of zooplankton community and water quality assessment before and after ecological restoration of Xinglong Lake. *Environmental Ecology, 5*(3), 75−80., [In Chinese].

Fu, K., Chai, X., Wan, L., Zhao, B., Xue, Y., & Tang, T. (2023b). Response of phytoplankton communities to ecological restoration of Xinglong Lake in Chengdu. *Environmental Ecology, 5*(1), 82−87., [In Chinese].

Galoie, M., Motamedi, A., Fan, J., Moudi, M. (2023). Prediction of water quality under the impacts of fine dust and sand storm events using an experimental model and multivariate regression analysis. Journal of Environmental Pollution 336 (11), 122462. doi: 10.1016/j.envpol.2023.122462

Galoie, M., Motamedi M., Fan J., Eslamian S. (2022). Iterative floodway modeling using HEC-RAS and GIS. In Flood handbook (pp. 389-406). CRC Press. eBook ISBN: 9780429463938

González del Tánago, M., García de Jalón, D., & Román, M. (2012). River restoration in Spain: Theoretical and practical approach in the context of the European Water Framework Directive. *Environmental Management, 50*, 123−139.

Herdendorf, C. E. (1982). Large lakes of the world. *Journal of Great Lakes Research, 8*, 106−113.

Hou, X., Wang, H., Guan, E., Yao., Xing., Li, F., & Qin, C. (2022). The concept and practice of ecological wetland construction in Tianfu New Area: Taking Xinglong Lake wetland park as an example. *Resources and Human Settlements, 2*, 49−53, [In Chinese].

Hounam, C. E. (1973). Comparison between pan and lake evaporation. *World Meteorological Organisation Technical Note, 126*, 52.

Kohler, M. A., Nordenson, T. J., & Fox, W. E. (1955). Evaporation from pans and lakes. *United States Weather Bureau Research Paper, 38*, 21.

Meybeck, M. (1995) Global distribution of lakes. In Lerman, A., Imboden, D.M., Gat, J.R. (Eds.). Physics and chemistry of lakes (pp. 1− 35). Springer-Verlag, Berlin.

O'Sullivan, P.E., Reynolds, C.S. (2005). The lakes handbook. Vol. 2: Lake restoration and rehabilitation (p. 579). Blackwell Publishing.

Ouyang, L., Jia, B., & Zan, X. (2018). Characteristics of planktonic community structure in spring and winter of Xinglong Lake and its indicative significance for water quality status. *Sichuan Environment, 37*(3), 91−97..

Peterson, R. H., & Martin-Robichaud, D. J. (1988). Community analysis of fish populations in headwater lakes of New Brunswick and Nova Scotia. *Proceedings of the Nova Scotia Institute of Science, 38*, 55−72.

Schindler, D. W., Curtis, P. J., Bayley, S. E., Beaty, K. G., & Stainton, M. P. (1997). Climate-induced changes in the dissolved organic carbon budgets of boreal lakes. *Biogeochemistry, 36*, 9−28. Available from https://doi.org/10.1023/A:1005792014547.

Servos, M. R., Muir, D. C. G., & Webster, G. R. (1989). The effect of dissolved organic matter on the bioavailability of polychlorinated dibenzo-p-dioxins. *Aquatic Toxicology, 14*, 169−184.

Shi, T., Chen, Y., Liu, Z., Zhang, H., Wang, H., Fan, H., & Ding, Y. (2022). *Numerical simulation and comprehensive evaluation of key water quality factors in urban shallow lakes: Xinglong Lake case study. Proceedings of the China water conservancy society.* Yellow River Water Conservancy Press: China Water Conservancy Society., [In Chinese].

Singh, K. P. (1982). Lake restoration methods and feasibility of water quality management in Lake of the Woods. *ISWS Contract Report No. CR*, 301.

Tan, Z., Ren, J., Zhang, R., Hu, C., & Feng, W. (2021). A study on the relationship between chlorophyll a index and nitrogen and phosphorus characteristics in Xinglong Lake. *Environmental Science and Technology, 44*(S2), 205−209, [In Chinese].

Tan, Z., Wang, R., Ren, J., Hu, C., & Feng, W. (2020). Remote sensing information extraction and nitrogen and phosphorus distribution characteristics of aquatic plants in Xinglong Lake. *Sichuan Water Conservancy, 41*(4), 3−8, [In Chinese].

Tian, C., 2020.Research on the landscape planning and design of artificial lakes in urban new area. Master thesis. Beijing Forestry University, p. 203.

Vollenweider, R. A. (1976). Advances in defining critical loading levels for phosphorus in lake eutrophication. *Memorie dell'Istituto Italiano di Idrobiologia, 33*, 53−83..

Wang, G. (2022). Planning map is transforming into realistic map by Xinglong Lake. *Sichuan Daily* (011), 2, [In Chinese].

Wikipedia, available on: https://en.wikipedia.org/wiki/Caspian_Sea.

Winter, T. C. (1976). Numerical simulation analysis of the interaction of lakes and ground water. *United States Geological Survey Professional Paper, 1001*, 45.

Winter, T.C. (1995). Hydrological processes and the water budget of lakes. In Lerman, A., Imboden, D.M., Gat, J.R. (Eds). Physics and Chemistry of Lakes (pp. 37−62). Springer-Verlag, Berlin.

Zamparas, M., & Zacsharias, I. (2014). Restoration of eutrophic freshwater by managing internal nutrient loads. A review. *Science of the Total Environment, 496*, 551−562.

Zaragüeta, M., & Acebes, P. (2017). Controlling eutrophication in a Mediterranean shallow reservoir by phosphorus loading reduction: The need for an integrated management approach. *Environmental Management, 59*, 635−651.

Zhang, H., Wang, H., Chen, Y., Liu, Z., Luo, B., Li, Y., Fan, H., Xie, H. (2022). Framework design and typical demonstration of park city water system based on digital twins. In Proceedings of the China Water Conservancy Society. China Water Conservancy Academic Conference, Vol. 7. Yellow River Water Conservancy Publishing House: China Water Conservancy Society, p. 6. [In Chinese]

Zhang, J. (2022). Exploration of ecological construction of urban parks. Master thesis. Sichuan Agricultural University. p. 71. [In Chinese]

Zhao, Z., & Lu, Y. (2023). Chengdu Xinglong Lake Zhongxin underwater bookstore. *Journal of Idea and Criticism*, 105−107. Available from https://doi.org/10.19953/j.at.2023.02.011, [In Chinese].

Chapter 16

Analysis of surface flows of Urmia Lake Basin: a review

Mina Mahdizadeh[1], Saeid Eslamian[2] and Yaser Sabzevari[2]

[1]Department of Soil Science and Engineering, College of Agriculture, Isfahan University of Technology, Isfahan, Iran, [2]Department of Water Sciences and Engineering, College of Agriculture, Isfahan University of Technology, Isfahan, Iran

16.1 Introduction

Providing food and a productive environment for the human population, animals, plants, microorganisms, and other organisms worldwide has caused the global demand for freshwater to increase rapidly. Although water is considered a renewable resource, its availability is limited in terms of the amount available per unit of time in each region. So, when managing water resources, all agricultural, social, and ecological systems should be considered (Pimentel david et al., 2004). Therefore, in hydrology, when studying the distribution and movement of water, man's role and investigating its action is inevitable (Davie, 2008). Indiscriminate use of surface water available on the earth's surface, such as rivers, lakes, lagoons, and groundwater, has caused water bodies to face irreparable problems (Alizadeh, 2001). Also, increasing water demand often leads to unsustainable water consumption and insufficient water remaining to protect the environment (Ahmadaali et al., 2018). Since the limitation of water resources with unequal distribution and increasing demand is one of the main challenges of managing water resources in the world in general and in Iran in particular, it is very important to investigate water ecosystems and the factors that have caused their destruction (Zarghami, 2011).

Today, one of the greatest important and discussed issues in Iran is the restoration of Lake Urmia, which has received the attention of the government and development programs in recent years (Moghimi, 2020). Lake Urmia has always been important in various aspects, such as environmental, economic, political, and social. Since Lake Urmia is one of Iran's most important and valuable water ecosystems, the decrease in the water level in Lake Urmia leads to an environmental disaster. It destroys the birds and creatures that live in that area. Therefore studying the hydrological basins of Urmia is of great importance.

Urmia Lake is one of the most valuable permanent super-saline lakes in the world, and it is the largest lake of this type in the Middle East, located in the northwest of Iran. Due to its unique natural features and ecology, including the presence of Artemia as a freshwater population, Urmia saltwater lake was introduced as a wetland of international importance by the Ramsar Convention in 1971 and as a biosphere reserve in 1976 (Soudi et al., 2017). As the second largest saline lake in the world, this lake has been destroyed during the last two decades, and this has caused many social and environmental effects (Shadkam et al., 2016). The shape and natural structure of the catchment area of Lake Urmia have changed over time with the creation of large structures such as dams. The effects of human factors on Lake Urmia have led to changes in the flow path, destruction of the natural drainage network and waterways, development of irrigation and water transfer channels, and changes in land use and vegetation downstream of the structures (Moghimi, 2020). These changes have had devastating effects on the region's ecosystem, such that the existence of long-term abnormal climatic conditions could eventually lead to severe droughts in Lake Urmia (Moghimi, 2020).

According to the available studies, the wetlands of 29 countries in the world have been included in the Montreux list, among which Iran, having six wetlands, has the second most endangered wetlands after Greece. Although the comprehensive management studies of Lake Urmia show that Lake Urmia will face a severe crisis for at least the next 20 years, nine large islands of this biome are completely connected to the land, and the continuation of this process endangers the region's life. It will drop, but Lake Urmia is not actively included in Montero's list (AslHashemi & Farahmand, 2021). Today, one of the most important and discussed issues in Iran is the restoration of Lake Urmia, which has received the attention of the government and

Hydrosystem Restoration Handbook. DOI: https://doi.org/10.1016/B978-0-443-29802-8.00016-9

development programs in recent years (Moghimi, 2020). Lake Urmia has always been important in various aspects, such as environmental, economic, political, and social. Since Lake Urmia is one of Iran's most important and valuable water ecosystems, the decrease in the water level in Lake Urmia leads to an environmental disaster. It destroys the birds and creatures that live in that area. Therefore it is very important to examine the studies conducted in the surface currents of the Urmia River basin. By examining the studies conducted in this field, several solutions can be taken to reduce the threat of destruction of Lake Urmia as much as possible.

16.2 Study area

16.2.1 Geographic features

The catchment of Urmia Lake, with an area of about 51,762 km², is located in the northwestern part of Iran between the geographical coordinates of 44 degrees and 14 minutes to 47 degrees and 53 minutes of east longitude and 35 degrees and 40 minutes to 38 degrees and 30 minutes of north latitude and Azarbaijan province including West and also parts of East Azarbaijan (24 km east of Urmia) and Kurdistan provinces (Hemmati et al., 2020) (Fig. 16.1). The extent of the plains of the Urmia Lake catchment area is 14,923 km², including the water area of the lake and its edges, and

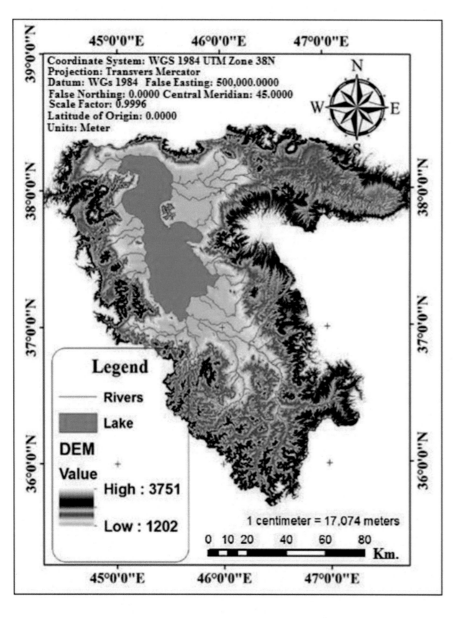

FIGURE 16.1 Geographical location of Lake Urmia catchment area (Moghimi, 2020). Map lines delineate study areas and do not necessarily depict accepted national boundaries.

the area of the height of the catchment area is also 36,839 km². According to this, the catchment area of Urmia Lake is a mountainous basin (Ministry of Energy, 2016) (Moghimi, 2020). Urmia Lake, with an average height of 1275.6 m above sea level, is the highest and most water-filled internal and permanent lake in the country and the 20th largest lake in the world. Fig. 16.2 shows sub-basins in the basin.

Urmia Lake is located in the center of the Urmia catchment basin and collects the whole basin's surface water. Many rivers enter Lake Urmia from the surrounding areas, such as Aji Chai, Qala Chai, Zarineh Rood, Simine Rood, Godarchai, Baranduzchai, Mahabad Chai, Lilan Chai, Nazlochai, Shahrchai, and Zola Chai (Ministry of Energy, 2016) (Moghimi, 2020).

Fig. 16.3 shows a view of the rivers around Urmia Lake (Alborzi, 2018). Seventeen permanent rivers and 12 seasonal rivers end in Lake Urmia (Shadkam et al., 2016).

16.2.2   Climate characteristics

The climate of Lake Urmia Bowl is mainland and unforgiving (Roshan et al., 2016) and is characterized by cold winters and dry and direct summers. Urmia meteorological stations show that the normal yearly temperature in this locale is 11.2°C with a least of −2.5°C in winter and a most extreme of 23.9°C in summer (Fig. 16.4) (Roshan et al., 2016; Sharifi et al., 2018).

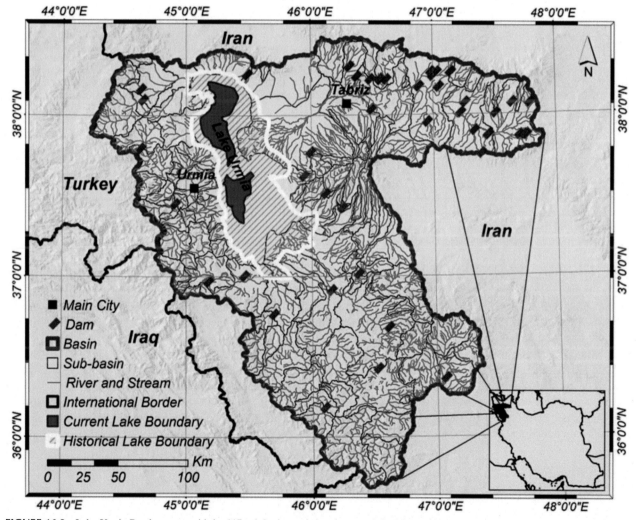

FIGURE 16.2   Lake Urmia Bowl appears with its 117 sub-basins, existing dams, and the lakes surface zone. The white boundary shows the lakes surface region in 1998 based on Landsat symbolism. The moment boundary portrays the current condition based on the 2017 Landsat symbolism (the dim blue zone appears to be the locale where water can be unquestionably identified from space) (Alborzi, 2018). Map lines delineate study areas and do not necessarily depict accepted national boundaries.

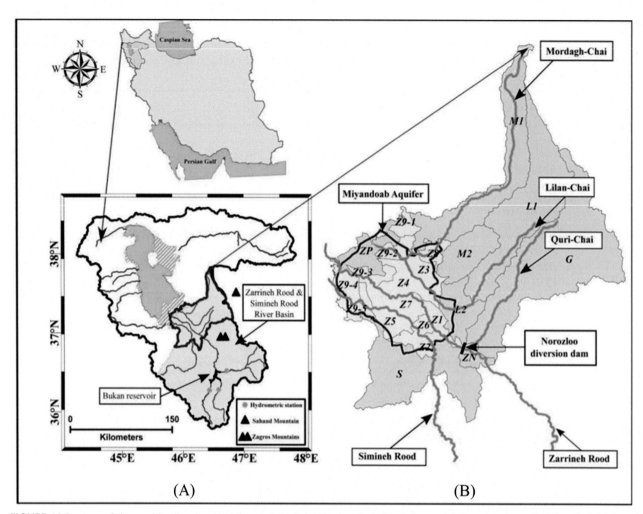

FIGURE 16.3 Area of the considered region (A) Miyandoab Plain within the Urmia bowl, Iran and (B) agrarian zones in Miyandoab Plain and Miyandoab aquifer. The inner and outside zones appear individually in yellow and green (Shadkam et al., 2016). Map lines delineate study areas and do not necessarily depict accepted national boundaries.

The most rainfall occurs in the months of winter and spring, and the maximum rainfall is usually recorded from March to June. This basin, which is located in a semiarid Mediterranean climate, has an average annual rainfall of 361 mm (Razmara et al., 2016; Sima et al., 2013). Average annual rainfall data in Urmia during the years 1999−2016, accessible through the Urmia weather station located west of the lake, shows that the average rainfall has decreased to 315 mm. Also, the amount of evaporation is estimated to be about 1 m per year (Razmara et al., 2016; Sima et al., 2013), which can increase up to 1.5 m per year due to strong seasonal winds blowing from the northwest−southeast direction (Kelts & Shahrabi, 1986). Since this lake is super saline, the average dissolved salts in it reach 200 g/L, compared to the normal salinity of seawater, which reaches about 35 g/L (Dalezios et al., 2018).

This is despite the fact that the lake's salinity has reached more than 350 g/L in recent years (Razmara et al., 2016). Lake Urmia is between 130 and 140 km long, and its average width is 40 km. Sixty-five percent of the Urmia basin is covered by mountainous areas, 25% by foothills and plains, and, on average, 10% by lakes.

16.2.3 Runoff

Since the world's quick financial advancement, populace development, and the presentation of more up-to-date and more commonsense innovations with an accentuation on clean vitality are of incredible significance to realize an economic environment and economy, the considers conducted amid a long time 1961−90 appear that the runoff 4.7% expanded at the 25% hazard level, 13.8% at the 50% chance level, and 18.4% at the 75% chance level (Fig. 16.4 and scenario A) (Razmara et al., 2016). It is additionally vital to reinforce territorial statistic strengths by considering family values and traditions,

(A)

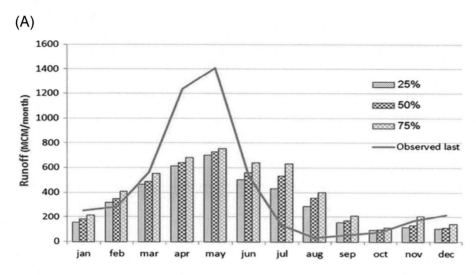

(B)

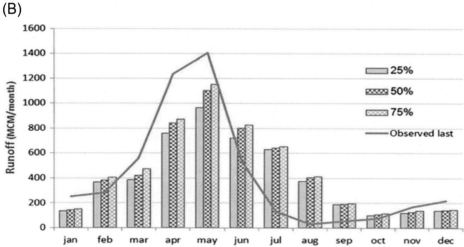

FIGURE **16.4** Average of observed and future runoff in (A) scenario and (B) scenario (Razmara et al., 2016).

populace development, and less reliance on financial improvement in this locale; consequently, the considers conducted between 1961−90 appear to diminish the normal surface runoff to the lake by 21% at the chance level of 25%, 13% at the chance level of 50%, and 0.3% at the chance level of 75% (Fig. 16.4 and situation B) (Razmara et al., 2016).

The yearly runoff within the Urmia catchment has diminished by 11.3% compared to the base period, with the biggest diminish in April and May. The diminish in precipitation in this range will lead to numerous dangers for a long time, 2050−21 (Razmara et al., 2016), since the need for runoff decreases the river's stream; hence, the show is worth for month-to-month normal runoff projections as appeared in Fig. 16.5.

16.2.4 Rivers

The northern part of this area is located in the alluvial plain of Maragheh-Baonab, and its middle and southern parts are located in the alluvial plain of Miandoab. Zarineh Rood, Mardichai, and Sufichai rivers are the most important rivers in this region The Zarineh Rood, with an area of 11,729 km^2 and a waterway of 218 km, and the Simineh Rood, with an area of 3488 km^2 and a length of the main waterway equal to 180 km, are the main surface currents of Lake Urmia. Among the rivers in the catchment area of Lake Urmia, Zarineh Rood alone supplies about 42% of the water entering this lake (Shemshaki & Karami, 2019; Yusuf et al., 2017).

Nowadays, due to the various dams built on these streams and their tributaries since 2000, which are the foremost critical providers of Lake Urmia, the rate of water saving of the streams and their tributaries has diminished (Fig. 16.6) (Nhu, 2020).

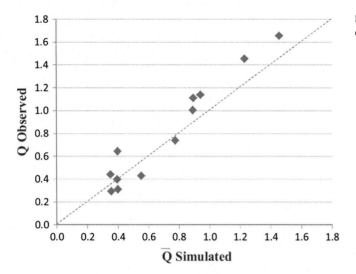

FIGURE 16.5 Average monthly runoff over 1961−90 (Razmara et al., 2016).

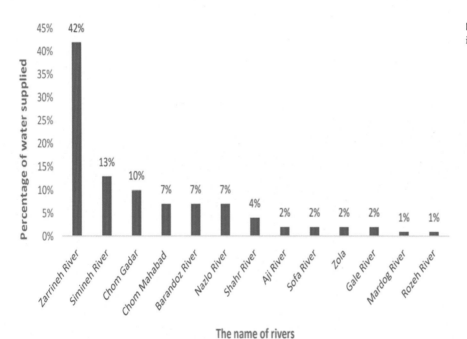

FIGURE 16.6 Main rivers' discharge into the Lake Urmia (Nhu, 2020).

16.2.5 Geology characteristics

From the geological point of view, the study area mainly includes Quaternary alluvial deposits. The grading of the harvested soil samples shows that most soils in the Maragheh-Bonab Plain include L, SP, and SM, and in the Miandoab Plain, including CL. The size of the sediment grains is relatively coarser on the coast of the Maragheh-Baonab study area and smaller on the coast of the Miandoab study area (Fig. 16.7) (Shemshaki & Karami, 2019).

Urmia Lake is located in a tectonic collision zone between the Eurasian and Arabian plates, and it is an area in the lowest depression of Azerbaijan. It covers a plateau with a height of more than 2000 m (AslHashemi & Farahmand, 2021). The activity of the Tabriz fault in this area has caused the rise of the northern part of this fault, and the part mentioned above has caused the formation of Lake Urmia by creating an obstacle against the flow of water.

In the Plio-Pleistocene, Lake Urmia extended to Tabriz and Maragheh. Unlike Islamic Island, which is made of Pliocene volcanic rocks, the deposits of the other islands are Lower Cretaceous flysch limestone or Miocene reef rocks (Qom Formation) (AslHashemi & Farahmand, 2021). The lake's geological location is quite diverse, with a drainage

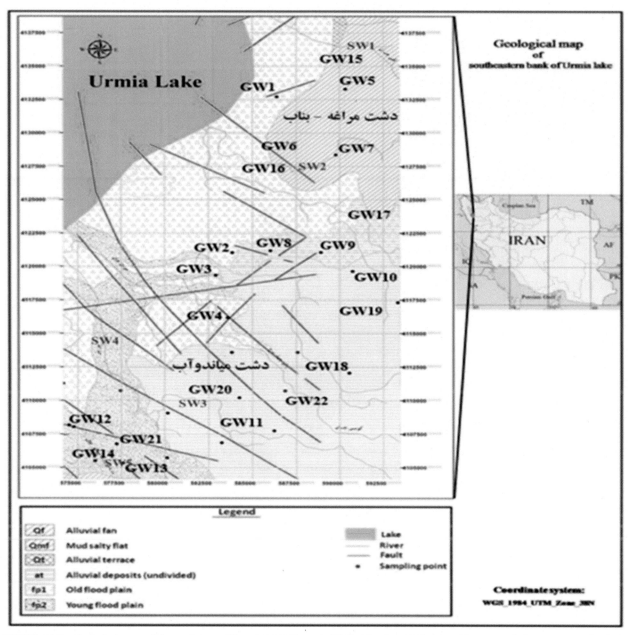

FIGURE 16.7 Geological map of the study area (Shemshaki & Karami, 2019). Map lines delineate study areas and do not necessarily depict accepted national boundaries.

basin composed of various rock formations of different ages, ranging from Precambrian metamorphic complexes to Quaternary mud deposits (Fig. 16.8) (Sharifi et al., 2018).

In old reports, Lake Urmia is considered a remnant of the Mediterranean Sea, and the climate of the Urmia region is classified as a continental-seasonal Mediterranean climate. However, from a geological point of view, this basin results from pressure fault systems such as the Tabriz and Zarineh Rood faults (AslHashemi & Farahmand, 2021; Rivas-Martínez et al., n.d.).

Lake Urmia's water level and area have decreased drastically since the late 1990s (Khazaei, 2019). The time series of the water level of Lake Urmia is shown in Figs. 16.9 and 16.10 (Dehghanipour et al., 2020). The investigated sediments from Lake Urmia show that during the lake's life, there were also dry climatic conditions. Most of the sediments seen in this area are chemical sediments. It is quartz, calcite, plagioclase, and kaolinite, which kaolinite indicates the

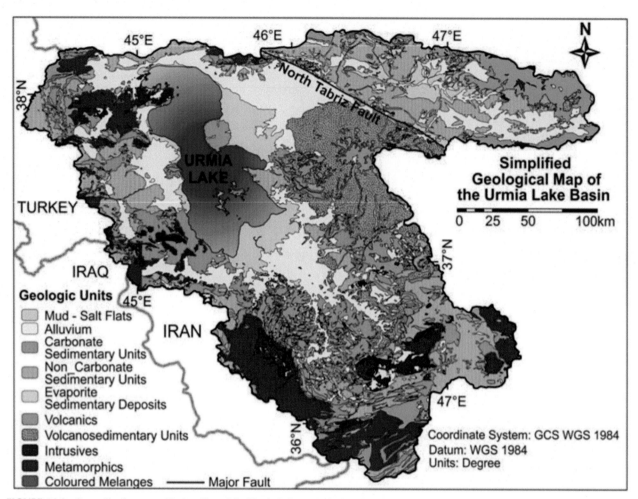

FIGURE 16.8 Streamlined topographical outline of the Urmia Lake watershed region. Shake units with comparable lithology are combined as one unit notwithstanding of their geographical ages (Sharifi et al., 2018). Map lines delineate study areas and do not necessarily depict accepted national boundaries.

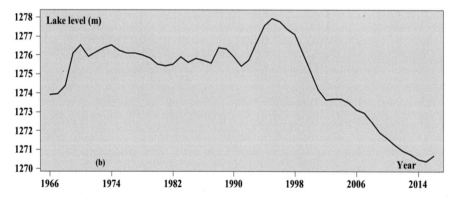

FIGURE 16.9 Changes in the water level in Lake Urmia, 1965−2018 (Dehghanipour et al., 2020).

temporary change of weather and salinity of the lake (AslHashemi & Farahmand, 2021). The seasonal biological cycle in Lake Urmia leads to aragonite precipitation in surface waters (Kelts & Hsü, 1978). Therefore the sediment most visible in this area is aragonite, which is seen as thin and regular or irregular blades because biological activities lead to surface changes of carbon dioxide (Kelts & Hsü, 1978).

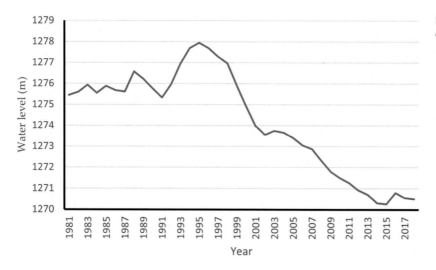

16.2.6 Vegetation cover

Available studies show about 564 plant species from 299 genera and 64 families in the ecological area of Lake Urmia. Most species are annual or perennial grasses. Of all the available species, 32.2% are annual, 9.1% are biennial, 47.1% are permanent, and 11.5% are shrubs and trees. A total of 17.8% of all species are salt-loving plants around the lake and salt marshes (Zarandian et al., 2023).

16.2.7 Living organisms

This lake is the habitat of various species (Razmara et al., 2016) and has diverse groups of bacterial species. Two types of bacteria, including *Clostridium perfringens* and *Enterococcus faecalis*, are among the most important bacterial species identified in the lake water and its sediments (Asem et al., 2014). These bacteria are present in the digestive system, and their observation in lake water and sediments indicates that urban sewage has entered Lake Urmia, which can be a serious threat to the living species in Lake Urmia.

This area includes 26 urban areas and more than a thousand rural areas. The most important human activities are agriculture and animal husbandry (Zarandian et al., 2023). Due to the increase in population, many industrial factories have been built, which, as mentioned, have entered Lake Urmia due to the improper management of urban sewage, which will cause many risks.

Parasites are an expansive group of eukaryotic living beings widely distributed throughout the world. They live in a differing range of extraordinary environments, from deserts to hypersaline situations (Asadi et al., 2011), and can be observed within the catchment range of Lake Urmia. Twelve organisms have been detailed in a few halophytes and glycophytes of the islands and from the western shores of Lake Urmia, the foremost vital species of which is Artemia.

Nowadays, unscheduled dam developments due to dishonorable administration, the event of dry spells (expanding temperature and diminishing precipitation), over-the-top misuse of groundwater by agriculturists, and unlawful and over-the-top improvement of wells are among the most common reasons for the diminishing water level of Lake Urmia (Zoljoodi and Didevarasl, 2014) (Nhu, 2020; Roshan et al., 2016). The lake's surface region in 1997 was evaluated to be 6100 km². However, since 1995, the surface range of Lake Urmia has, for the most part, diminished. In Eminent 2011, based on today's information, it was assessed to be 2366 km² (Fig. 16.11) (Jalili et al., 2016).

16.3 Review studies

Fathian et al. (2015) studied the evaluation of land cover changes under the eastern basins of Lake Urmia through classification and statistics. The results show that in 1976 and 2011, respectively, 41.4% and 27.2% of the area was covered by pasture, and the investigations show that due to the development of agricultural land, this area decreased during the specified years. So, agricultural land, garden land, and rain-fed land have increased by 412%, 333%, and 672%, respectively, during the last 35 years. The analysis of the obtained statistics shows the increase in temperature throughout the region and the trend of specific precipitations. Also, the studies shown in the downstream stations well represent the

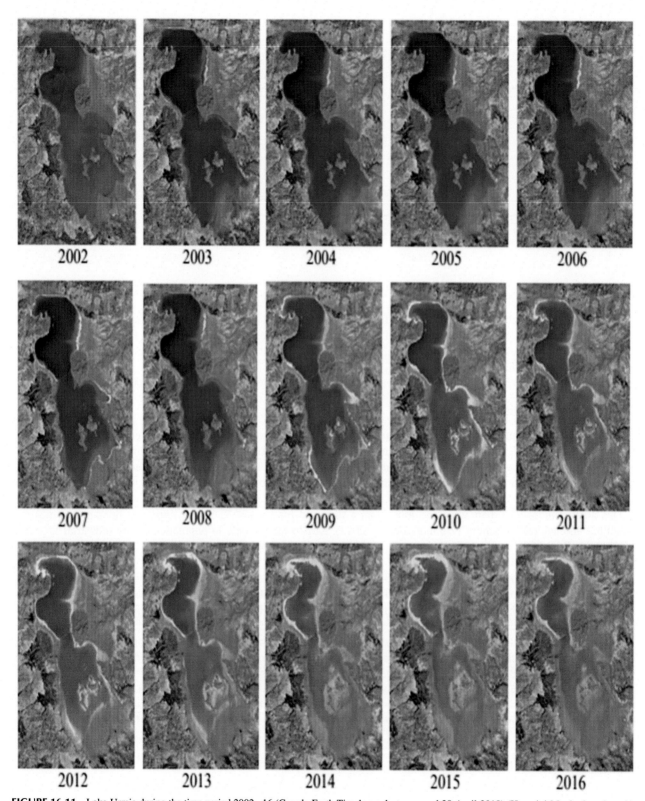

FIGURE 16.11 Lake Urmia during the time period 2002−16 (Google Earth Timelapse, last accessed 28 April 2018) (Hosseini-Moghari et al., n.d.).

overall flow reduction in this area. Finally, the correlation between changes in watercourse flow with simultaneous changes in climatic variables and land cover showed that river flow is more sensitive to land cover changes than temperature (Fathian et al., 2016a).

This has been aimed to examine the impact of climate change on drought in the Lake Urmia basin. Drought events have been analyzed from 2011 to 2100 using the Standardized Precipitation Index (SPI) and the Standardized Precipitation Evapotranspiration Index (SPEI), comparing them with data from 1976 to 2005. Using precipitation and temperature data from the CanESM2 model under different Representative Concentration Pathway (RCP) scenarios, drought frequency and duration are projected to increase, especially under the pessimistic RCP8.5 scenario. The SPEI, which considers both precipitation and potential evapotranspiration (PET), proved to be more effective than the SPI in assessing drought in a changing climate (Ahmadebrahimpour et al., 2019).

Fazli Fard et al. (2019) investigated the reasons for the decline in the lake's water level by studying the trends of two factors involved in its nourishment, namely, precipitation and river discharge. Accordingly, long-term time series of variables were prepared throughout the basin on a seasonal and annual scale, and their temporal trends were analyzed using the Mann−Kendall (MK) and seasonal Kendall tests. The results indicated that the annual precipitation variable shows an increasing trend in some areas and a decreasing trend in others. Therefore, statistically, the significant decline in the water yield of the rivers leading to the lake, and consequently the decrease in the lake's water level, cannot be attributed to the reduction in precipitation. As a result, other factors such as the increase in the evapotranspiration process due to atmospheric warming, excessive extraction of groundwater, land use changes, or even water storage in dam reservoirs and reduction in inflows to the lake reservoir may have caused the decline in the lake's water level (Fazli Fard et al., 2019).

Safari and Zarghami (2012) conducted a study using a multicriteria decision-making model based on distance, along with simple weighted aggregation methods, adaptive planning, and TOPSIS method, to allocate water from the Lake Urmia basin among stakeholders considering social, economic, and environmental criteria. They determined the optimal share of surface water for each province. According to the results of the adaptive planning model, the annual long-term average water share of the Lake Urmia basin was estimated at 7.2 billion cubic meters per year, with the share of West Azerbaijan province equal to 1.804, East Azerbaijan province 1.312, and Kurdistan province 0.984 (Safari & Zarghami, 2012).

An approach has been embraced to examine the interaction impact of arrive utilize and arrive cover on water surrender utilizing the Johnson Neyman (JN) strategy. The soil and water appraisal apparatus (Soil and Water Assessment Tool [SWAT]) was utilized in Urmia Lake Bowl to appraise water abdicate after effective calibration and approval of the demonstration by stream. With the help of the used models, they found that in each sub-basin, the influence of soil layer on WYLD water reduction is significant. The critical decrease in the water level of Lake Urmia started in 1995, which showed that part of this contraction was most likely due to the decrease in water inflow in a 4-year time delay. (Sakizadeh et al., 2023).

The water level of Lake Urmia has been diminishing for the past two decades, and as a result, this range has become a critical center for logical investigation. Thus Habibi et al. (2021) made strides in understanding the relationship between the decreasing the lake level and the alteration of the nearby dry spell conditions, with the point of showing a comparative examination of dry spells within the Lake Urmia bowl by combining distinctive dry spell records among the three common dry spell for examining and utilizing investigation amid the long time 1981−2018, which incorporates Standard Precipitation Record (SPI), Standard Precipitation Evapotranspiration File (SPEI), and Standard Rain and Snow Soften List (SMRA). In spite of the fact that precipitation is a critical marker of water accessibility, temperature is additionally a key factor, since it determines the rate of vanishing and snow melting. The examination shows a descending slant for the Standardized Precipitation Vanishing Record (SPEI) and the Standardized Snowmelt and Precipitation List (SMRA) (but not for the SPI Standardized Precipitation Record), demonstrating that both dissipation and snowmelt worsen in dry seasons. Moreover, it appears that the decrease in the lake's water level from 2010 to 2018 was not due to changes within the components of the water adjustment but due to unsustainable water administration (Habibi et al., 2021).

Since Urmia Lake Bowl has confronted natural, social, and financial challenges for a long time, this circumstance will likely compound by the impact of climate change. Hence, two models have been utilized: the Water Evaluation And Planning System (WEAP; Stockholm Natural Established, Stockholm, Sweden) and the low emissions analysis platform (LEAP (Stockholm Environmental Institute, Boston, MA, USA)) are coordinated to simulate changes in water, vitality, nourishment, and the environment over these 20 years. Two climate scenarios and nine approach scenarios are combined to evaluate maintainable improvement by employing a multicriteria choice approach. Economic improvement will be accomplished by seeking challenging objectives in agrarian, consumable water, vitality, and mechanical segments. In this situation, Lake Urmia's water level will reach its environmental water level in 2040. On the other hand, social, specialized, and political challenges are considered impediments to actualizing the objectives of this situation. In expansion, industry development and structure alteration have the foremost effect on feasible improvement accomplishment (Nasrollahi et al., 2021a).

The slant of hydrological and climatic time arrangement information of the Lake Urmia bowl in Iran utilizing four distinctive MK approach adaptations has been examined. This inquiry distinguished the slant of hydrological and climatic information in a month-to-month and yearly time scales for 25 temperature stations, 35 precipitation, and 35 waterway stream estimation stations chosen from the Urmia Lake Bowl. Moreover, in this inquiry, MK and Pearson's tests were utilized to examine the relationship between temperature, precipitation, and stream patterns. The results show factually critical upward and descending patterns in yearly and month-to-month hydrological and climatic factors. Upward patterns in temperature, not at all like waterway streams, are much bigger than downtrends, but for precipitation, the slant conduct is diverse on month-to-month and yearly time scales. Moreover, the relationship between stream and climate factors shows that the waterway stream within the Urmia Lake Bowl is more responsive to temperature changes than precipitation. (Fathian et al., 2016b).

The commitment of human water has been measured in lessening the influx to Lake Urmia and the diminishing lake water volume, groundwater, and the overall water capacity within the whole bowl of Lake Urmia over a long time from 2003 to 2013. For this reason, the Water Crevice Worldwide Hydrology Show was physically calibrated particularly for the bowl against different in situ and spaceborne information, and the best-performing calibration variation was run with or without considering water use. Observation information includes remote-sensing-based time arrangement of yearly flooded range within the bowl from moderate resolution imaging spectroradiometer (MODIS), month-to-month add-up to water capacity irregularity from Elegance satellites and month-to-month lake volume. Existing considerations show that the elements of lakes and groundwater can be well mimicked, as they were in case the calibration against the groundwater level alters the division of human water utilization from groundwater and surface water. According to our studies, human utilization of water was the reason for 50% of the entire water misfortune within the bowl, measuring around 10 km^3 within the period of 2003−13, accounting for 40% of the water misfortune. Lake Urmia was around 8 km^3 and up to 90% of the groundwater misfortune. Moreover, the lake's influx was 40% lower than it would have been without human water utilization. An audit of the accessible proof shows that, indeed, without human water utilization, Lake Urmia does not compensate for 30 cases of critical diminish within the lake's water volume caused by the 2008 dry spell. These discoveries can be valuable for supporting water administration within the bowl and particularly for Urmia Lake reclamation ventures (Hosseini-Moghari et al., 2018).

Shadkam et al. (2016) proposed a water administration arrangement based on lessening water utilization within the agrarian division to reestablish the lake, which incorporates a 40% diminishment in irrigation water utilization rather than the past arrangement pointing at creating supplies and water systems. Be that as it may, none of these water administration plans, which have expansive socioeconomic impacts, have been evaluated under future climate and water accessibility changes. By embracing a strategy of natural stream prerequisites (environmental flow requirements [EFRs]) for hypersaline lakes, they evaluated that yearly water is required to protect Urmia Lake. The variable penetration capacity (variable invasion capacity [VIC]) hydrological constrained with bias-corrected climate show yields for both the most reduced (RCP2.6) and most noteworthy (RCP8.5) greenhouse gas concentration scenarios to appraise future water accessibility and impacts of water administration methodologies. It shows a 10% decrease in future water accessibility within the bowl under RCP2.6 and 27% under RCP8.5. The result showed that in the event that future climate change is profoundly constrained (RCP2.6), an influx can be sufficient to meet the EFRs by actualizing the decreased water system arrangement. In any case, under a faster climate situation (RCP8.5), lessening water system utilization will not be sufficient to spare the lake, and more exceptional measures are required. Moreover, long-haul water administration plans in this locale do not appear solid and economical in the face of climate change (Shadkam et al., n.d.).

Radmanesh et al. (2022) utilized the TOPSIS technique to analyze the impact of climate on the water level and shrinkage of the Urmia Lake catchment, selecting the top 10 public circulation models from 23 available models from 1951 to 2005. They merged 10 general circulation models (GCMs) using the KNN method and quantified their uncertainties. Future period data (2028−79) was generated with the LARS-WG model. Results indicate that temperature increases across all seasons. Under the RCP4.5 scenario, spring and autumn precipitation decreases by 10.4% and 27.8%, respectively, while summer and winter see 18.2% and 3.4% increases, respectively. Similarly, the RCP8.5 scenario predicts spring, autumn, and winter precipitation decreases of 11.4%, 22.7%, and 4.8%, respectively, and a 26.5% increase in summer. SPI and SPEI were employed to calculate short-, medium-, and long-term meteorological droughts in baseline and future periods, revealing increased drought occurrences and peaks alongside decreased durations in the future. Overall, SPEI exhibits a stronger correlation than SPI with changes in Lake Urmia's water level (Radmanesh et al., 2022).

The water budget of Lake Urmia and the severity of drought in Lake Urmia from 1985 to 2010 has been investigated. A new hypothesis to quantify the human and climatic effects of Lake Urmia's decrease has been presented. The observations showed that human effects and activities can be much more important than climatic effects. Human influences and climatic factors affect 80% and 20% of Lake Urmia's drying. The investigation showed that although it was previously assumed that the groundwater output from Lake Urmia is insignificant, there was a significant leakage of groundwater from Lake Urmia. Major changes in the reduction of the water level of Lake Urmia have been observed since 1998. Therefore the first step in the restoration of Lake Urmia can be a review of surface water management and the exploitation of dams and groundwater resources. The next step is to investigate and classify agricultural products grown in the region in terms of water consumption and teach the local people about proper irrigation methods (Alizade Govarchin Ghale et al., 2018).

Shadkam et al. (2016) evaluated the relative commitment of climate alter and water asset improvement, which incorporates the development of stores and the development of water regions, to changes within the input of Lake Urmia during 1960−2010, utilizing the VIC. It appears that the decrease in influx, for the most part, takes place after the observed decrease in precipitation, in spite of the fact that the inconsistency in influx is more prominent than the alteration in precipitation. The audit of the utilization of water for the water system has placed more weight on the basin's water supply and has caused a 40% decline in streams over a long time. On the other hand, stores have played a noteworthy part in getting to water during moderately dry season for a long time. It has altogether anticipated a decrease in the influx of the Urmia Lake Bowl. Therefore, with the extension of the water system framework within the bowl, the existence of stores has, in a roundabout way, made a difference in diminishing the water input within the bowl.

Also, the results show that the annual input to the Urmia Lake Basin has decreased by 48% during 1960−2010. The evidence indicates that about three-fifths of these changes were caused by climatic changes and about two-fifths by the development of water resources. The results of this study show that in order to prevent further drying of the Lake Urmia, in addition to national programs to reduce irrigation water consumption, there is a need for international programs to deal with climate change (Shadkam et al., n.d.).

Since the downward trend of the water level of the Urmia Lake is a serious problem in the northwest of Iran and has had negative effects on agriculture and industry, three statistical tests MK, Sen's T, and Spearman rho have been used to estimate annual and seasonal time series trends of temperature, precipitation, and stream flow in 95 stations throughout the basin. The results showed a significant increase in temperature throughout the basin and an area-specific precipitation trend. Finally, the review of existing studies shows that the river flow in the Urmia Lake Basin is more sensitive to temperature changes than precipitation. The decrease in the water level of the lake can be related to both the increase in temperature in the basin and the improvement of overexploitation of water resources (Fathian et al., 2015).

Yarahmadi and Rostamizad (2019) investigated and analyzed hydrological droughts in the north of Lake Urmia, especially the Darian Chai River over 31 years, using the threshold level method with NIZOWKA 2003 software. The characteristics of hydrological droughts include time of occurrence, duration, and intensity. They have calculated the minimum discharge observed during the drought event. Frequency analysis based on partial series and by fitting different distribution functions has been done to investigate the probability of occurrence of drought events, as well as their intensity and duration. The results showed that 38 hydrological dry periods occurred in this river, and more than 60% had a duration of more than 200 days. In terms of time distribution, about 71% (27 occurrences) of dry spells started in spring. The largest drought event was 577 days, and the second largest drought period, with a duration of 365 days, occurred with a delay of 2 months. The deficit of surface flow caused by dry periods in this river equals 117 million cubic meters. This issue, its generalization to other rivers of the watershed of Lake Urmia, and its effect on reducing the lake's water level and drying it up are very important (Yarahmadi & Rostamizad, 2019).

The main cause of the drying up crisis of the Lake Urmia Basin is the continuous reduction of streams to the lake due to excessive use of water for agriculture during the last two decades. It is necessary to determine the environmental water requirements for each of these rivers. Yasi and Rezaei (2022), to determine the environmental flows of rivers in the Lake Urmia Basin, present a hydrological method that is suitable for rapid assessment of environmental flows of at least these rivers since specific bio-river data are not available. This research considered nine different eco-hydrological methods to estimate the EFs for the second-largest river in the Urmia Lake Basin, the Simineh River. The ecological flow needs were investigated in four different reaches of the river, upstream, and downstream of the Simineh Dam site. The results indicate that the method of "Flow Duration Curve Shifting" is well adapted to the natural river flow regime. In order to improve the river's environmental status by one step, a range of 20%−30% of the mean annual flow (MAF) is to be allocated in four different reaches of the river. The environmental water release from the Simineh Dam is to be revised and increased from the prescribed value of 10% to about 23% of MAF. This revision is guaranteed by reducing the dam height and reservoir capacity (Yasi & Rezaei, 2022).

Okhravi and Eslamian (2017) believe that Lake Urmia has one of the most severe water problems that Iranians have experienced, which is not only due to frequent droughts and aggressive use of water upstream, water diversion, and storage but also due to human effects. The shrinking of the lake and the construction of a controversial bridge over the lake have had significant consequences for the lake's valuable ecosystem and the region's economy. Therefore this chapter shows how endangering of Lake Urmia is based on the results of water-related scientific studies and provides complete information for the management of the lake, its ecosystem, and its watershed, which may be of immediate help to revive it. They showed that the situation of Urmia Lake is critical and undeniable: the lake is in a very dire state, and if effective restoration measures are not carried out properly and in time, it will turn into a disaster of a huge scale. Also, the necessity and importance of the corrective actions related to the hemispheres to restore the ecological, limno-logical, recreational, esthetic, and climatic characteristics of Lake Urmia cannot be overstated. Therefore there is an urgent need for Lake Urmia's rapid and effective development. Therefore, with a large reduction in the share of agricultural water from 90% to 50% through fallow agricultural lands, using appropriate irrigation methods and increasing irrigation efficiency, purchasing 40% of surface water rights in the irrigation network around Lake Urmia, preventing the planting of crops, water and illegal extraction from rivers and groundwaters throughout the lake, revision of 13 large dams in the basin in order to release 30%−40% of the river flow to the lake, and physical activities to flow the river to the lake with minimal environmental effects can be sustainable and immediate solutions to save Urmia Lake (Okhravi and Eslamian, 2017).

Emami and Koch (2019) simulated the effects of climate change on water availability in the Zarineh Rood Basin, which is the resource of Urmia Lake in western Iran, with the Bukan Dam under different climate scenarios until 2029 using the SWAT hydrologic model. After that, meteorological variables are predicted from MPI-ESM-LR-GCM (rainfall) and CanESM2-GCM (temperature) models under RCP2.6, RCP4.5, and RCP8.5 climate scenarios and scaled using quantile mapping (QM) methods. QM is reduced for error correction and SDSM. In this chapter, of the two types of QM models used, the empirical-CDF-QM model specifically reduced the errors of precipitation forecasts from the original GCM. Then, SWAT was calibrated with Zarineh Rood flow during the years (1981−2011) using the SWAT-CUP model. SWAT simulations for the future period of 2012−2029 show that the projected climate changes for all RCPs lead to a decrease in inflow to the Bukan dam and the entire Zarineh Rood sub-basin, the main reason being a decrease in future precipitation of 23%−35%, which causes a concomitant decrease in flow. The groundwater goes to the main channel. However, future runoff coefficients show increases of 3%, 2%, and 1%, as the −2% to −26% decrease in surface runoff is offset by the decrease in precipitation. This study showed that in the near future, due to the progress of agriculture and changes in water needs, Zarineh Rood River will face a water shortage because the water efficiency will decrease from −17% to −39%. This issue requires proper planning and better management of water resources (Emami & Koch, 2019).

Simineh Rood relies heavily on water resources for agricultural purposes in Urmia Lake. However, the hydrological system of the Simineh Rood Basin is highly susceptible primarily due to its diverse topographical features, limited data availability, and the complex nature of the local climate. In a study, Abghari and Erfanian (2023) simulated the monthly flow of Simineh Rood using the SWAT model, and the effects of climate change on the monthly flow of this river were evaluated. Future climate scenarios for 2011−2030 were generated using HadCM3 meteorological models under A2, B1, and A1B scenarios. Results of LARS WG showed that minimum and maximum temperature for the Simineh Rood watershed, the output of HadCM3 in A1B, B1, and A2 scenarios was reduced, and the desired meteorological parameters were predicted, from these predicted values as input for SWAT model was used. Since the main focus of this study was on the impact of climate change scenarios, fixed land use has been assumed. However, by optimizing irrigation efficiency with new methods, appropriate measures can be taken to save water consumption in Simineh Rood. This issue is very important because the available results show that the decrease of up to 25% of discharge in the Urmia Lake catchment due to climate change, to the values of 398, 394, and 440 million under scenarios A1B, B1, and A2, respectively, leads to a possible reduction. Attention is paid to the average annual input to the lake. Therefore, since Simineh Rood supplies 11% of the water of the Urmia Lake catchment, the necessary measures to preserve the water resources of this river are very important (Abghari & Erfanian, 2023).

Narimani et al. (2023) used long-term observational data to evaluate the stability and reliability of base flow estimation to determine nondeterministic quantities on base flow separation in the Urmia Lake Basin. This research compared seven basic flow separation methods with the mass balance filter as a reference method. Also, two indices, including the average annual base flow coefficient and baseflow index, were used to evaluate the effect of weather conditions on base flow separation. This study investigated two sources of uncertainty in digital filter methods, which include the constant value of stagnation and the approximation of groundwater supply. This research indicates that Eckhart's method is a better estimate of base flow in both wet and dry years and also shows acceptable accuracy during peak flows, even during multiple events. Therefore the uncertainty between the filter methods showed that the Lynie and Holick algorithm and Eckhardt's method are more sensitive to the value of the recession constant (α). In addition, Eckhart's method showed a more reliable estimate of the ratio of groundwater recharge to flow (Narimani et al., 2023).

Considering the importance of Urmia Lake, Hesari and Zeinalzadeh (2018) evaluated human activities in the process of surface and groundwater changes in Urmia Plain. For this purpose, based on the statistics of nine hydrometric stations (1978–2011), information from 74 piezometers and satellite images, the changes in inflow and outflow, aquifer discharge, and land use changes in Urmia Plain were studied. The test used for trend analysis was the nonparametric MK test, considering the effect of autocorrelation. This research is an attempt at human activities in the form of changes in inflow and outflow, aquifer cells, and land use in Urmia Plain. For this purpose, the statistics and information related to the inlet and outlet flow of nine stations located in the Urmia plain, piezometric information of the plain, and MSS, TM, ETM + , and SPOT satellite images were used. By demarcating the sub-basins of Nazlochai, Rozechai, Shahrchai, and Barandozchai rivers and in Urmia Plain, which overlooks Urmia Lake, and by determining the inlet and outlet stations of the plain, the process of changes in discharge in and out of the plains, land use, and aquifer discharge were investigated separately. The available evidence shows that the inflow to the Urmia Plain is -0.3, and the outflow from it is -1.6. Therefore the inflow trend into the Urmia Plain and its outflow is negative, indicating the decreasing trend of inflow into Lake Urmia. A comparison of the average results of two periods before and after 1996 (the year Urmia Lake began to dry up) showed that the average discharge entering the Urmia Plain decreased in the period of 1374–1390 compared to the period of 1978–95. Also, with time, the amount of discharge from the Urmia aquifer has increased, and the volume of the aquifer has gradually decreased. Satellite images show that the area of wet and rain-fed land has increased over time, and irrigated agriculture and horticulture have significantly prospered. The increase in the amount of inflow to the Urmia Plain and the decrease in the amount of outflow from it, together with the results of groundwater discharge, show that human activities reducing the inflows to Lake Urmia and, as a result, reducing its level along with climatic factors such as reducing the changes in the frequency of precipitation in the basin have played a role (Hessari & Kamran, 2020; Salehi Bavil et al., 2018).

Taheri et al., SEBAL assessed. Moreover, by estimating the precipitation conveyance within the bowl, the agrarian water system water needs pattern was calculated utilizing the arrival utilize outline within a long time 1995, 2010, and 2014 in all rural lands and in its seven primary sub-basins where ET changed from 369 to 1000 mm. The most extreme water system water utilization within the Urmia sub-basin was 535 million cubic meters in 2010 and 469 million cubic meters within the Miandoab sub-basin in 2014, the normal of which was 2108 million cubic meters within the bowl. The most extreme and least dissipation and transpiration rates in all rural lands were 765 and 555 mm in 2010 and 1995, respectively. This approach gives a basic but valuable evaluation to portray the design of water system water utilization in sub-basins utilizing negligible ground information and recognizing water administration techniques in bowl horticulture and climate alteration (Taheri et al., 2019).

Alborzi et al. (2018) set up a system for extricating energetic and climate-aware natural streams to drying up lakes, taking into consideration climatic and human conditions. In this inquiry, they give a quantitative evaluation of the basin's water assets by employing a wealthy dataset of hydrological highlights, water needs, and withdrawals, as well as water administration framework, that is, store capacity and operational approaches, which shows that Urmia Lake has come to the turning point within the early 2000s. The lake level at that point fizzled to return to its assigned environmental limit (1274 m over ocean level) amid a generally ordinary water period promptly taking after the dry spell recorded during (1998–2002). The choice of this breakthrough was due to the unmistakable increment within the hydrological capacity of the bowl, which was caused by human dry spells within the confront of extraordinary climatic stresses. An audit of existing investigations shows that the fast reduction in water levels after the dry spell recorded during 1998–2002, which drove a 48% diminish in runoff, coincided with an inexact 25% increment in surface water withdrawal, particularly within the agrarian sector, which proceeded for a long time. In this research, the natural flow for different climatic conditions (dry, wet and close to normal) was shown along with three water harvesting scenarios, and by accepting the convincing implementation of the 40% reduction proposal in the current water harvesting, the natural inputs from 2900 It varies from million cubic meters per year in dry conditions to 5400 million cubic meters per year in wet periods with a normal of 4100 million cubic meters per year in wet periods.(Alborzi, 2018).

Jalili et al. (2016), to examine the impact of hydroclimate inconstancy on the lake water level, considered two tele-connections for investigation, which included the Southern Wavering File and the North Atlantic Swaying. Moreover, unearthly examination was utilized to distinguish recurrence components and the relationship between inaccessible associations and lake water level changes. It appears that the common occasional behavior of Lake Urmia cannot clarify the diminishing and increasing water levels from 1994 to 1999. A nonparametric slant investigation examined the human impact on the lake water level. It appears that the lake's water level had a positive slant for a long time, from 1995 to 1966, but this drift is not noteworthy at the 95% certainty level. The drift from 1995 to 2009 was negative and critical at the 99% certainty level. It appears that the recent decline in the lake watershed's water level was due to human exercises instead of climate alteration (Jalili et al., 2016).

In their study, Esmailzadeh et al. (2023) sought to identify the primary actors in the revival efforts, how the policies to date have succeeded or failed, and where the government of Iran needs to go from here. Unfortunately, based on the available evidence, several recent efforts to revive the lake have had no effect, and the situation has become more dire than before (AghaKouchak, 2015).

For instance, farmers in the locality utilize additional water resources to enhance the productivity of their lands. Within this region, the majority of farmers sustain meager incomes, being heavily dependent on agricultural activities. Consequently, a significant number of them have opted to expand their cultivated acreage and cultivate crops such as apples and sugar beets, which offer greater profitability—but thirstier—crops (Schmidt et al., 2021; Y. & Molan-Nejad, 2017; Ženko & Menga, 2019).

To support their larger ranches and thirstier crops, they built more than 88,000 wells within the bowl, generally 40,000 of which are unlawful, all drawing extra groundwater (Ženko & Menga, 2019). A comparison made by common water asset administration within the Urmia Lake Bowl is another calculation that has caused tall water utilization in this range. In this manner, due to regional water administration, there's furious competition between the territories of Iran, West Azerbaijan, East Azarbaijan, and Kurdistan for the utilization and advancement of more water (Farajzadeh, 2019). Urmia Lake Bowl farmland's normal water system effectiveness is less than 35% (Najafi et al., n.d.). In the meantime, the suitable level for water system effectiveness is more than 40% (Rai et al., 2017). Rural efficiency in this locale is amazingly high, with each cubic meter of water creating a 500 g abdicate, compared with 1500 g within the rest of the world (Hassanzadeh et al., 2012). In expansion, rural arrival has been separated into smaller bundles (on normal 1.7 ha per agriculturist), which has expanded water utilization, particularly groundwater (Rajabi Hashijan, n.d.).

This chapter showed that despite the government's 12-year effort, significant financial expenses, and the participation of three prominent Iranian government organizations, the government has failed to improve the condition of Urmia Lake. From the point of view of these organizations, the low efficiency of irrigation, the low efficiency of water transfer, and the lack of water in the basin of Urmia Lake are the main reasons for its dryness. Based on a review of previous research, this chapter shows that many other socioeconomic and political problems, for example, water development policies and projects, low farm income, small ownership, local economies based on more water consumption, and political and social factors, also have a negative impact. Therefore the problems mentioned above have affected Lake Urmia (Esmailzadeh et al., 2023).

Shadkam (2017) assessed the most common reasons for the diminished influx by utilizing the VIC hydrology show, counting supplies, and water system modules. It appeared that climate alteration was the most important figure in lessening the influx. By the way, the advancement of water assets, particularly the utilization of water for water systems, has played a vital part in decreasing influx. After that, the influx of Lake Urmia was assessed under two scenarios: climate change and water system water in the future. At that point, the VIC was assessed with the climate demonstrate yields rectified for the least (RCP2.6) and most noteworthy (RCP8.5) GHG concentration scenarios for future water accessibility. It appears that water assets programs are not safe from climate change. In other words, in the event that future climate changes are restricted due to fast relief measures (RCP2.6), the unused procedure to decrease water system water utilization can offer assistance to protect Urmia Lake. This proposition showed that an economical approach to protecting Lake Urmia ought to incorporate request administration (considering financial complexities) and adaptable supply administration techniques (managing vulnerabilities in climate change) in a collaborative approach. This approach should be taken into account to manage the scenarios of reducing the input of Lake Urmia, either due to climate changes or improving water assets, in order to manage critical changes in the current and future framework (Shadkam, 2017).

An article has been done with the aim of how land use change affects surface flows and incoming flows to the catchment of Lake Urmia in the Mardaq Chai basin. In this study, future changes using the Dyna-CLUE model have been mapped and modeled. In this research, land use changes until 2030 under four scenarios is predicted: continuation of the current trend of water use, 40% water reduction, and two other scenarios with a 40% reduction in water consumption and improvement of irrigation efficiency up to 50% and 85%. Between 1993 and 2015, 21% of the studied area was converted to orchards and cropland, mostly at the expense of pasture. However, after reducing water harvesting, the analysis showed that orchards would have to be reduced by between 27% and 40%. The rain-fed land is projected to experience a major increase in all scenarios, especially in the case of reduced water abstraction, where it will increase by 217%. In order to achieve sustainable management of water resources, land use plays a major role and will lead to different land use in this type of semiarid area in the future (Shirmohammadi, 2020).

Using climate, hydrology, and vegetation data available during the years 1981−2015, the dryness of Urmia Lake has been analyzed and explained based on the changes in hydro-climate and vegetation observed in the watershed: Urmia Lake and exploratory statistical methods. The analysis calculates the relationships between changes in atmospheric climate (precipitation and temperature) and changes in hydrology (soil moisture and water levels) and vegetation (including crops and other vegetation). The results show that precipitation, temperature, and soil moisture cannot justify the sharp decrease in water level since 2000. Instead, the increase in agricultural activities and vegetation in the watershed is well correlated with the changes in the lake water level, which shows that agricultural and vegetation activities are human-driven VC, and due to the expansion of irrigation, human factors are one of the most important factors that have caused the drying of Urmia Lake. In particular, increased transpiration from crops causes increased irrigation in this region. As a result, the runoff to the basin is reduced, which ultimately causes a drop in the water level of the catchment area of Lake Urmia (Khazaei, 2019).

16.4   Results and discussion

This study reviewed the effects of two important factors, climate change and human effects, on the surface currents of Lake Urmia's catchment area. Urmia Lake Basin is very important due to its climatic, economic, ecological, and environmental importance, and climate changes and human effects are among the factors that can endanger the safety and health of this basin.

Assessing climate effects is important in water resources planning (Sheikh & Bahremand, n.d.). Since most of the rains in the Urmia Lake region occur in the winter and spring seasons and regulate the base flow of rivers, the downward trend of precipitation in a considerable number of stations in these seasons can significantly reduce the water flow of rivers. Relevant studies show that temperatures have increased significantly across the region (Fathian et al., 2016a). Under global warming and climate change, the important role of PET was emphasized. The available results show that the drought stress index (SPEI) will perform better than the conventional drought stress index (SPI) in drought studies in a changing climate (Ahmadebrahimpour et al., 2019). Most studies have analyzed monthly or annual precipitation data in a specific period. Although monthly data are usually sufficient to show the general state of seasonal patterns of climate elements, the natural change of seasons is not necessarily consistent with the period of months.

On the other hand, daily values have a greater impact on the experiences and health of plants, animals, and humans than monthly values. Most studies in Urmia Lake have focused on monthly and annual rainfall time series. The results of the previous analyses in the area of Lake Urmia show that there is no definite trend in annual rainfall in most stations. In some others, a decreasing or increasing trend is observed (Fathian, 2012; Fathian & Morid, 2014). A review of existing studies shows that the frequency of low rainfall, which is prone to more evaporation losses, has been increasing during the 30-year statistical period, while the frequency of rainfall above 10 and 15 mm, which can affect the resulting runoff, shows a decreasing trend in the basin. Fig. 16.12 shows the spatial changes of increasing and decreasing rainfall frequency in the studied areas based on the MK test over 30 years (based on Thiessen polygons) (Salehi Bavil et al., 2018).

The results of recreation in a few studies have shown that climate change causes a diminishing of the inputs to the lake during the high-water season, particularly in April and June. It appears that, on the off chance that climate change does not exist, the influx would be adequate to meet the natural needs of the lake for all water asset scenarios. Salt lakes, particularly hypersaline lakes, are exceptionally sensitive to any slight alteration in any component of their hydrological budget, particularly vanishing. Hence, it is not shocking that the lake reacts rapidly to any changes in the climate. (Shadkam et al., n.d.). Investigation of the normal yearly precipitation observed within the basin appeared to have a diminishing slant between 1960 and 2010, from 390 to 330 mm per year over the final 50 years. Moreover, precipitation within the bowl has diminished by more than 1.12 mm per year amid the ponder period, which demonstrates extreme climate changes (Fig. 16.13).

Investigations showed that the trend of normalized flow was similar to precipitation but with a more obvious decreasing trend and higher interannual variability (Fig. 16.14).

The total natural flow to the lake has decreased by 1.5×10^9 m^3 in the last 50 years. The 10-year average annual natural inflow was higher than the environmental flow requirements (EFRs) for the entire study period (Fig. 16.15). However, during the dry period, 1995–2001, the natural inflow into the lake was generally lower than the EFRs (Shadkam et al., n.d.).

The scientific literature debates whether Lake Urmia's water decline was mainly caused by atmospheric climate change or human activities in the watershed landscape. Water consumption data and its availability in the basin are very important for the evaluation, management, and development of water resources projects, especially in arid and semiarid regions. Therefore human activities and natural factors affecting these changes have also been evaluated (Taheri et al., 2019).

Lake Urmia's water adjustment has been significantly affected by anthropogenic impacts such as controlled inflows into the lake. The progressively negative drift within the lake water level over the years after 1995 is not clarified by

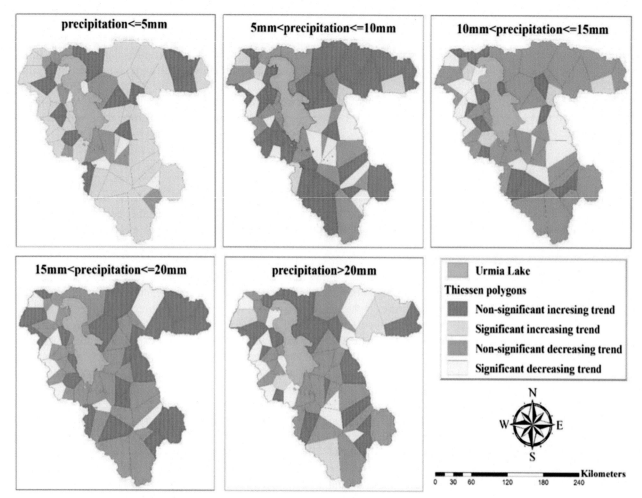

FIGURE 16.12   Spatial varieties of the precipitation recurrence slant in Urmia Lake Bowl (Salehi Bavil et al., 2018).

common lake water level inconstancy, and there might have been man-made causes for the condition that the lake is encountering nowadays. In that event, to arrange to bring the lake back to life once more, all measures of control and direction ought to be expelled from influx into the lake and agricultural exercises or any other movement that will result in negative water adjustment within the Urmia Lake Bowl ought to be essentially restricted. In this manner, it looks coherent to spend their best endeavors on the lake-regulated streams to compensate for the climate inconstancy and alter over the span of a long time to decades. Building numerous dams on the rivers flowing into the lake and overseeing the lake water adjustment on a year-by-year premise may be a drastically unsafe administration that seems to drive the lake to where it is today and, indeed, totally dehydrate the lake in the near future. In brief, the fumble of water assets within the lake bowl jeopardizes the long-term presence of the lake (Jalili et al., 2016).

The arrival cover changes of the eastern sub-basins of Lake Urmia and their impact on the hydrological administrations of the sub-basins have been analyzed. The arrival cover changes within the study area appeared to be in the early 1970s when most parts of the eastern sub-basin of Lake Urmia were secured by field arrival (PL). In any case, the arrival cover comes about appeared that PL, for the most part, diminished between 1976 and 2011, whereas rural arrival expanded essentially. So, PL diminished from 41.4% in 1976 to 27.2% in 2011. Agrarian income expanded from 2.9% in 1976 to 18.3% in 2011. Rural exercises and the development rate of rural arrival extension in 2002 overwhelm the considered zone. Over the final three decades, the populace has expanded, driving the overexploitation of assets and causing corruption as the request for nourishment increases (Fathian et al., 2016a). In common, the alteration in arriving utilization and vegetation that people do nowadays due to the increment in populace and utilization of assets can have unsalvageable impacts. The results show that despite the critical extension of watered rural areas within the watershed, the yearly stream of the watershed has diminished altogether over time (Fathian et al., 2016a).

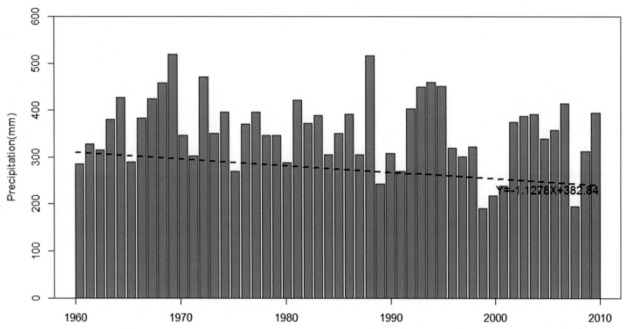

FIGURE 16.13   Watched cruel yearly precipitation over the bowl gotten from 146 stations (1960–2010). The dashed straight lines show related direct relapses (Shadkam et al., n.d.).

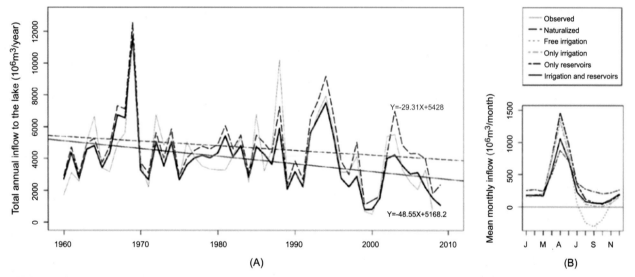

FIGURE 16.14   Re-enacting and watching add up to the yearly influx into the lake. The influx was reenacted for naturalized conditions and counting water system and store. The dashed straight lines demonstrate related direct relapses for naturalized and water system and store run (Shadkam et al., n.d.).

The government of Iran has embraced various measures within the past decade to restore the lake, but it has had a small victory (Esmailzadeh et al., 2023). In this way, a total assessment of Iran's water approaches can give a modern understanding of existing issues and offer assistance in recognizing future arrangements (Fathian & Morid, 2014). Saline lakes are confronting phenomenal natural challenges on a worldwide scale as rising temperatures, diminished precipitation, and abuse exhaust their water levels. The Incredible Salt Lake in Utah, the Salton Ocean in California, and the Aral Ocean on the border of Kazakhstan and Uzbekistan are three cases of saline lakes that have experienced extreme drying that debilitates not only the lake itself but, moreover, the well-being of the encompassing communities (Esmailzadeh et al., 2023).

Over the past 50 years, more than 50 dams with 2 billion cubic meters of supply volume and 21 huge water system systems have been built within the Urmia Lake Bowl (I. W. R. M. Company, 2020). These ventures have tripled the

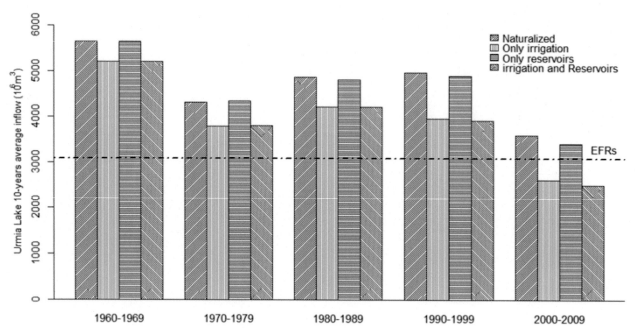

FIGURE 16.15 The 10-year average inflow to the lake, dash lines represent the Environmental Flow Requirements, 3085×10^6 m^3, for Urmia basin calculated by Abbaspour and Nazaridoust (2007) (Shadkam et al., n.d.).

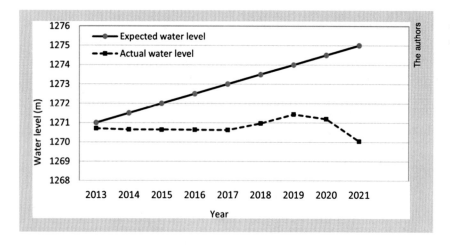

FIGURE 16.16 The actual and expected water levels of Urmia Lake (Esmailzadeh et al., 2023).

sum of allowed agrarian arrival within the final 40 years (Shadkam, 2017) and made a change of over about 2000 square kilometers of rain-fed farmland to watered farmland, all of which contributed to the drying up of the lake (Schmidt et al., 2021). Seriously, rural advancement has made nourishment generation in this basin mainly territorial (Feizizadeh et al., 2022). This bowl, too, makes a difference in supplying nourishment in other parts of the nation. In 2021, 9% of all rural crops and 13.5% of all agricultural crops in Iran have been delivered in this bowl (Esmailzadeh et al., 2023). In the interim, as it were, 7% of Iran's populace lives in this bowl, which incorporates, as it were, 3% of the country's add-up range (Nasrollahi et al., 2021b). Two major development ventures incorporate the development venture of exchanging water from the Zarineh Rood to the city of Tabriz. Due to fast populace development and high water utilization, the city confronted water shortage (Bozorg-Haddad et al., 2020). Since 1999, 200 million cubic meters of water have been exchanged every year from the Zarrineh Rud Waterway to Tabriz to address this water shortage (Schmidt et al., 2021).

The moment venture that adversely affected Lake Urmia was the development of a street that separated the lake into two sub-lakes, north and south (Eimanifar & Mohebbi, 2007). This thruway avoids water trade between the north and south of the lake, causing the water to warm more rapidly, hence expanding dissipation (Schmidt et al., 2021). Fig. 16.16 shows the actual and expected water levels of Urmia Lake.

Understanding and distinguishing atmospheric-climatic and human-scenery drivers of major changes in the water level of lakes (drying) is a research priority for water resources management and planning (Clites et al., 2014). The review of studies shows that the effect of climatic factors on the drying of Lake Urmia is much less than human effects. Human influences and climatic factors have an effect of 80% and 20%, respectively, on the drying of Lake Urmia

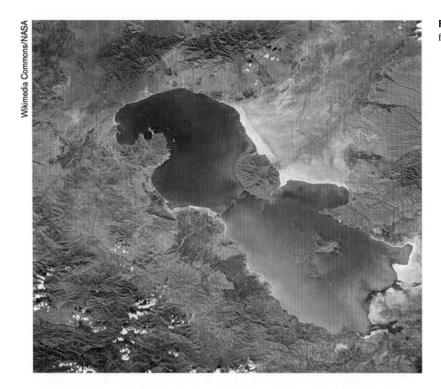

FIGURE 16.17 Urmia Lake satellite images from 2016 (Esmailzadeh et al., 2023).

FIGURE 16.18 Urmia Lake satellite images from 2016 (Esmailzadeh et al., 2023).

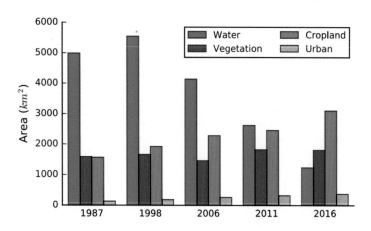

FIGURE 16.19 Graphical representation of land cover change in Urmia Lake basin from 1987 to 2016 (Chaudhari et al., 2018).

(Alizade Govarchin Ghale et al., 2018). Therefore proper management and specific planning for the revival of Lake Urmia are very important and should be paid attention to. The failure of the previous administrations shows that this matter should be taken into account in advance because Lake Urmia plays a vital role in Iran. Figures 16.17 and 16.18 show the Urmia Lake satellite images from 2016.

As shown in Fig. 16.19, both Cropland and Urban classes show a strong positive trend from 1987 to 2016, which can be directly related to human activities in the basin. Urban areas in the watershed have increased by 180%, with a 98% increase in agricultural land (Chaudhari et al., 2018).

16.5 Conclusion

Urmia Lake is one of the most important water sources in West Asia. The extreme salinity of the lake water has formed a special ecosystem. The stress on the lake has disturbed the living conditions of Artemia, one of the valuable organisms of this ecosystem that feeds on the lake's algae (Radmanesh et al., 2022). Urmia is one of the largest brackish waters in the Middle East, which now has only 10% of its volume and has turned into a desert within 15 years (Nabavi, 2017). During the last two decades, this lake's water level and area have decreased drastically. This tension has been imposed on the lake for two main reasons. The first factor is human activities, which include the increasing use of the lake's water supply resources for the development of industry and agriculture, the inappropriate pattern of cultivation, and the construction of the highway dividing the lake into two northern and southern parts. The second factor, less obvious than the first, is climate change and the resulting droughts (Radmanesh et al., 2022).

The analysis of land cover change indicates that human activities in the Urmia Lake basin from 1987 to 2016 have increased significantly, especially in agricultural development. Some previous studies have estimated the change in agricultural and water areas in the Lake Urmia region. However, due to the difference in the method used for estimation, a direct comparison (pixel by pixel) cannot be made (Chaudhari et al., 2018). Unauthorized harvesting of river water is common in this basin. In this way, 88,000 wells have been created in this area, more than half of which are illegal (Ženko & Menga, 2019). The competition created by the provincial water resources management in the Urmia Lake basin is another factor that has caused high water consumption in this area. Therefore, due to regional water management, there is fierce competition between the provinces of West Azerbaijan, East Azerbaijan, and Kurdistan for the use and development of more water (Farajzadeh, 2019). Planting plants and products that require extensive irrigation is another factor that has brought Urmia Lake into crisis.

Although climatic variables have always impacted the water level of Urmia Lake, in recent decades, a significant decrease in rainfall and an increase in temperature have significantly accelerated the decrease of the lake's water (Shadkam, 2017). Today, this area's lack of proper rainfall and increased evaporation have threatened Lake Urmia more than in the past. The average annual rainfall in the Urmia Lake basin has decreased by 17% from 1967 to 2006. The river's flow is more sensitive to temperature than precipitation, so the rise in temperature followed by the increase in evaporation causes irreparable effects. It is suggested to reduce the effects of the agricultural sector in reducing the water level of Lake Urmia, to push the region's farmers to plant low-consumption crops, increase irrigation efficiency, and integrate land for better water management in the study area. Also, using artificial wetlands and plant treatment is one of the methods that can help revive Lake Urmia. Collecting floodwaters and preventing the construction of dams can be other solutions to restore the lake.

References

Abghari, H., & Erfanian, M. (2023) Quantifying the effects of climate change on Simineh River discharge in Lake Urmia Basin. Available from https://ijonfest.gedik.edu.tr/.

AghaKouchak, A., et al. (2015). Aral Sea syndrome desiccates Lake Urmia: call for action. *Journal of Great Lakes Research*, *41*(1)), 307−311. Available from https://doi.org/10.1016/j.jglr.2014.12.007.

Ahmadaali, J., Barani, G. A., Qaderi, K., & Hessari, B. (2018). Analysis of the effects of water management strategies and climate change on the environmental and agricultural sustainability of Urmia Lake Basin, Iran. *Water*, *10*(2), 1092. Available from https://doi.org/10.3390/w10020160.

Ahmadebrahimpour, E., Aminnejad, B., & Khalili, K. (2019). Assessing future drought conditions under a changing climate: a case study of the Lake Urmia basin in Iran. *Water Supply*, *19*(6)), 1851−1861. Available from https://doi.org/10.2166/ws.2019.062.

Alborzi, A., et al. (2018). Climate-informed environmental inflows to revive a drying lake facing meteorological and anthropogenic droughts. *Environmental Research Letters*, *13*(8). Available from https://doi.org/10.1088/1748-9326/aad246.

Alizade Govarchin Ghale, Y., Altunkaynak, A., & Unal, A. (2018). Investigation anthropogenic impacts and climate factors on drying up of Urmia Lake using water budget and drought analysis. *Water Resources Management*, *32*(1), 325−337. Available from https://doi.org/10.1007/s11269-017-1812-5.

Alizadeh, A. (2001). *Principles of Applied Hydrology*. Mashhad: Astan Quds Razavi.

Asadi, M., Dehghan, G., Zarrini, G., & Soltani, N. (2011). Taxonomic survey of cyanobacteria of Urmia Lake (NW Iran) and their adjacent ecosystems based on morphological and molecular methods. *Rostaniha*, *12*(2), 153−163. Available from https://doi.org/10.22092/botany.2012.101408.

Asem, A., Eimanifar, A., Djamali, M., De Los Rios, P., & Wink, M. (2014). Biodiversity of the hypersaline Urmia Lake National Park (NW Iran). *Diversity (Basel)*, *6*, 102−132. Available from https://doi.org/10.3390/d6020102.

AslHashemi, A., & Farahmand, N. (2021). Examining the ecological characteristics of Lake Urmia and its restoration strategies. *Application of Chemistry in the Environment*, *12*(54), 49−60.

Bozorg-Haddad, X. C. O., Abutalebi, M., & Loáiciga, H. A. (2020). Assessment of potential of intraregional conflicts by developing a transferability index for inter-basin water transfers, and their impacts on the water resources. *Environmental Monitoring and Assessment*, *192*(1), 40. Available from https://doi.org/10.1007/s10661-019-8011-1.

Chaudhari, S., Felfelani, F., Shin, S., & Pokhrel, Y. (2018). Climate and anthropogenic contributions to the desiccation of the second largest saline lake in the twentieth century. *Journal of Hydrology*, *560*, 342−353. Available from https://doi.org/10.1016/j.jhydrol.2018.03.034.

Clites, A. D., Smith, A. H., Hunter, J. P., & Gronewold, T. S. (2014). Visualizing relationships between hydrology, climate, and water level fluctuations on Earth's largest system of lakes. *Journal of Great Lakes Research*, *40*, 807−811. Available from https://doi.org/10.1016/J.JGLR.2014.05.

Dalezios, N. R., Angelakis, A. N., & Eslamian, S. S. (2018). Water scarcity management. Part 1: methodological framework. *International Journal of Global Environmental Issues*, *17*(1), 1−40. Available from https://doi.org/10.1504/IJGENVI.2018.090629.

Davie, T. (2008). *Fundamentals of hydrology*. Routledge: Routledge Fundamentals of Physical Geography Series Available. Available from https://books.google.com/books?id = x0HfA6HJvogC.

Dehghanipour, A. H., Panahi, D. M., Mousavi, H., Kalantari, Z., & Tajrishy, M. (2020). Effects of water level decline in Lake Urmia, Iran, on local climate conditions. *Water*, *12*(8), 2153. Available from https://doi.org/10.3390/W12082153.

Eimanifar, A., & Mohebbi, F. (2007). Urmia Lake (Northwest Iran): a brief review. *Saline Systems*, *3*(1), 1−8. Available from https://doi.org/10.1186/1746-1448-3-5.

Emami, F., & Koch, M. (2019). Modeling the impact of climate change on water availability in the Zarrine River Basin and inflow to the Boukan Dam, Iran. *Climate*, *7*(4). Available from https://doi.org/10.3390/cli7040051.

Esmailzadeh, S., Rizi, A. P., & Mianabadi, H. (2023). Evaluation of the water policies of the Urmia Lake Basin: has the government accurately identified the problem? *Environment*, *65*(6), 18−34. Available from https://doi.org/10.1080/00139157.2023.2245741.

Farajzadeh, M. (2019). MSc thesis, Tarbiat Modares University. *Analytical Conflict Mapping in the Complex Water System of Urmia Lake Basin*.

Fathian, F, & M.S. (2012). *Study of climate and hydrologic trends in Lake Urmia watershed using non-parametric*. Tehran, Iran.

Fathian, F., Dehghan, Z., Bazrkar, M. H., & Eslamian, S. (2016b). Trends in hydrological and climatic variables affected by four variations of the Mann-Kendall approach in Urmia Lake basin, Iran. *Hydrological Sciences Journal*, *61*(5), 892−904. Available from https://doi.org/10.1080/02626667.2014.932911.

Fathian, F., Dehghan, Z., & Eslamian, S. (2016a). Evaluating the impact of changes in land cover and climate variability on streamflow trends (case study: eastern subbasins of Lake Urmia, Iran). *International Journal of Hydrology Science and Technology*, *6*(1), 1−26. Available from https://doi.org/10.1504/IJHST.2016.073881.

Fathian, K. E., & Morid, S. (2014). Identification of trends in hydrological and climatic variables in Urmia Lake basin, Iran. *Iranian water and soil research journal, volume: 43, number:, 3.

Fathian, F., Morid, S., & Kahya, E. (2015). Identification of trends in hydrological and climatic variables in Urmia Lake basin, Iran. *Theoretical and Applied Climatology*, *119*(3−4), 443−464. Available from https://doi.org/10.1007/s00704-014-1120-4.

Fazli Fard, P., Sheikh, V., Sadoddin, A., & Hessari, B. (2019). Study of trends in precipitation and stream flow of the Lake Urmia Basin during the past four decades. *Water and Soil Science*, *29*(4), 27−41, . Available. Available from https://water-soil.tabrizu.ac.ir/article_10007.html.

Feizizadeh, A. S. B., Lakes, T., Omarzadeh, D., & Blaschke, S. K. T. (2022). Scenario-based analysis of the impacts of lake drying on food production in the Lake Urmia Basin of northern Iran. *Scientific Reports*, *12*(1). Available from https://doi.org/10.1038/s41598-022-10159-2.

Habibi, M., Babaeian, I., & Schöner, W. (2021). Changing causes of drought in the Urmia Lake basin—increasing influence of evaporation and disappearing snow cover. *Water*, *13*(22), 3273. Available from https://doi.org/10.3390/w13223273.

Hassanzadeh, E., Zarghami, M., & Hassanzadeh, Y. (2012). Determining the main factors in declining the Urmia Lake level by using system dynamics modeling. *Water Resources Management, 26*(1), 129−145. Available from https://doi.org/10.1007/s11269-011-9909-8.

Hemmati, M., Ahmadi, H., & Sajad-Ahmad, H. (2020). 3D numerical simulation of wind direction effect on flow circulation and salinity patterns at Lake Urmia. *Irrigation and Water Engineering, 10*(3), 22−33. Available from https://doi.org/10.22125/iwe.2020.107016.

Hessari, B., & Kamran, K. (2020). Investigation on human activities effects on water resources trend of Urmia Plain. *Watershed Engineering and Management, 12*(2), 415−427. Available from https://doi.org/10.22092/ijwmse.2019.124426.1580.

Hosseini-Moghari, S.-M., Araghinejad, S., Tourian, M.J., Ebrahimi, K., Döll, P. (2018). Quantifying the impacts of human water use and climate variations on recent drying of Lake Urmia basin: the value of different sets of spaceborne and in-situ data for calibrating a hydrological model. Hydrology and Earth System Sciences Discussions, 1−29. Available from https://doi.org/10.5194/hess-2018-318.

I. W. R. M. Company (November 3, 2020) List of Dams in Iran. http://daminfo.wrm.ir/fa/dam/tabularview.

Jalili, S., Hamidi, S. A., & Namdar Ghanbari, R. (2016). Climate variability and anthropogenic effects on Lake Urmia water level fluctuations, north-western Iran. *Hydrological Sciences Journal, 61*(10), 1759−1769. Available from https://doi.org/10.1080/02626667.2015.1036757.

Kelts, K., & Hsü, K.J. (1978). Freshwater carbonate sedimentation. In: Lerman A. (Eds.) Lakes. Springer, New York, NY. Available from https://doi.org/10.1007/978-1-4757-1152-3_9.

Kelts, K., & Shahrabi, M. (1986). Holocene sedimentology of hypersaline Lake Urmia, Northwestern Iran. *Palaeogeography, Palaeoclimatology, Palaeoecology, 54*(1), 105−130. Available from https://doi.org/10.1016/0031-0182(86)90120-3.

Khazaei, B., et al. (2019). Climatic or regionally induced by humans? Tracing hydro-climatic and land-use changes to better understand the Lake Urmia tragedy. *Journal of Hydrology, 569*, 203−217. Available from https://doi.org/10.1016/j.jhydrol.2018.12.004.

Moghimi,H. (2020) Reliability analysis of stream network density changes in Urmia Lake catchment area using FORM Probabilistic Method. Iran—Water Resource Research 16(3), 188−197.

Nabavi, E. (2017) (Ground)water governance and legal development in Iran, 1906−2016. Middle East Law and Governance 9(1), 43−70. Available from https://doi.org/10.1163/18763375-00901005.

Najafi, A., Samadi, A., & Rasouli, K. (2013) Environmental crisis in Lake Urmia, Iran: a systematic review of causes, negative consequences and possible solutions. In: Proceedings of the 6ht International Perspective on Water Resources & the Environment (IPWE), Izmir, Turkey. Available from https://doi.org/10.13140/RG.2.1.4737.0088.

Narimani, R., Jun, C., Nezhad, S. M., Bateni, S. M., Lee, J., & Baik, J. (2023). The role of climate conditions and groundwater on baseflow separation in Urmia Lake Basin, Iran. *Journal of Hydrology: Regional Studies, 47*, 101383. Available from https://doi.org/10.1016/j.ejrh.2023.101383.

Nasrollahi, H., Shirazizadeh, R., Shirmohammadi, R., Pourali, O., & Amidpour, M. (2021a). Unraveling the water-energy-food-environment nexus for climate change adaptation in Iran: Urmia Lake Basin case study. *Water, 13*(9), 1282. Available from https://doi.org/10.3390/w13091282.

Nasrollahi, R. S. H., Shirazizadeh, R., & Pourali, M. A. O. (2021b). Unraveling the water-energy-food-environment nexus for climate change adaptation in Iran: Urmia Lake Basin case study. *Water, 13*(9), 1−25. Available from https://doi.org/10.3390/w13091282.

Nhu, V. H., et al. (2020). Monitoring and assessment of water level fluctuations of the Lake Urmia and its environmental consequences using multi-temporal landsat 7 etm + images. *International Journal of Environmental Research and Public Health, 17*(12), 1−18. Available from https://doi.org/10.3390/ijerph17124210.

Okhravi, S., Eslamian S. (2017). Drought in Lake Urmia. Available: https://www.researchgate.net/publication/323174495.

Pimentel, D., Berger, B., Filiberto, D., Newton, M., Wolfe, B., Karabinakis, E., Clark, S., Poon, E., Abbett, E., & Nandgopal, S. (2004). Water resources: agricultural and environmental issues. *BioScience, 54*(10), 909−918.

Radmanesh, F., Esmaeili-Gisavandani, H., & Lotfirad, M. (2022). Climate change impacts on the shrinkage of Lake Urmia. *Journal of Water and Climate Change, 13*(6), 2255−2277. Available from https://doi.org/10.2166/wcc.2022.300.

Rajabi Hashijan, M. (n.d.) The need to restorate Lake Urmia, the causes of drought and threats [in Persion].

Rai, U. A., Singh, R. K., & Upadhyay, A. (2017). *Scheme irrigation efficiency. Planning and Evaluation of Irrigation Projects* (pp. 525−538). Elsevier.

Razmara, P., Motiee, H., Massah Bavani, A., Saghafian, B., & Torabi, S. (2016). Risk assessment of climate change impacts on runoff in Urmia Lake Basin, Iran. *Journal of Hydrologic Engineering, 21*(8). Available from https://doi.org/10.1061/(asce)he.1943-5584.0001379.

Rivas-Martínez, S., et al. (n.d.). ITINERA GEOBOTANICA es una publicación periódica de la Asociación Editorial Board (Comisión editorial) Technicals Editors (Responsables de la edición).

Roshan, G., Samakosh, J. M., & Orosa, J. A. (2016). The impacts of drying of Lake Urmia on changes of degree day index of the surrounding cities by meteorological modelling. *Environmental Earth Sciences, 75*(20), 1387. Available from https://doi.org/10.1007/s12665-016-6200-6.

Safari, N., & Zarghami, M. (2012). Allocating the surface water resources of the Urmia Lake Basin to the stakeholder provinces by distance based decision making methods. *Water and Soil Science, 23*, 135−149.

Sakizadeh, M., Milewski, A., & Sattari, M. T. (2023). Analysis of long-term trend of stream flow and interaction effect of land use and land cover on water yield by SWAT model and statistical learning in part of Urmia Lake Basin, northwest of Iran. *Water, 15*(4), 690. Available from https://doi.org/10.3390/w15040690.

Salehi Bavil, S., Zeinalzadeh, K., & Hessari, B. (2018). The changes in the frequency of daily precipitation in Urmia Lake basin, Iran. *Theoretical and Applied Climatology, 133*(1−2), 205−214. Available from https://doi.org/10.1007/s00704-017-2177-7.

Shadkam, S. (2017). Preserving Urmia Lake in a changing world: reconciling anthropogenic and climate drivers by hydrological modelling and policy assessment. PhD thesis, Wageningen.

Shadkam, S., Ludwig, F., Van Oel, P., Kirmit, Ç., Kabat. P. (n.d.). Impacts of climate change and water resources development on the declining inflow into Iran's Urmia Lake. Available: http://www.sage.wisc.edu/.

Schmidt, M., Gonda, R., & Transiskus, S. (2021). Environmental degradation at Lake Urmia (Iran): exploring the causes and their impacts on rural livelihoods. *GeoJournal, 86*(5), 2149−2163. Available from https://doi.org/10.1007/s10708-020-10180-w.

Shadkam, S., Ludwig, F., van Vliet, M. T. H., Pastor, A., & Kabat, P. (2016). Preserving the world second largest hypersaline lake under future irrigation and climate change. *Science of the Total Environment, 559*, 317−325. Available from https://doi.org/10.1016/j.scitotenv.2016.03.190.

Sharifi, A., Shah-Hosseini, M., Pourmand, A., Esfahaninejad, M., & Haeri-Ardakani, O. (2018). *The vanishing of Urmia Lake: a geolimnological perspective on the hydrological imbalance of the world's second largest hypersaline lake,* . *Handbook of Environmental Chemistry* (Part F1). Springer. Available from https://doi.org/10.1007/698_2018_359.

Sheikh, V., & Bahremand, A. (n.d.). Investigation of the hydrological processes in the Atrak Basin.

Shemshaki, A., & Karami, G. H. (2019). Hydrogeological, hydrogeochemical and isotopic study of coastal aquifer in southeastern bank of Urmia Lake. *Scientific Quarterly Journal of Iranian Association of Engineering Geology, 11*(4), 109−121, . Available. Available from https://www.jir-aeg.ir/article_88497.html.

Shirmohammadi, B., et al. (2020). Scenario analysis for integrated water resources management under future land use change in the Urmia Lake region, Iran. *Land Use Policy, 90*, 104299. Available from https://doi.org/10.1016/j.landusepol.2019.104299.

Sima, S., Ahmadalipour, A., & Tajrishy, M. (2013). Mapping surface temperature in a hyper-saline lake and investigating the effect of temperature distribution on the lake evaporation. *Remote Sensing of the Environment, 136*, 374−385. Available from https://doi.org/10.1016/j.rse.2013.05.014.

Soudi, M., Ahmadi, H., Yasi, M., & Hamidi, S. A. (2017). Sustainable restoration of the Urmia Lake: History, threats, opportunities and challenges. *European Water, 60*, 341−347.

Taheri, M., Emadzadeh, M., Gholizadeh, M., Tajrishi, M., Ahmadi, M., & Moradi, M. (2019). Investigating the temporal and spatial variations of water consumption in Urmia Lake River Basin considering the climate and anthropogenic effects on the agriculture in the basin. *Agricultural Water Management, 213*, 782−791. Available from https://doi.org/10.1016/j.agwat.2018.11.013.

Y., J., & Molan-Nejad, L. (2017). Assessing attitudes of farmers to participate in the process of preserving and restoring Urmia Lake and its related factors in Miandoab Township. *Iranian Agricultural Extension and Education Journal, 13*(1), 47−58, In Persian.

Yarahmadi, J., & Rostamizad, G. (2019). The analysis of the hydrological droughts in the northern part of Lake Urmia. *Hydrogeomorphology, 6*(19), 79−100. Available. Available from https://hyd.tabrizu.ac.ir/article_9313.html.

Yasi, M., & Rezaei, N. (2022). Eco-hydro desktop assessment for determining rivers environmental flows in Urmia Lake Basin (case study: Simineh River). *Sigma Journal of Engineering and Natural Sciences, 40*(2), 243−251. Available from https://doi.org/10.14744/sigma.2022.00029.

Yusuf, S., Sina, B., Vahid, V., & Javad, K. (2017). *Hydrological drought monitoring in the watersheds of Zarineh Rood and Simineh Rood rivers. Proceedings of the 16th Iranian Hydraulic Conference.* Mohaghegh Ardabili University.

Zarandian, A., Mousazadeh, R., & Mehrian, M. R. (2023). Assessment of the hazards of vegetation loss in the eastern basin of the Lake Urmia based on the ecosystem services modeling approach under the conceptual framework of DPSIR. *Journal of Environmental Research, 14*(27), 219−251. Available from https://doi.org/10.22034/eiap.2023.179862.

Zarghami, M. (2011). Effective watershed management: case study of Urmia Lake, Iran. *Lake and Reservoir Management, 27*(1), 87−94. Available from https://doi.org/10.1080/07438141.2010.541327.

Ženko, M., & Menga, F. (2019). Linking water scarcity to mental health: hydro-social interruptions in the Lake Urmia Basin, Iran. *Water, 11*(5), 1092. Available from https://doi.org/10.3390/w11051092.

Chapter 17

Stream flow restoration case studies in the Middle East and North Africa countries

Saeid Eslamian[1], Mousa Maleki[1] and Hemraj Ramdas Kumavat[2]

[1]Department of Water Sciences and Engineering, College of Agriculture, Isfahan University of Technology, Isfahan, Iran, [2]Department of Civil Engineering, R C Patel Institute of Technology, Shirpur, Maharashtra, India

17.1 Introduction

Streamflow recharge (SFR), also known as managed aquifer recharge (MAR) or artificial groundwater recharge, refers to the process of intentionally directing water into an aquifer through artificial means, either underground or on the surface, at a rate that exceeds natural recharge. The primary objective of SFR projects is to enable the infiltration of excess surface water runoff and seasonal floodwaters into aquifers through the use of engineeed structures and infrastructure. This helps to replenish groundwater reserves, improve dry season base flows that support rivers and riparian ecosystems, support stream—aquifer interactions, and increase the overall availability and reliability of water resources, both underground and on the surface (Andualem et al., 2021; Wu et al., 2021).

There are a few key techniques commonly employed for SFR. Subsurface dams involve subsurface barriers constructed underground to block or redirect shallow aquifers and encourage the infiltration and storage of water. Infiltration basins and trenches utilize unlined earthen structures to directly percolate surface waters underground. Recharge wells convey surface water via pipes inserted vertically down into aquifers. Sand dams are structures built across riverbeds to temporarily pond and infiltrate floodwaters. Check dams are small, often rubble or earthen barriers installed across drainage lines to impound water and promote infiltration through the beds and banks. Traditional techniques such as qanats in Iran and foggara systems in North Africa also function as long, underground conduits to channel seasonal rains into aquifer recharge zones. Forest and rangeland management practices that help conserve soil and increase moisture retention can also contribute to diffuse natural recharge (Arnold & Allen, 1999; Gaaloul et al., 2021).

The Middle East and North Africa (MENA) region faces severe water scarcity issues that act as a major developmental constraint. MENA encompasses 17 countries and over 400 million people spread across diverse climatic and geographic zones, yet the region as a whole has some of the lowest per capita renewable freshwater resources availability globally at under 1000 cubic meters annually on average. Over half of the total land area receives less than 250 mm of annual rainfall, qualifying as hyperarid according to The United Nations' definitions. Population growth rates around 2%—3% per annum and rapid urbanization trends are exacerbating already high water stress levels, with projections that two-thirds of MENA's population will inhabit extreme arid and desert fringe zones by 2050 (Scanlon et al., 2007; Tropp & Jagerskog, 2006).

Surface water resources are limited and unevenly distributed across the region. Notable rivers such as the Nile, Tigris, and Euphrates flow through multiple countries and depend on flows originating from outside MENA boundaries, making them politically and environmentally vulnerable. Internal renewable water supplies are characterized by flash flood-prone wadis and ephemeral streams that often run dry for most of the year. Groundwater reserves provide a larger relative contribution, storing an estimated 35,000 cubic km of nonrenewable fossil aquifers and renewable rechargeable aquifers. However, depletion rates of nonrenewable aquifers from unsustainable extraction already exceed natural replenishment by over 25% in countries like Saudi Arabia, Libya, and Yemen (Foster & Chilton, 2003; Gleeson et al., 2012; Haddadin, 2001; Konikow & Kendy, 2005).

Hydrosystem Restoration Handbook. DOI: https://doi.org/10.1016/B978-0-443-29802-8.00017-0

Water scarcity severely impacts socioeconomic development and environmental security across MENA. The agricultural sector, which accounts for over 80% of water demand region-wide, suffers from low and declining productivity on nonirrigated rain-fed lands. Municipal and industrial water needs are increasingly difficult and expensive to meet in many urban centers and economic activity hubs. Ecosystems and biodiversity are degraded by flow reductions and modifications within arid freshwater systems. Climate change projections indicate rising temperatures and more erratic rainfall patterns, and longer more frequent droughts will further exacerbate the region's water challenges in the coming decades. A particular threat is reduced mountain snowpack and glaciers in water tower countries that feed cross-boundary surface flows relied on by downstream nations. The socioeconomic toll of water scarcity is reflected in high desalination energy consumption and costs to countries like Saudi Arabia and Qatar. Well-being indicators are impacted, with more than 50% of rural populations in Yemen, Djibouti, and Libya lacking access to safe drinking water (Hoekstra & Chapagain, 2007; Ibrahim & Mensah, 2017; Shah, 2009; Taylor et al., 2013).

Within the pressing context of MENA's water crisis, SFR techniques hold promise as a distributed, community-based solution able to augment limited local freshwater resources on a sustainable basis. To date, a number of small-scale SFR pilot and demonstration projects have emerged across the arid region in countries experiencing some of the most acute impacts of water scarcity. However, there has been limited documentation and cross-analysis of implemented case studies to draw lessons regarding drivers of success, limitations, and potential for upscaling approaches (Alam et al., 2021; Gleick, 1998; Hanjra & Qureshi, 2010).

The aim of this introductory chapter is to begin addressing this gap by presenting case studies from three representative MENA nations—Oman, Morocco, and Jordan—where innovative SFR methods have been tested under diverse climatic settings and socioeconomic conditions. These countries each face serious but distinct water challenges and have distinctive hydrogeological characteristics. Comparing SFR models adapted to local hydrological and societal contexts can provide insight into how natural infrastructure techniques can be tailored and replicated to supplement the water security of other water-stressed communities across similar arid systems within the region and beyond. The objective is to analyze technical, institutional, economic, environmental, and social aspects of implemented projects to identify factors enabling or constraining outcomes. It is hoped the chapter will help inform the development of an evidence base regarding strengths and weaknesses of different recharge approaches, facilitating future expansion of such sustainable freshwater augmentation strategies (Alam et al., 2021; Gleick & Palaniappan, 2010).

17.2 Case study 1: Wadi restoration, Oman

17.2.1 Description of degraded wadi hydrological system

The wadi hydrological system in Oman has been experiencing significant degradation due to various factors such as climate change, increasing water demand, and unsustainable land use practices. The once-thriving wadis, which are seasonal rivers or watercourses, have suffered from reduced streamflow, erosion, and loss of biodiversity. This case study focuses on the restoration efforts carried out in a specific wadi in Oman to address these challenges and rejuvenate the hydrological system (Sadoff & Grey, 2002). Fig. 17.1 shows a graphical overview of a wadi in Oman.

Table 17.1 describes names of Oman wadis and their features.

17.2.2 Steamflow recharge techniques piloted

To restore the degraded wadi hydrological system, several SFR techniques were piloted. These techniques aimed to increase water infiltration and recharge the groundwater, ultimately replenishing the streamflow during both wet and dry seasons. The following SFR techniques were implemented.

17.2.2.1 Subsurface dams

Subsurface dams were constructed strategically along the wadi to impede the flow of water and allow for increased water infiltration. These dams are designed to slow down the movement of water, creating temporary water storage underground. As a result, the groundwater recharge is enhanced, leading to augmented streamflow during low-flow periods.

17.2.2.2 Sand dams

Sand dams were implemented in specific sections of the wadi to capture and store water during the rainy season. These structures consist of a reinforced concrete wall built across the wadi, which allows sediment to accumulate behind it.

FIGURE 17.1 Graphical overview of a wadi in Oman (author).

TABLE 17.1 Brief description of Oman wadis (author).

Wadi name	Location	Features	Activities
Wadi Shab	Ash Sharqiyah	Stunning natural beauty with turquoise pools, waterfalls, and narrow canyons.	Hiking, swimming, cliff jumping, and cave exploration.
Wadi Bani Khalid	Ash Sharqiyah	Lush palm-lined oasis with crystal-clear pools and natural springs.	Swimming, picnicking, and exploring the scenic surroundings.
Wadi Al Arbaeen	Muscat	Deep canyons with towering cliffs and pools of emerald green water.	Hiking, canyoning, swimming, and photography.
Wadi Tiwi	Ash Sharqiyah	Scenic Wadi with steep cliffs, terraced plantations, and small villages.	Hiking and exploring the traditional Omani villages.
Wadi Al Hoqain	Al Batinah	Picturesque Wadi with rocky landscapes, water pools, and agricultural terraces.	Hiking, photography, and experiencing traditional Omani culture.
Wadi Darbat	Dhofar	Lush green Wadi with cascading waterfalls, caves, and abundant wildlife.	Sightseeing, birdwatching, and enjoying the natural beauty.
Wadi Al Jizzi	Al Batinah	Serene Wadi with date palm plantations, freshwater pools, and scenic mountain views.	Picnicking, camping, and enjoying the peaceful surroundings.
Wadi Shuwaymiyah	Dhofar	Dramatic wadi with towering cliffs, natural pools, and a seasonal waterfall.	Hiking, swimming, and enjoying the scenic landscapes.
Wadi Ghul	Al Dakhiliyah	Deep canyon known as the "Omani Grand Canyon" with rugged cliffs and breathtaking views.	Hiking, trekking, and enjoying panoramic vistas.
Wadi Bani Awf	Al Dakhiliyah	Challenging wadi known for its rugged terrain, narrow gorges, and adventurous trails.	Off-road driving, hiking, rock climbing, and canyoning.

The trapped sediment acts as a natural filter, allowing water to percolate into the ground, recharging the aquifers and contributing to sustained streamflow during dry periods.

17.2.3   Monitoring of streamflow response pre- and postintervention

To assess the effectiveness of the restoration efforts, a comprehensive monitoring program was initiated to measure streamflow response before and after the implementation of SFR techniques. The monitoring involved the installation of stream gauges and water level sensors at strategic locations along the wadi. Prior to the intervention, baseline data was collected over an extended period to establish the degraded condition of the hydrological system. This included measurements of streamflow, water levels, and sediment transport rates during different seasons and flow conditions. Following the implementation of SFR techniques, continuous monitoring was carried out to evaluate the impact of the restoration efforts. Parameters such as streamflow volume, water table levels, and sediment deposition rates were measured regularly. Additionally, ecological monitoring was conducted to assess changes in vegetation cover, wildlife habitat, and overall ecosystem health.

17.2.4   Discussion of environmental and socioeconomic benefits

The restoration of the wadi hydrological system in Oman has yielded significant environmental and socioeconomic benefits. The implementation of SFR techniques has resulted in the following positive outcomes.

17.2.4.1   Environmental benefits

- Increased streamflow: The restoration efforts have led to enhanced streamflow, particularly during dry periods, providing a more reliable water source for both natural ecosystems and human populations.
- Biodiversity conservation: The revived wadi ecosystem has created favorable conditions for the recovery of native flora and fauna. Increased water availability has supported the growth of vegetation, contributing to improved habitat quality and biodiversity.
- Reduced erosion: The implementation of subsurface dams and sand dams has helped in reducing erosion and sediment transport downstream, thus protecting the ecological integrity of the wadi and adjacent areas.

17.2.4.2   Socioeconomic benefits

- Water availability for agriculture: The increased streamflow and groundwater recharge have facilitated agricultural activities in the region, supporting local food production and improving livelihoods.
- Enhanced water security: The restoration efforts have contributed to increased water availability, reducing dependency on external water sources and enhancing water security for communities residing in the vicinity of the wadi.
- Recreation and tourism opportunities: The revitalized wadi hydrological system has created attractive landscapes and water features, drawing tourists and contributing to the local economy through recreational activities such as hiking, camping, and wildlife observation.

In conclusion, the case study of wadi restoration in Oman highlights the effectiveness of SFR techniques in rejuvenating degraded hydrological systems. The successful implementation of subsurface dams and sand dams has resulted in increased streamflow, improved ecosystem health, and socioeconomic benefits for the local communities. These restoration efforts serve as a valuable example for similar initiatives in MENA countries, emphasizing the importance of sustainable water management and conservation practices in arid regions.

17.3   Case study 2: Aquifer recharge from floods, Morocco

17.3.1   Existing floodwater harvesting infrastructure

In Morocco, the hydrological system heavily relies on seasonal floods for water supply and groundwater recharge. Over the years, the country has developed floodwater harvesting infrastructure to capture and utilize the abundant water during flood events. This case study focuses on the existing floodwater harvesting infrastructure in Morocco and its role in aquifer recharge. The infrastructure includes various elements such as cisterns and ponds strategically placed in flood-prone areas. Cisterns are underground storage structures designed to collect and store floodwater, while ponds are surface storage reservoirs. These structures are typically located in river basins or areas prone to flash floods, where they can capture and retain large volumes of floodwater (Döll et al., 2009; Falkenmark & Molden, 2008). An example of a flood water harvester channel in Morocco is depicted in Fig. 17.2.

FIGURE 17.2   A flood water harvester channel in Morroco (author).

17.3.2   Improvement/expansion of structures to optimize groundwater recharge

To optimize the recharge of aquifers from floodwater, ongoing efforts have been made to improve and expand the existing floodwater harvesting structures in Morocco. These improvements aim to enhance the capture and infiltration of floodwater into the groundwater system. Some of the key measures undertaken are discussed in the following subsections.

17.3.2.1   Increased storage capacity

Existing cisterns and ponds have been expanded in size to accommodate larger volumes of floodwater. This expansion allows for a more efficient capture and storage of floodwater, maximizing the recharge potential.

17.3.2.2   Enhanced infiltration mechanisms

Structural modifications have been made to facilitate the infiltration of captured floodwater into the underlying aquifers. This includes the construction of infiltration basins, trenches, or infiltration galleries adjacent to the storage structures. These features promote the percolation of floodwater into the ground, replenishing the aquifers.

17.3.2.3   Sediment trapping and filtering

To optimize aquifer recharge, sediment trapping and filtering techniques have been incorporated into the floodwater harvesting infrastructure. Sedimentation basins or settling ponds are constructed to capture and retain sediment carried by the floodwater. This allows for sediment deposition and subsequent filtration of the water before it infiltrates into the aquifers, minimizing the risk of aquifer clogging.

17.3.3   Community water supply/agricultural impacts

The improvement and expansion of floodwater harvesting infrastructure in Morocco have had significant impacts on community water supply and agricultural practices.

17.3.3.1 Community water supply

The optimized aquifer recharge from floodwater provides a reliable and sustainable source of water for communities. Increased groundwater levels and improved water quality have led to enhanced access to potable water, reducing the dependency on other water sources. This has improved the overall water security and livelihoods of local communities.

17.3.3.2 Agricultural impacts

The availability of replenished aquifers has facilitated agricultural activities in the region. Farmers can access groundwater for irrigation, enabling crop cultivation even during dry periods. This has led to increased agricultural productivity, diversification of crops, and improved food security. The use of floodwater as a source of irrigation has also reduced dependence on surface water sources, which may be limited or heavily regulated (Postel, 1999; Rosegrant et al., 2009; Siebert et al., 2010).

17.3.4 Lessons for integrated flood management

The case study of aquifer recharge from floods in Morocco offers valuable lessons for integrated flood management strategies in other regions. Some key takeaways are given in the following subsections.

17.3.4.1 Synergies between flood mitigation and water resource management

The floodwater harvesting infrastructure serves a dual purpose of flood mitigation and aquifer recharge. By capturing and storing floodwater, the infrastructure reduces the risk of flood-related damages while simultaneously replenishing groundwater resources.

17.3.4.2 Importance of adaptive infrastructure

The continuous improvement and expansion of floodwater harvesting structures demonstrate the importance of adaptive infrastructure that can accommodate changing hydrological conditions. Flexibility in design and capacity allows for better optimization of aquifer recharge.

17.3.4.3 Integration of multiple stakeholders

Successful implementation of integrated flood management requires collaboration between various stakeholders, including government agencies, local communities, and water resource management authorities. Engaging stakeholders throughout the planning, implementation, and monitoring phases ensures effective floodwater harvesting and aquifer recharge strategies.

In closing, this evaluation of MAR from flood events in Morocco underscores the advantages of harnessing overflow for underground water inventory replenishment. The adaptive infrastructure already in situ, coupled with boosts and extensions, has favorably influenced local water provision and crop cultivation methods. Additionally, this pilot study furnishes important understandings with respect to integrated flood regulation schemes, highlighting the significance of flexible construction and engagement among diverse stakeholders. The takeaways gleaned can educate and direct analogous undertakings elsewhere, contributing to sustainable water administration and habituation to climatic unpredictability.

17.4   Case study 3: Managed aquifer recharge, Jordan

17.4.1   Artificial recharge of strategic aquifers

In Jordan, a country facing water scarcity challenges, MAR has been implemented as a strategy to replenish strategic aquifers. MAR involves the deliberate infiltration of water from various sources, such as treated wastewater and surface runoff, into underground aquifers. This case study focuses on the artificial recharge efforts in Jordan and their significance in sustaining water resources. To address water scarcity, treated wastewater is utilized as a valuable resource for MAR. The treated wastewater, which meets specific quality standards, is carefully discharged into targeted recharge areas. Additionally, surface runoff from rainfall events is captured and directed towards recharge basins or infiltration galleries, allowing the water to seep into the aquifers (Pescod, 1992). A graphical design of a channel for aquifer recharge purpose is represented in Fig. 17.3.

FIGURE 17.3 A graphical design of a channel for aquifer recharge purpose (author).

17.4.2 Conjunctive surface−groundwater management model

To optimize the efficiency of MAR in Jordan, a conjunctive surface−groundwater management model has been tested. This model integrates surface water and groundwater management approaches to achieve sustainable water resource utilization. The conjunctive management approach involves the coordinated operation of surface reservoirs and groundwater abstraction wells. During periods of high surface water availability, excess water is intentionally recharged into the aquifers, effectively storing it for future use. Conversely, during periods of water scarcity, groundwater pumping is increased, reducing reliance on surface water sources. The conjunctive surface−groundwater management model allows for flexibility and adaptability in water allocation, balancing the demand and supply in an integrated manner.

17.4.3 Water quality/quantity monitoring results

Continuous monitoring of water quality and quantity is vital to ensure the success and sustainability of MAR in Jordan. Various parameters are regularly monitored to assess the effectiveness and impacts of the recharge efforts.

17.4.3.1 Water quality monitoring

Water quality parameters such as pH, electrical conductivity, and concentrations of major ions and contaminants are measured to ensure compliance with water quality standards. Monitoring activities also include the assessment of treated wastewater quality before recharge and the monitoring of groundwater quality after recharge to evaluate any changes.

17.4.3.2 Water quantity monitoring

Monitoring the quantity of water recharged into the aquifers is crucial for assessing the effectiveness of the MAR system. Flow meters and water level sensors are installed at recharge points to measure the volume of water infiltrated. Groundwater monitoring wells are used to track changes in water levels and recharge rates.

17.4.4 Sustainability considerations

MAR in Jordan takes into account various sustainability considerations to ensure the long-term viability of the recharge efforts.

17.4.4.1 Water resource availability

The selection of recharge areas considers the availability and capacity of the aquifers to accommodate the recharged water. Sustainable water resource management practices are implemented to prevent overexploitation and ensure the continued availability of water for future generations.

17.4.4.2 Water treatment and quality

Stringent water treatment processes are employed to ensure that the recharged water meets the required quality standards. Treatment technologies are continuously evaluated and improved to minimize potential risks to the aquifers and safeguard water resources.

17.4.4.3 Environmental impact assessment

Environmental impact assessments are conducted to evaluate the potential ecological and hydrological impacts of MAR. This assessment helps identify and mitigate any adverse effects on groundwater-dependent ecosystems and adjacent surface water bodies.

17.4.4.4 Stakeholder engagement

Engaging stakeholders, including local communities, water management authorities, and relevant organizations, is crucial for the success of MAR. Public awareness campaigns, education, and participatory decision-making processes ensure the acceptance and sustainable management of the recharge initiatives (Eyitayo et al., 2023; Falkenmark & Rockström, 2004).

In summary, the Jordan case study on managed aquifer replenishment emphasizes the importance of leveraging treated wastewater and surface runoff for replenishing strategic underground water sources. The integrated approach combining surface and groundwater administration, along with meticulous tracking efforts, contributes to sustainable water asset administration. Considering water quality, quantity, and long-term sustainability ensures that MAR projects remain viable long-term solutions. Lessons from this case study can steer comparable water administration strategies in additional areas challenged by scarce water availability, promoting efficient and sustainable water resource utilization.

17.5 Cross-case analysis

In this section of the chapter, a cross-case analysis will be conducted to compare and analyze the MAR projects discussed in the previous case studies. The analysis will focus on the hydrogeological conditions and different approaches used for surface runoff (SFR) in each case. Furthermore, it will identify key drivers of project success across technical, institutional, and social factors. The replication potential of these approaches to other aquifer systems in the MENA region will also be assessed. Finally, recommendations for scaling up SFR projects across the region will be provided.

17.5.1 Comparative hydrogeological conditions and streamflow recharge approaches

The cross-case analysis will begin by comparing the hydrogeological conditions in each case study. This includes factors such as aquifer characteristics, recharge rates, and groundwater availability. The different approaches used for surface runoff (SFR) in each case will also be analyzed, considering factors such as collection methods, storage systems, and infiltration techniques. By comparing these aspects, a comprehensive understanding of the variations and similarities between the cases can be achieved.

17.5.2 Key drivers of project success

The analysis will then identify the key drivers of project success across technical, institutional, and social factors. Technical factors may include the effectiveness of recharge methods, water quality monitoring systems, and management practices. Institutional factors may refer to the presence of supportive policies, coordination among stakeholders, and capacity building efforts. Social factors may encompass community engagement, awareness campaigns, and the involvement of local institutions. By examining the success factors in each case, valuable insights can be gained to inform future project planning and implementation.

17.5.3 Replication potential to other Middle East and North Africa aquifer systems

One of the objectives of the cross-case analysis is to assess the replication potential of the MAR and SFR approaches to other aquifer systems in the MENA region. This assessment will consider factors such as hydrogeological similarities, availability of water sources, and the suitability of the proposed methods within different contexts. By evaluating the transferability of these approaches, policymakers and stakeholders can identify opportunities for implementing similar projects in other regions facing water scarcity challenges.

17.5.4 Recommendations for scaling up streamflow recharge across the region

Based on the findings from the cross-case analysis, recommendations will be provided for scaling up SFR projects across the MENA region. These recommendations may include policy suggestions, technical guidelines, capacity building initiatives, and strategies for stakeholder engagement. The aim is to provide actionable insights and guidance for policymakers and practitioners interested in implementing and expanding SFR initiatives to enhance water resource sustainability in the region.

In conclusion, the cross-case analysis will provide a comprehensive evaluation of the hydrogeological conditions, SFR approaches, success factors, replication potential, and recommendations for scaling up SFR projects across the MENA region. By leveraging the lessons learned from the case studies, decision-makers can promote sustainable water management practices and address water scarcity challenges effectively.

17.6   Conclusions

In light of the extensive analysis of the MAR projects presented in the case studies, it is evident that innovative hydrological interventions hold significant promise in alleviating water scarcity challenges across the Middle East and North Africa (MENA) region. The successful implementation of various SFR techniques in Oman, Morocco, and Jordan underscores the potential of these natural infrastructural solutions to bolster water resources and improve socioenvironmental conditions. The case studies demonstrate that tailoring SFR techniques to local hydrogeological and societal contexts can yield substantial environmental and socioeconomic benefits. Techniques such as subsurface dams, sand dams, and improved floodwater harvesting infrastructure have proven effective in enhancing streamflow, conserving biodiversity, and supporting local livelihoods. Moreover, the adoption of a conjunctive surface−groundwater management model in Jordan exemplifies the importance of integrated approaches in optimizing water resource utilization.

Crucially, the cross-case analysis illuminates key drivers of project success. It highlights the significance of factors such as hydrogeological synergies with SFR methods, robust institutional frameworks, and active community participation. These elements are essential in ensuring the sustainability and longevity of SFR initiatives. Furthermore, the assessment of replication potential underscores the transferability of these approaches to other aquifer systems within the MENA region. By identifying shared hydrogeological characteristics and similar contexts, policymakers can leverage the successes of these case studies to address water stress in other areas.

In conclusion, the case studies and subsequent cross-case analysis provide valuable insights and recommendations for scaling up SFR projects across the MENA region. By prioritizing adaptive, community-centered approaches and leveraging existing infrastructural resources, the region can take significant strides toward mitigating water scarcity and ensuring sustainable water management for generations to come.

References

Alam, S., Borthakur, A., Ravi, S., Gebremichael, M., & Mohanty, S. K. (2021). Managed aquifer recharge implementation criteria to achieve water sustainability. *Science of The Total Environment, 768*, 144992.

Andualem, T. G., Demeke, G. G., Ahmed, I., Dar, M. A., & Yibeltal, M. (2021). Groundwater recharge estimation using empirical methods from rainfall and streamflow records. *Journal of Hydrology: Regional Studies, 37*, 100917.

Arnold, J. G., & Allen, P. M. (1999). Automated methods for estimating baseflow and ground water recharge from streamflow records 1. *JAWRA Journal of the American Water Resources Association, 35*(2), 411−424.

Döll, P., Fiedler, K., & Zhang, J. (2009). Global-scale analysis of river flow alterations due to water withdrawals and reservoirs. *Hydrology and Earth System Sciences, 13*(12), 2413−2432.

Eyitayo, S.I., Watson, M.C., Kolawole, O. (2023). Produced water management and utilization: Challenges and future directions. SPE Production & Operations 38(11), 1−16.

Falkenmark, M., & Molden, D. (2008). Wake up to realities of river basin closure. *International Journal of Water Resources Development, 24*(2), 201−215.

Falkenmark, M., & Rockström, J. (2004). *Balancing water for humans and nature: The new approach in ecohydrology.* Earthscan.

Foster, S. S. D., & Chilton, P. J. (2003). Groundwater: The processes and global significance of aquifer degradation. *Philosophical Transactions of the Royal Society of London. Series B: Biological Sciences, 358*(1440), 1957−1972.

Gaaloul, N., Eslamian, S., & Katlane, R. (2021). Tunisian experiences of traditional water harvesting, conservation, and recharge. In Eslamian, S., Eslamian, F., (Eds.). Handbook of Water Harvesting and Conservation. Vol. 2: Case studies and application examples (pp. 171−198). Wiley, New Jersey, USA.

Gleeson, T., Alley, W. M., Allen, D. M., Sophocleous, M. A., Zhou, Y., Taniguchi, M., & VanderSteen, J. (2012). Towards sustainable groundwater use: Setting long-term goals, backcasting, and managing adaptively. *Groundwater, 50*(1), 19−26.

Gleick, P. H. (1998). *The world's water 1998−1999: The biennial report on freshwater resources.* Island Press.

Gleick, P. H., & Palaniappan, M. (2010). Peak water limits to freshwater withdrawal and use. *Proceedings of the National Academy of Sciences of the United States of America, 107*(25), 11155−11162.

Haddadin, M. J. (2001). Water scarcity impacts and potential conflicts in the MENA region. *Water International, 26*(4), 460−470.

Hanjra, M. A., & Qureshi, M. E. (2010). Global water crisis and future food security in an era of climate change. *Food Policy, 35*(5), 365−377.

Hoekstra, A. Y., & Chapagain, A. K. (2007). Water footprints of nations: Water use by people as a function of their consumption pattern. *Integrated Assessment of Water Resources and Global Change: A North−South Analysis,* 35−48.

Ibrahim, B., & Mensah, H. (2017). Linking environmental water scarcity and options for adaptation in the MENA Region. *Journal of Water Resource and Protection, 9*(4), 378.

Konikow, L. F., & Kendy, E. (2005). Groundwater depletion: A global problem. *Hydrogeology Journal, 13,* 317−320.

Pescod, M. B. (1992). *Wastewater treatment and use in agriculture: FAO irrigation and drainage paper 47.* Rome: Food and Agriculture Organization of the United Nations.

Postel, S. (1999). *Pillar of sand: Can the irrigation miracle last.* WW Norton & Company.

Rosegrant, M. W., Ringler, C., & Zhu, T. (2009). Water for agriculture: maintaining food security under growing scarcity. *Annual Review of Environment and Resources, 34,* 205−222.

Sadoff, C. W., & Grey, D. (2002). Beyond the river: The benefits of cooperation on international rivers. *Water Policy, 4*(5), 389−403.

Scanlon, B. R., Jolly, I., Sophocleous, M., & Zhang, L. (2007). Global impacts of conversions from natural to agricultural ecosystems on water resources: Quantity versus quality. *Water Resources Research, 43*(3).

Shah, T. (2009). Climate change and groundwater: India's opportunities for mitigation and adaptation. *Environmental Research Letters, 4*(3), 5005.

Siebert, S., Burke, J., Faures, J. M., Frenken, K., Hoogeveen, J., Döll, P., & Portmann, F. T. (2010). Groundwater use for irrigation: A global inventory. *Hydrology and Earth System Sciences, 14*(10), 1863−1880.

Taylor, R. G., Scanlon, B., Döll, P., Rodell, M., Van Beek, R., Wada, Y., Longuevergne, L., Leblanc, M., Famiglietti, J. S., Edmunds, M., & Konikow, L. (2013). Ground water and climate change. *Nature Climate Change, 3*(4), 322−329.

Tropp, H., Jagerskog, A. (2006). Water scarcity challenges in the Middle East and North Africa (MENA). Human Development Report, pp.1−26.

Wu, H., Huang, Q., Fu, C., Song, F., Liu, J., & Li, J. (2021). Stable isotope signatures of river and lake water from Poyang Lake, China: Implications for river−lake interactions. *Journal of Hydrology, 592,* 125619.

Chapter 18

Streamflow recharge: case studies in Zayandeh Roud River, Iran

Fatemeh Dadvand, Yaser Sabzevari and Saeid Eslamian

Department of Water Sciences and Engineering, College of Agriculture, Isfahan University of Technology, Isfahan, Iran

18.1 Introduction

Rivers can be identified as the most extensive sources of freshwater that are available to human beings. The aforementioned freshwater sources play an immensely significant role in the ecology, hydrology, and environment of diverse regions. Consequently, they are constantly susceptible to the modifications under the influence of social actions and weather changes. In contrast, the organization of water resources encompasses the administration of the variability in the climate system. As per the findings of climate research, a reduction in rain and a rise in air temperature, wind speed, solar radiation, evaporation, transpiration, and dust have augmented the likelihood of the intensification or reduction of the aforementioned phenomenon. Therefore the efficient administration of the existing water resources necessitates the management of river water consumption. Multiple solutions are being contemplated to address this issue. For instance, long-term predictions of rainfall and river flow have the probability to enhance the water resource administration system (Araghinejad et al., 2006). In recent years, experiential climate data have unambiguously demonstrated a warming trend in numerous regions of the world, which has resulted in a broad spectrum of the climate impacts (Lenderink & Van Meijgaard, 2008; Haerter & Berg, 2009; Berg et al., 2013; Zhang et al., 2011).

Iran is situated in the dry and semidry belt of the Earth, a region that is characterized by notable climatic anomalies. As an effect of organization issues, Iran has been plagued by various calamities reaching from the shrinking of an extensive number of lakes and rivers to land subsiding, floods, and droughts. It is noteworthy that Iran's annual precipitation level is significantly lower than the global average, while its evaporation rate is remarkably higher. Precipitation levels in Iran vary from as low as 50 mm in the eastern and central regions to as high as 1730 mm on the southwestern shores of the Caspian Sea (Kaboli et al., 2021). The spatiotemporal analysis of precipitation in Iran indicates a transformation in the characteristics of precipitation throughout the country (Alizadeh, 2011). The western and northwestern regions experience the highest amount of precipitation, while the central regions, east, and southwest record less rainfall. The trend analysis reveals that the incidence of drought has augmented in most parts of Iran, although it has been significantly reduced in some stations in the northern part of the country. In addition, the results indicate that Iran's rainfall patterns adopt an ascending cluster model, which exhibits the positive spatial autocorrelation in some areas and negative spatial autocorrelation in others (Ghaedi., 2021; Saemian et al., 2021).

Iran, a country with an annual rainfall that is merely one-third of the world's average, faces a challenging situation due to its evaporation rate, which is approximately thrice the world's average. This country is recognized as the foremost groundwater miner in the world. Unfortunately, between the years 2015 and 2018, nearly six major floods took place in areas that were not anticipated, situated in arid and semiarid regions of the country. Note, Iran is segmented into six main hydrological basins, namely, the Central Plateau, the Persian Gulf, and Oman Sea, the Caspian Sea, Urmia, Qara Qom, and the eastern border basin. These basins are then subdivided into 30 primary basins with varying sub-basins. Among these basins, the main basin of Gavkhoni, with its sub-basin of Zayandeh Roud, holds a particular significance. It is found in the central plateau of Iran's basin (Torfe et al., 2017), and the Zayandeh Roud, is the primary water source in the middle of Iran's plateau. It plays a vital role in this hot and dry region, and it is a crucial river in the center of Iran that supports the development of agriculture, national water source, and general financial activities of Isfahan province.

Hydrosystem Restoration Handbook. DOI: https://doi.org/10.1016/B978-0-443-29802-8.00018-2

The basin in question has experienced a surge in immigration, largely due to favorable job opportunities resulting from the development of industry, accessibility of land and water resources related to adjacent areas, and tall populace growth rates, which have been driven by industrial activities and immigration from surrounding regions. Consequently, there has been an increase in demand for water and competition within the basin. Improper management and climate change have also put the Zayandeh Roud River at risk of disappearing. The scarcity of water not only results in difficulties in water supply but also causes soil uniformity. Additionally, the agrarian segment, particularly in the downstream regions of the Zayandeh Roud River sink, influences the quality of water in the river. A potential solution to address this issue is the analysis of water levels in farms, which is crucial for the proper management and expansion of water resources. Given the limited availability of freshwater resources and multiple uses of existing resources, it is also vital to consider reappearance movements resulting from seepage, infiltration, and surface runoff, which have usually been perceived as "losses" at the farm and system level (Shafiee & Safamehr, 2011).

As per the definition provided by the US Environmental Protection Agency, a stream that flows during specific times of the year is referred to as a seasonal stream (Berg et al., 2013). Additionally, the US Geological Survey defines a permanent stream as one that is always present. This can be corroborated by Moreno (1990). In the catchment area of the Central Plateau of Iran, the Zayandeh Roud River and Gavkhoni International Wetland are considered to be the most significant river and wetlands, respectively. These bodies of water possess a permanent irrigation regime, as noted by Ventra and Clarke (2018) as well as Wilber (2014). However, the River discharge, located within the study subgroups, has undergone a transition from permanent to seasonal as a result of human influence (Enteshari et al., 2020, 2014). The continuous development of cities, agricultural sectors and factories in the vicinity of Zayandeh Rood is an important issue that must be paid attention to. The basin of this river has been divided into different study areas, which include from the Plasjan basin to the Saman basin. Since then, it has become seasonal under the Koshkroud and Morghab basins, as illustrated by research conducted by Hajian and Hajian burden on local water directors to cater to the needs of the agricultural sector, which eats more than 73% of the available water resources. The basin supplied by the surface reservoir has an average annual requirement of agricultural water that amounts to 1088 mm^3, with a standard deviation of 73 mm^3. Moreover, there is an annual use of 416 mm^3 for national and manufacturing purposes in the basin. During droughts, water tank provision is solely designated for agrarian needs, as emphasized by Araghinejad (2011).

The Zayandeh Roud River is a geographical feature that spans approximately 350 km from the Zagros Mountain range's western region near Isfahan city, flowing eastward to its endpoint at the Gavkhoni swamp. However, it is crucial to note that the water levels of the Zayandeh Roud River have fluctuated throughout different seasons and times. During late winter and early spring, this river frequently experiences flooding, with the maximum instantaneous discharge in Varzane bridge station in April 1949 reaching 400 m^3 s^{-1}. It is noteworthy that Zayandeh Roud and its estuary, Gavkhoni, underwent a natural process until 1954. As a result, the river has historically flowed up to the town of Zarinshahr in Isfahan's provinces. However, since 2006, it has lost its permanent flow along the entire route. The primary reasons for the drying up of Zayandeh Roud are the several decades of drought in the upper reaches and the Zagros Mountains. The height of snow in springs such as ZardKouh and Chelgerd, which used to exceed 2 m even in the summer, decreased to less than 1 m in the spring of 2016. In addition to the decline in the discharge of the sources of Zayandeh Roud, the discharge of tunnels that transfer Karun water to Zayandeh Roud has also been reduced. However, one of the primary reasons for this river's drought is human intervention in the minor scopes of the Zayandeh Roud River. These interventions and mismanagement in the Zayandeh Roud basin include changing 180,000 ha of resource fallows into gardens in the upper reaches of the Zayandeh Roud River in the Chaharmahal and Bakhtiari province and Fereidan region. Also, structural dams and reservoirs for the expansion of agriculture are located upstream of the watershed on the water transfer route. Improper cultivation patterns in the downstream, such as rice cultivation in Isfahan province and the cities of Lanjan, Zarin Shahr and Bagh Bahadran, have caused the Zayandeh River to dry up for long periods of time (Saedpanah et al., 2021).

Between the years 2000 and 2010, it has been observed that a staggering average of 2 billion square meters of water was utilized within the Zayandeh Roud catchment for a multitude of purposes, including farming, trade, drinking water supply, and transfer to neighboring provinces. This amount of water usage exceeded the available water supply from both the dam and groundwater. Furthermore, the condition is compounded by the fact that there have been significant variations in the water output from the Zayandeh Roud tank, which has varied from 533 to 1720 million cubic meters per year, as highlighted in the research conducted by Molle et al. (2009).

Given the aforementioned circumstances, it is possible to comprehend the sheer importance and the corresponding position of the Zayandeh Roud River as an ecological, economic, and environmental resource. Therefore, it is imperative to undertake a multitude of studies concerning replenishment of the flow of this river. Additionally, it is necessary to identify the root causes for the damage and provide effective solutions in this field. To that end, this research delves into various studies, including books, journal articles, and conferences that have been held in the area, and thoroughly reviews and discusses their findings.

Given the importance of the Zayandeh Roud, it is imperative to monitor its hydrological changes in the region. Hence, this study aims to examine the variations in the flow of the Zayandeh Roud basin.

18.2 Study area

The Zayandeh Roud basin, situated in the center of Iran, holds immense significance due to its vast expanse of around 26,917 km². The river system, being the primary river of Central Iran, poses a strategic and intricate area concerning water capitals (Gohari et al., 2013b). The geographical boundaries of the Zayandeh Roud basin stretch from Namak Lake and Kavir Siah watersheds in the east, to the Dez and Karun watersheds in the west and southwest, and to the Shahreza watershed in the south. The highest point of the basin is situated at an altitude of 3974 m above sea level, while the lowest altitude of the basin is found in the Gavakhuni swamp, standing at an elevation of 1450 m above sea level. The river emerges from an altitude of 4221 m above sea level in the Chaharmahal and Bakhtiari province.

The Kouhrang tunnel in the Chaharmahal and Bakhtiari province serve as the primary source of the Zayandeh Roud. Furthermore, the river system receives the contribution of the Plasjan river, Langan, and Morghab from Isfahan province, eventually flowing into Isfahan province. Several measures have been implemented to transfer Kouhrang water to the central plateau of Iran, with the first tunnel in the Kouhrang area being noteworthy. The average discharge of 1400 MCM, comprising 650 MCM of natural flow and 750 MCM of interbasin transfer flow, has made the Zayandeh Roud River a vital source of fulfilling diverse consumer needs. The basin has observed swift socioeconomic progress and growth, leading to an increase in population around the river in recent decades (Madani and Mariño, 2009; Gohari et al., 2017).

The ecological, economic, and environmental significance of the Zayandeh Roud River cannot be understated. Therefore, it is imperative to conduct extensive research to study the feeding of the river flow, identify the causes of damage, and provide sustainable solutions for the future. This research encompasses an array of studies, including books, journal articles, and conferences held in this particular domain, that are explored and deliberated upon.

Numerous initiatives, both completed and ongoing, have been undertaken or are currently being examined to transfer water from the extensive Karun catchment. As per the approvals received during the fourth meeting of the Coordinative Body for the Combined Administration of Zayandeh Roud Water Resources, the resources and utilization of the Zayandeh Roud watershed have been allocated in the following manner: 176 MCM shall be dedicated to the environment, 419 and 655 MCM for rainfall of Isfahan, 30 MCM for drinking purposes, 32 MCM for industrial usage, and 237 MCM from Kouhrang 1 and 237 MCM from Kouhrang 2 shall be assigned to the Chaharmahal and Bakhtiari province. Additionally, 404 MCM will be allocated from Kouhrang 3 for drinking purposes in Isfahan, 49 MCM from Langan spring and Khadangestan for drinking purposes in Kashan, and 98 MCM from Behesht Abad will be dedicated to drinking purposes in the Yazd province.

The Zayandeh Roud River is heavily reliant on the yearly accumulation of snow in the Zagros Mountains, making it exceedingly vulnerable to fluctuations in the climate. Long-term climatic data reveal that during the winter season, temperature averages of 0.4°C are typically observed in high-altitude areas, whereas the eastern lowland regions experience average temperatures of 11°C. Moreover, during the summer season, temperature averages exceeding 7°C and 24°C have been recorded in the western and eastern regions, respectively. Temperature exhibits considerable variation across the year, with July experiencing a peak of 35°C and January witnessing a dip to 5°C. These climatic patterns have significant implications for various ecological and socioeconomic systems and warrant further investigation (Fig. 18.1).

The Zayandeh Roud Dam, which is situated in the Zayandeh Roud basin, is the solitary dam of significant magnitude in the region. According to the research conducted by Norouzi and Mohammadi (2016), it was determined that the volumetric capacity of the water resources contained within the Zayandeh Roud Dam was 2500 MCM (Fig. 18.2).

18.2.1 Upstream and downstream Zayandeh Roud Dam

The region surrounding the Upstream Zayandeh Roud Dam is unique in its climatical and hydrological situations when compared to extra areas within the study area. This is primarily due to its high altitude, which results in snowfall and

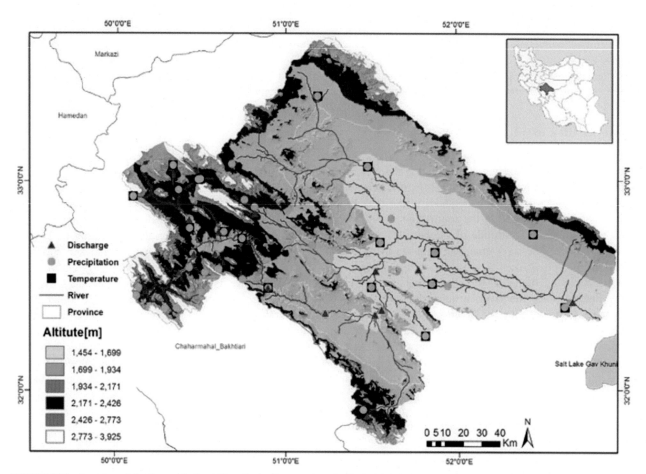

FIGURE 18.1 Location of the Zayandeh Roud River basin (Faramarzi et al., 2017). Map lines delineate study areas and do not necessarily depict accepted national boundaries.

gathering during the fall and winter seasons, as well as snow melt during springtime. The precipitation in high-altitude areas is approximately 400 mm. The total drainage area for this region is roughly 4100 km², and it serves as the main source of water source for the Zayandeh Roud basin. The flow regime and water yield are significantly impacted by water transported from Karun and Dez basins. Kouhrang Channels (Tunnel 1 and Tunnel 2) were constructed in 1950, and they redirect some of the Kouhrang's water toward the Zayandeh Roud River. The annual yield of these water transfer projects ranges from 250 to 300 MCM.

In the vicinity situated downstream of the Zayandeh Roud Dam, the flow regime is subjected to stringent regulation owing to the presence of the Chadegan Dam and Zayandeh Roud Regulatory Dam Station and subsequently affected by diverse water diversion schemes. These schemes encompass contemporary and conventional irrigation networks, diversion dams along the river, and water collection wells that are distributed along the river banks, exemplified by Felman wells. The natural climatic conditions in this area exhibit a variation ranging from 300 mm during the initial year close to the Chadegan Dam to below 75 mm during the first year near the Gavkhoni swamp.

18.3 Overview of the research done on stream feeding in the Zayandeh Roud basin

Marani-Barzani et al. (2017) assess aridity in the Zayandeh Roud basin using a geographical information system (GIS) and three aridity models. In this chapter, GIS was applied for the valuation of the Zayandeh Roud basin using climatical information composed from 11 stations located in the basin.

Molden and Sakthivadivel (1999) put forward a comprehensive framework for water organization within the upstream sub-basin of the Zayandeh Roud basin. This framework categorizes the various components of water balance into distinct water use categories, each of which reflects the different consequences arising from human interferences in

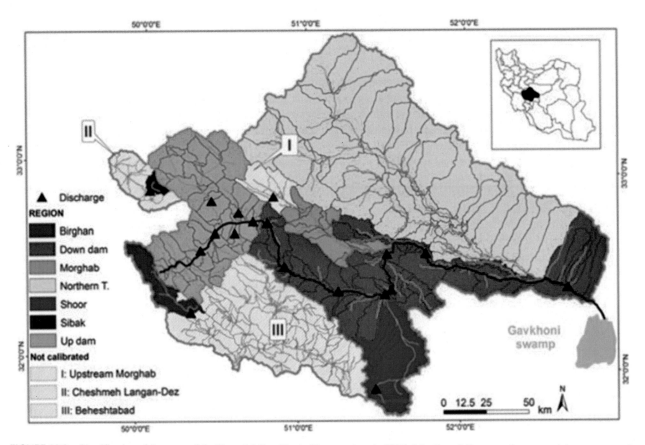

FIGURE 18.2 Classification of the areas of the Zayandeh Roud basin (Faramarzi et al., 2017). Map lines delineate study areas and do not necessarily depict accepted national boundaries.

the hydrological set. The catchment area under consideration is dominated by irrigation, with this particular activity accounting for almost 90% of the surface and groundwater resources. It is a natural phenomenon that in years of low rainfall and arid conditions, the demand for water in agriculture increases significantly.

Abou Zaki et al. (2020) researched and analyzed the consequences of alterations in land use and the incidence of metrological and hydrological droughts through the use of groundwater information sourced from 30 shafts, the standardized precipitation index, and the river flow drought index. The data and changes in the wetland were assessed using the normalized difference water index values and the reduction in water mass in the basin with gravity recovery and data from the weather trial gravity recovery and climate experiment (GRACE). The outcomes of the study revealed that hydrological droughts escalated after the implementation of large irrigation projects in the Zayandeh Roud basin. The depletion of groundwater and drought in the river were shown to be related to human activities.

Nazemi et al. (2019) conducted a complete investigation of various agricultural water governance scenarios. Specifically, they focused on the Zayandeh Roud basin in Iran. To identify crucial adaptation strategies, an examining and seminal scenario planning approach was utilized. From these strategies, a small set of scenarios that are coherent, acceptable, and diverse was developed. The authors also analyzed forthcoming scenarios of the Zayandeh Roud watershed in Iran in 2040.

Raber (2017) considered several key parameters to develop an agricultural transformation strategy for the Zayandeh Roud watershed. These included the place and amount of educated and irrigated areas, types of crops and orchards, irrigation water sources and applied irrigation methods, crop calendar, and specific crop data for calculating crop water requirements. The collected data was analyzed to provide the basis for the aforementioned strategy.

Kaltofen et al. (2017) employed a choice support tool for combined water resources running to demonstrate the complex use and organization of processed water in the Zayandeh Roud basin. Their objective was to assist in the water organization procedure. The authors utilized the MIKE basin to model consumption and organization procedures for water in the Zayandeh Roud basin.

Torfe et al. (2017) proposed a new water supply plan for the Zayandeh Roud watershed in the central plateau of Iran. However, the proposed plan is founded on a petition known as Sheikh Bahai. The findings of their study indicate that the Zayandeh Roud watershed has a surging demand for water. Thus, the traditional water distribution system requires fundamental amendments.

Ababei and Sohrabi (2009) evaluated the efficacy of the Soil and Water Assessment Tool (SWAT) within the Zayandeh Roud watershed, with the aim of simulating river discharge values. Their results reveal that SWAT can serve as a suitable tool for simulating river discharge values. Additionally, the study affirms that the SWAT model is adept at simulating river discharge values, with acceptable performance based on statistical analysis.

Faramarzi et al. (2017) employed SWAT along with the Sequential Uncertainty Fitting version 2 (SUFI2) program to calibrate and validate a hydrological model of the Zayandeh Roud watershed. The model was calibrated at the sub-basin level, with a monthly time step, and an uncertainty vein analysis was conducted for explicit quantification of the hydrological mechanisms of water resources. The study utilized these techniques to assess the potential of the Zayandeh Roud watershed for water availability.

Similarly, Modares and Eslamian (2006) discuss monthly flow modeling within the Zayandeh Roud River, Iran, using multiplicative seasonal autoregressive combined affecting normal models. The study concluded that the multiplicative seasonal autoregressive integrated moving average (ARIMA) model is appropriate for flow modeling, with the capability to predict and forecast streamflow.

Shafiee and Safamehr (2011) investigated the water resource system of Zayandeh Roud dam sediments in Isfahan province, Iran. The study evaluated various aspects of the Zayandeh Roud dam lake exploitation and the sediment—water resource system using increasing and decreasing area methods. The research results revealed that the capacity had decreased to 150 MCM due to drought in recent years. This occurrence had a detrimental influence on various water resources in the region, including the drying up of the Goukhoni swamp and reduced access to agricultural and industrial water.

Bahreinimotlagh et al. (2019) conducted incessant nursing of flow in common watersheds through the use of a progressive underwater aural tomography system with a case study of the Zayandeh Roud River. The study proposed a new equation for the precise selection of accurate angles for the width of different rivers, thereby reducing the measurement error of FAT, a method for quantitative flow measurement. The research findings indicated that the maximum possible error of FAT was less than 15%.

Safavi et al. (2015) conducted a study on expert-based knowledge modeling for the planning and integrated management of water resources in the Zayandeh Roud River basin. The research led to the construction of an integrated water resources management (IWRM) planning model that is feasible with limited data availability. Furthermore, the Water Evaluation and Planning (WEAP) model of Zayandeh Roud accurately predicted the future performance of the water supply.

Byzedi et al. (2013) performed an analysis of river flow during a drought using the cut level method (at 70% level) with daily flow in 54 stations in the southwest of Iran. The research findings indicated that the area of the watershed was the most significant factor that had a high correlation with the amount of drought deficiency.

In the research conducted by Zareian (2021), the optimal allocation of water at varying levels of climate change was analyzed to minimize water shortage in dry areas, specifically in the case study of the Zayandeh Roud basin in Iran. The analysis showed a reduction of 21%—38% in the volume of input to the Zayandeh Roud Reservoir, which was attributed to climate change predictions on the river.

A geostatistics-based method with a resident reversion method was proposed by Araghinejad et al. (2006) to forecast seasonal streamflow using ocean—atmosphere signals and basin hydrological conditions as predictors. The GBPF method, which incorporates nonlinear relationships and multiple predictors, allows for long-term streamflow forecasts to be made.

Saedpanah et al. (2021) investigated the relationship between land use variations and water quality within the Zayandeh Roud basin in Isfahan, Iran, at the spatial scale of the entire basin, sub-basin, and defined buffers (10 and 15 km). The study found that land use changes influence water quality parameters, with the sub-basin scale showing the strongest correlation between land use parameters and water quality.

The financial effects of weather variation on water resources and farming of the Zayandeh Roud basin in Iran were investigated by Aghapour Sabbaghi et al. (2020) for the years 2040 and 2070 in three main phases. In the first phase, two universal flow models (HadCM2 and CGCM3T63), an ANN, and physical models were utilized to investigate the dual impact of water resources.

Framework elements investigation has been utilized to ponder and oversee the Zayandeh Roud Waterway Bowl in Iran. The bowl is characterized by complex intelligence between physical, social, financial, and political subsystems. The need for total information around these subsystems has driven disappointments in tending to water deficiencies within the bowl. Madani and Mariño (2009) with a stude, built based on causal circle graphs of the issue, show that transbasin diversion is not efficient alone and populace control can be more compelling in tending to the water emergency of the Zayandeh Roud Stream Bowl (ZRB). Framework flow recreation permits assessment of territorial arrangements. Strategists can make choices to amend the current framework.

Weather variation is predictable to worsen the water system's challenges, and adaptation strategies are needed to minimize its impacts. Supply-oriented plans alone are not effective, but together with water claim organizations, they can alleviate weather change-related water stress.

Alinezhad et al. (2019) utilized a framework flow demonstration to assess climate alter adjustment procedures for Iran's Zayandeh Roud water framework. In this chapter, a probabilistic multimodel outfit situation is utilized to characterize instabilities in climate alter projections for the ponder period (2015−44), and the Zayandeh Roud Watershed Administration and Supportability Show is run beneath an outfit situation with different instability levels to assess the impacts of climate alter on the Zayandeh Roud water system and to recognize successful adjustment methodologies to play down these effects. Gavkhoni Bog will be seriously corrupted without natural flows. Supply-arranged methodologies alone are not successful. Stakeholder analysis and socially organized investigation have been utilized to get control elements within the stream basin's administration framework. The results showed that partners with the same control and intrigued may play distinctive parts, highlighting the significance of control relations in administration forms.

Enteshari et al. (2020) showed that the grounded theory approach has also been used to develop dynamic hypotheses for the basin, addressing challenges in system dynamics modeling and identifying key problems in the natural and human systems. The proposed strategy must have demonstrated palatable in appearing most issues of the natural and human activities within the Zayandeh Roud.

−   Grounded theory approach effectively addresses challenges in system dynamics modeling.
−   Main problems in the Zayandeh Roud basin were identified: population growth, water consumption, industrial/agricultural growth, unfair water allocation, and unemployment.

In this chapter, the authors provided an overview of the hydrology and water use in the Zayandeh Roud basin based on the data available over the 11 years from 1988 to 1998. Frequent entry and discharge in the Zayandeh Rood basin worsens the effects of drought despite the limited storage in this basin.

Grundmann et al. (2020) studied Participatory Development of Strategies for the Transformation of Agriculture in the Zayandeh Roud River basin. The authors discussed the combined management of water and land resources in Zayandeh Roud basin.

During this study, the challenges of water organization in the Zayandeh Roud basin and unpredictable irrigation water sources for agriculture were investigated.

Zareian (2021) investigated ideal water allotment at diverse levels of climate alter to play down water deficiency in bone-dry locales in ZRB, Iran, which predicts a diminish in influx volume to the Zayandeh Roud Store by 21% to 38% due to climate alteration, which can be diminished with water request control.

The Zayandeh Roud River basin in Iran is facing water stress and limited water source for agrarian creation. The basin has experienced water scarcity for the past 50 years, and the current water supply falls behind the growing demand in the urban, industrial, and agricultural sectors. Zamani et al. (2019) investigated other situations of timing and summary water supply in terms of their influences on agricultural land use, farm revenue, and food crop production in the Zayandeh Roud River basin in central Iran. The study investigates the impacts of limiting and timing water supply on agricultural production in the Zayandeh Roud River basin in Iran, Timing of water supply affects agricultural production, and limiting water supply in peak demand months is less adverse.

The Zayandeh Roud watershed was modeled utilizing the SWAT to mimic streamflow information for water asset administration by Faramarzi et al. (2017). In that study, the soil and water evaluation apparatus (SWAT) was utilized in combination with the SUFI2 program to calibrate and approve a hydrologic demonstration of the Zayandeh Roud watershed at the sub-basin level with vulnerability examination to unequivocally measure hydrological components of water assets on a month to month time step. The calibration moved forward bowl wide bR2 from 0.27 to 0.44 in upstream locale and from 0.39 to 0.46 in downstream catchment. The calibrated SWAT was utilized to create every day naturalized stream information for water administration.

A hybrid data-driven model for hydrological estimation has been developed in several studies. The models combine different techniques such as ANNs, ensemble empirical mode decomposition, and explainable artificial intelligence to improve the accuracy of streamflow forecasting. These hybrid models have shown better performance compared to traditional methods and single models.

Araghinejad et al. (2018) evaluated the performance of SWAT in the Zayandeh Roud watershed to act out river stream amount ethics, and the outcomes presented that SWAT could be a good instrument for simulating the stream amount value of the river.

The Zayandeh Roud basin is the focus of a groundwater model using FEFLOW software to describe seepage water volume to groundwater levels.

Javadinejad et al. (2019) utilized the water assessment and arranging (WEAP) program (Multi-Attribute Choice Investigation Show) for arranging water assets and administration at the watershed level, considering the multidimensionality of water asset administration approaches and prioritization of water requests.

Aghapour Sabbaghi et al. (2020) showed that climate alterations can have critical impacts on horticulture and water assets, as demonstrated in the Zayandeh Roud waterway bowl in Iran. This study presents a neural network (ANN) of rainfall-runoff, and crop water production functions were linked to evaluate the biophysical effects of water assets and different modified yields.

18.4 Summary of results obtained from past studies

Streamflow recharge case studies in the Zayandeh Roud River were conducted to evaluate the dynamic behavior of flow time series and the effect of time scale on flow rate. The chaotic behavior of the flow rate was investigated using the correlation dimension and the Lyapunov exponent test, indicating a chaotic flow rate on a daily scale at multiple hydrometric stations. Additionally, the role of recharge and catchment storage in streamflow drought sensitivity was examined using recharge stress tests. The tests revealed varying responses across catchments, suggesting different storage properties and recovery times from drought. Land use changes in the Zayandeh Roud basin were found to significantly impact hydrology and water quality, with an increase in average runoff volume and depth over time. The results can be used to monitor changes in land use and control the depth and volume of runoff. These studies provide insights into streamflow forecasting and the management of low flows during dry periods, highlighting the potential of managed aquifer recharge to augment streamflow.

The Zayanderud Basin in Iran is facing significant impacts from irrigation and drought. Irrigation accounts for completed 90% of water customs in the basin, leading to a depletion of groundwater and a diminution in the flow rate of the Zayanderud River (Abou Zaki et al., 2020). This has resulted in hydrological droughts and the drying of the Gavkhoni Wetland, which receives the river water. The depletion of groundwater and the occurrence of hydrological droughts are directly linked to human activities (Molle et al., 2008). The river basin has also experienced an increase in contamination during drought periods, as wastewater and effluents from various sources are discharged into the river. The vulnerability of the Zayanderud Basin to drought has been assessed using multiple indicators, including precipitation, water demand, and groundwater balance. Overall, the impacts of irrigation and drought on the river, groundwater, and wetland highlight the need for sustainable water management in the basin (Fig. 18.3).

SWAT has been used in the Zayandeh Roud watershed for various purposes such as calibrating and validating hydrologic models, simulating natural historical and future streamflow data, and assessing the effects of fertilizers on agricultural pollutants in the river (Javadinejad et al., 2019; Faramarzi et al., 2017; Ahmadzadeh et al., 2016). The SWAT model has been applied to measure hydrological components of water resources, evaluate the influence of changing irrigation systems on water productivity, and simulate the flow rate values of the river (Ababei & Sohrabi, 2009). The model has been calibrated and validated using discharge data from hydrometric stations, and its performance in simulating the hydrologic cycle and river flow rate values is acceptable. The results of these studies have provided valuable insights for water resource management, water allocation, and improving environmental conditions in the Zayandeh Roud watershed.

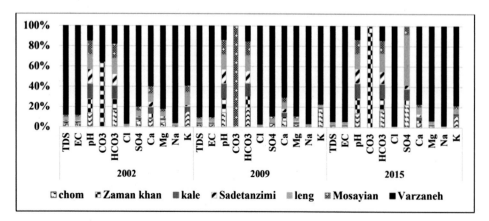

FIGURE 18.3 Surface water quality trend in the Zayandeh-Rud basin.

18.5 Conclusion

This study, by reviewing the various studies conducted in the Zayandeh Roud basin, monitored the status of water resources in this basin. For this purpose, various sources were used, including conference articles, journals, the Internet, and books. The outcomes of this study indicated a change in the climatic parameters of the basin, especially temperature and precipitation. Therefore the annual temperature in the east of the basin has increased and the rainfall in the west of the basin has decreased. Hydrological studies showed the deterioration of the Zayandeh Roud basin with the passage of time. Its effects can be seen in the region today. For example, the water quality of this basin changes over time in such a way that the highest amount of water pollution is in the summer season and the lowest amount is in the autumn season. One of the most important reasons for these changes is the growth of industries and less attention to the environment and the increase of greenhouse gases in this basin.

References

Ababei B., Sohrabi, T. (2009) Assessing the performance of SWAT model in Zayandeh-Roud Watershed. Journal of Soil and Water Conservation Researches 16(3):41−58. [in Persian]

Abou Zaki, N., Torabi Haghighi, A., Rossi, P. M., Tourian, M. J., Bakhshaee, A., & Kløve, B. (2020). Evaluating impacts of irrigation and drought on river, groundwater and a terminal wetland in the Ayanderud Basin, Iran. *Water, 12*(5), 1302. Available from https://doi.org/10.3390/w12051302.

Aghapour Sabbaghi, M., Mohammadreza, N., Araghinejad, S., & Soufizadeh, S. (2020). Economic impacts of climate change on water resources and agriculture in Zayandeh Roud river basin in Iran. *Agricultural Water Management, 241*, 106323. Available from https://doi.org/10.1016/J.AGWAT.2020.106323, [in Persian].

Ahmadzadeh, H., Morid, S., Delavar, M., & Srinivasan, R. (2016). Using the SWAT model to assess the impacts of changing irrigation from surface to pressurized systems on water productivity and water saving in the Zarrineh Rud catchment. *Agricultural Water Management, 175*, 15−28. Available from https://doi.org/10.1016/j.agwat.2015.10.026.

Alinezhad, A., Gohari, A., Eslamian, S., & Saberi, Z. (2019). Uncertainty Analysis of Climate Change Impacts on Streamflow Extremes in Zayandeh-Rud River by Bayesian Model Averaging. *jwss, 23*(4), 393−407. Available from http://jstnar.iut.ac.ir/article-1-3882-fa.html.

Alizadeh, A. (2011). *Principles of Applied hydrology*. Mashhad: Mashhad Ferdowsi University Press.

Araghinejad, S. (2011). An approach for probabilistic hydrological drought forecasting. *Water Resources Management, 25*, 191−200.

Araghinejad, S., Burn, D. H., & Karamouz, M. (2006). Long-lead probabilistic forecasting of streamflow using ocean-atmospheric and hydrological predictors. *Water Resources Research, 42*(3).

Araghinejad, S., Fayaz, N., & Hosseini-Moghari, S. M. (2018). Development of a hybrid data driven model for hydrological estimation. *Water Resources Management, 32*, 3737−3750.

Bahreinimotlagh, M., Kawanisi, K., Al Sawaf, M. B., Roozbahani, R., Eftekhari, M., & Khoshuie, A. K. (2019). Continuous streamflow monitoring in shared watersheds using advanced underwater acoustic tomography system: A case study on Zayanderud River. *Environmental Monitoring and Assessment, 191*(11). Available from https://doi.org/10.1007/s10661-019-7830-4.

Berg, P., Moseley, C., & Haerter, J. O. (2013). Strong increase in convective precipitation in response to higher temperatures. *Nature Geoscience, 6*(3), 181−185.

Byzedi, M., Saghafian, B., Mohammadi, K., & Siosemarde, M. (2013). Regional analysis of streamflow drought: A case study in southwestern Iran. *Environmental Earth Sciences, 71*(6), 2955−2972. Available from https://doi.org/10.1007/s12665-013-2674-7.

Enteshari, S., Safavi, H. R., & van der Zaag, P. (2020). Simulating the interactions between the water and the socio-economic system in a stressed endorheic basin. *Hydrological Sciences Journal, 65*(13), 2159−2174.

Faramarzi, M., Besalatpour, A. A., & Kaltofen, M. (2017). *Application of the hydrological model SWAT in the Zayandeh Roud Catchment. Reviving the Dying Giant: Integrated Water Resource Management in the Zayandeh Rud Basin, Iran* (pp. 219−240). Springer. Available from https://doi.org/10.1007/978-3-319-54922-4_14.

Ghaedi, S. (2021). Anomalies of precipitation and drought in objectively derived climate regions of Iran. *Hungarian Geographical Bulletin, 70*(2), 163−174. Available from https://doi.org/10.15201/HUNGEOBULL.70.2.5.

Gohari, A., Eslamian, S., Mirchi, A., Abedi-Koupaei, J., Bavani, A. M., & Madani, K. (2013b). Water transfer as a solution to water shortage: A fix that can backfire. *Journal of Hydrology, 491*, 23−39.

Gohari, A., Mirchi, A., & Madani, K. (2017). System dynamics evaluation of climate change adaptation strategies for water resources management in central Iran. *Water Resources Management, 31*, 1413−1434.

Grundmann P., Mohammad Naser R., Libra J.A., Lena H., Simone K., Omid Z., Mohammad Z. (2020) Participatory development of strategies for the transformation of agriculture in the Zayandeh Roud River Basin. In Standing up to Climate Change: Creating Prospects for a Sustainable Future in Rural Iran (pp. 265−279). Springer. doi: 10.1007/978-3-030-50684-1_12

Haerter, J. O., & Berg, P. (2009). Unexpected rise in extreme precipitation caused by a shift in rain type? *Nature Geoscience, 2*(6), 372−373.

Javadinejad, S., Ostad-Ali-Askari, K., & Eslamian, S. (2019). Application of multi-index decision analysis to management scenarios considering climate change prediction in the Zayandeh Roud River Basin. *Water Conservation Science and Engineering 4(1)*. Available from https://doi.org/10.1007/s41101-019-00068-3.

Kaboli, S., Hekmatzadeh, A. A., Darabi, H., & Haghighi, A. T. (2021). Variation in physical characteristics of rainfall in Iran, determined using daily rainfall concentration index and monthly rainfall percentage index. *Theoretical and Applied Climatology, 144*, 507−520.

Kaltofen, M., Müller, F., & Zabel, A. (2017). *Application of MIKE Basin in the Zayandeh Roud Catchment. Reviving the Dying Giant: Integrated Water Resource Management in the Zayandeh Rud Basin* (pp. 253−268). Springer. Available from http://doi.org/10.1007/978-3-319-54922-4_16.

Lenderink, G., & Van Meijgaard, E. (2008). Increase in hourly precipitation extremes beyond expectations from temperature changes. *Nature Geoscience, 1*(8), 511−514.

Madani, K., & Mariño, M. A. (2009). System dynamics analysis for managing Iran's Zayandeh-Roud river basin. *Water Resources Management, 23*, 2163−2187.

Marani-Barzani, M., Eslamian, S., Amoushahi-Khouzani, M., Gandomkar, A., Rajaei-Rizi, F., Kazemi, M., & Askari1, Z. (2017). Assessment of aridity using geographical information system in Zayandeh-Roud Basin, Isfahan, Iran. *International Journal of Mining Science, 3*(2), 49−61.

Modares R., Eslamian S.S. (2006) Streamflow time series modeling of Zayandeh Roud river. Iranian Journal of Science and Technology Transaction B: Engineering 30(4), 567−570.

Molden, D. J., & Sakthivadivel, R. (1999). Water accounting to assess use and productivity of water. *International Journal of Water Resources Development [Special Double Issue: Research from the International Water Management Institute (IWMI)], 15*(1/2), 55−71.

Molle, F., Ghazi, I., & Murray-Rust, H. (2009). Buying respite: Esfahan and the Zayandeh Roud River basin, Iran. In F. Molle, & P. Wester (Eds.), *River Basin Trajectories: Societies, Environments and Development. Comprehensive Assessment of Water Management in Agriculture, Seris 8.* Wallingford: CABI.

Molle, F., Hoogesteger, J., & Mamanpoush, A. (2008). Macro- and micro-level impacts of droughts: The case of the Zayandeh Roud river basin, Iran. *Irrigation and Drainage: Journal of the International Commission on Irrigation and Drainage, 57*(2), 219−227.

Moreno, E. M. (1990). Qanat, Kariz and Khattara: Traditional water systems in the Middle East and North Africa (Book Review). *Al-Qantara, 11*(1), 286.

Nazemi, N., Foley, R. W., Louis, G., & Keeler, L. W. (2019). Divergent agricultural water governance scenarios: The case of Zayanderud basin, Iran. *Agricultural Water Management, 229*, 105921. Available from https://doi.org/10.1016/j.agwat.2019.105921.

Norouzi, A., & Mohammadi, Z. (2016). Survey of hydrological drought and its effects on lenjan Agricultural Region. *Spatial Planning, 6*(2), 97−116.

Raber, W. (2017). *Current and future agricultural water use in the Zayandeh Roud Catchment. In In Reviving the Dying Giant: Integrated Water Resource Management in the Zayandeh Roud Catchment* (pp. 81−94). Springer. Available from https://doi.org/10.1007/978-3-319-54922-4_610.1007/978-3-319-54922-4_6.

Saedpanah, M., Reisi, M., & Nadoushan, M. A. (2021). The effect of land use changes on water quality (case study: Zayandeh-Roud Basin, Isfahan, Iran). *Pollution, 7*(4), 895−904. Available from https://doi.org/10.22059/POLL.2021.324387.1100.

Saemian, P., Hosseini-Moghari, S. M., Fatehi, I., Shoarinezhad, V., Modiri, E., Tourian, M. J., & Sneeuw, N. (2021). Comprehensive evaluation of precipitation datasets over Iran. *Journal of Hydrology, 603*, 127054.

Safavi, H. R., Golmohammadi, M. H., & Sandoval-Solis, S. (2015). Expert knowledge-based modeling for integrated water resources planning and management in the Zayandeh Roud River Basin. *Journal of Hydrology, 528*, 773−789. Available from https://doi.org/10.1016/j.jhydrol.2015.07.014.

Shafiee, A. H., & Safamehr, M. (2011). Study of sediments water resources system of Zayanderud Dam through area increment and area reduction methods, Isfahan Province, Iran. *Procedia Earth and Planetary Science, 4*, 29−38. Available from https://doi.org/10.1016/j.proeps.2011.11.004.

Torfe M.A., Mirmohammad Sadeghi M., Mohajeri S. (2017) Water management in the Zayandeh Roud Basin: Past, present and future. In Reviving the Dying Giant: Integrated Water Resource Management in the Zayandeh Roud Catchment, Iran (pp. 33−47). Springer.

Ventra, D., & Clarke, L. E. (Eds.), (2018). *Geology and geomorphology of alluvial and fluvial fans: Terrestrial and planetary perspectives.* Geological Society of London, July.

Wilber, D. N. (2014). *Iran, Past and Present: From Monarchy to Islamic Republic* (Vol. 529). Princeton University Press.

Zamani, O., Grundmann, P., Libra, J. A., & Nikouei, A. (2019). Limiting and timing water supply for agricultural production: The case of the Zayandeh Roud River Basin, Iran. *Agricultural Water Management, 222*, 322−335.

Zareian, M. J. (2021). Optimal water allocation at different levels of climate change to minimize water shortage in arid regions (Case Study: Zayandeh-Roud River Basin, Iran). *Journal of Hydro-Environment Research, 35*, 13−30. Available from https://doi.org/10.1016/j.jher.2021.01.004.

Zhang, X., Alexander, L., Hegerl, G. C., Jones, P., Tank, A. K., Peterson, T. C., & Zwiers, F. W. (2011). Indices for monitoring changes in extremes based on daily temperature and precipitation data. *Wiley Interdisciplinary Reviews: Climate Change, 2*(6), 851−870.

Further reading

Alinezhad, A., Gohari, A., Eslamian, S., & Saberi, Z. (2021). A probabilistic Bayesian framework to deal with the uncertainty in hydro-climate projection of Zayandeh-Roud River Basin. *Theoretical and Applied Climatology, 144*(3−4), 847−860. Available from https://doi.org/10.1007/s00704-021-03575-3.

Alireza, M. B., & Saeed, M. (2006). The effects of climate change on the flow of Zayandeh Roud river in Isfahan. *Water and Soil Sciences (Isfahan University of Technology), 9*(4), 17−28.

Barati, A. A., Pour, M. D., & Sardooei, M. A. (2023). Water crisis in Iran: A system dynamics approach on water, energy, food, land and climate (WEFLC) nexus. *Science of the Total Environment, 882*, 163549.

Gohari, A., Eslamian, S., Mirchi, A., Abedi-Koupaei, J., Bavani, A. M., & Madani, K. (2013a). Water transfer as a solution to water shortage: A fix that can backfire. *Journal of Hydrology, 491*, 23−39.

Hajian, N., Hajian P. (2013) Databases of Zayandeh Roud. [In Persian]

hossein Shafiee, A., & Safamehr, M. (2011). Study of sediments water resources system of Zayanderud Dam through area increment and area reduction methods, Isfahan Province, Iran. *Procedia Earth and Planetary Science, 4*, 29−38.

Hosseini, A., & Seyyed, H. (2009). The relationship between Sheikh Baha'i's scroll and the traditional water distribution of Zayandeh Roud? *Geography and Environmental Studies, 1*(2), 5−14, [in Persian].

Karami, M., & Asadi, M. (2021). Investigating the inter-annual precipitation changes of Iran. *Journal of Water and Climate Change, 12*(3), 879−894. Available from https://doi.org/10.2166/WCC.2020.205, [in Persian].

Murray-Rust, H., Sally, H., Salemi, H. R., & Mamanpoush, A. (2000). An overview of the hydrology of the Zayandeh Roud Basin. *IAERI-IWMI Research Reports, 3*, 52−54.

Perry C.J. (1996) The IIMI water balance framework: A model for project level analysis. IWMI Research Report No. 5. Colombo, Sri Lanka: International Irrigation Management Institute.

Saif A. (2015) Investigation of the process of formation of successive deltas of Zayandeh Roud using remote sensing technique. In Proceedings of the 24th Earth Sciences Meeting. Organization of Geology and Mineral Explorations of the Country.

Salehian, S., & Rahmani Fazli, A. (2018). Environmental consequences of water resources instability in the Zayandeh-Roud Basin. *Physical Geography Research Quarterly, 50*(2), 391−406. Available from https://doi.org/10.22059/jphgr.2018.226191.1006997.

Salemi, H. R., & Heydari, N. (2006). Assessment of water resources and uses in Zayandeh Roud watershed [Technical report]. *Water Resources Science and Engineering Association, 2*(1).

Salemi, H. R., Mamanpoush, A., Miranzadeh, M., Akbari, M., Torabi, M., Toomanian, N., Murray-Rust, H., Droogers, P., Sally, H., & Gieske, A. (2000). Water management for sustainable irrigated agriculture in the Zayandeh Roud basin, Esfahan Province, Iran. *IAERI-EARC-IWMI Research Report, 1*.

Seyed Ghasemi S. (2006) Prediction of river flow changes under the influence of climate change (case study: Zaindeh Roud basin, Master's thesis), Faculty of Civil Engineering, Sharif University of Technology.

Sorkhabi O.M. (2021). Spatial and temporal analysis of Iran precipitation. doi: 10.21203/RS.3.RS-509123/V2. [in Persian].

Verhram, G. (1991). Historical geography of Zayandeh Roud. *Geographical Research, 17*, 124−141.

Chapter 19

Community-based conservation initiatives for urban lakes: case studies from Bengaluru city, India

Dipak Mandal and S. Manasi

Centre for Research in Urban Affairs (CRUA), Institute for Social and Economic Change (ISEC), Bengaluru, Karnataka, India

19.1   Introduction

The rapid urbanization of Bengaluru, driven by a thriving Information Technology (IT) industry and increasing migration, poses a significant and ongoing threat to the city's lakes. Once renowned for its numerous water bodies, Bengaluru has witnessed the gradual reduction of lakes due to encroachments, illegal dumping, and unplanned construction. This loss of natural water resources has led to severe water scarcity, disrupted ecosystems, and heightened vulnerability to urban flooding during heavy rainfall. Despite efforts to restore and protect these lakes, balancing urban development and environmental preservation remains a pressing challenge for the city's authorities and environmentalists (Ahmadi & Eslamian, 2022).

The lakes in Bengaluru have a rich history that dates back centuries. Initially developed as irrigation tanks during the reign of the Kempegowda dynasty in the 16th century, these water bodies served as water storage and supply for agriculture and drinking. Over time, they became an integral part of the city's landscape, contributing to its charm and becoming essential landmarks.

However, the rapid urbanization and population growth experienced by Bengaluru in recent decades has posed significant challenges to these lakes. Encroachment, pollution, and untreated sewage have threatened their ecological balance and deteriorated water quality. Some lakes have even faced severe issues such as frothing, algal blooms, and diminishing water levels, highlighting the need for urgent conservation efforts.

Source: Survey of India, Topographic Maps (published in 1973) and Ramachandra and Kumar (2008).

Recognizing the importance of these lakes as ecological assets, ongoing initiatives have been to restore and rejuvenate them. The government, various environmental organizations, and citizen groups have been working toward reviving these water bodies by desilting, installing sewage treatment plants (STPs), and afforestation around the lake perimeters. Such efforts aim to restore the lakes' ecological balance, enhance their biodiversity, and provide sustainable water resources for the city.

In addition to their ecological significance, Bengaluru's lakes offer recreational opportunities and act as important cultural and social spaces. People visit these lakes for boating, birdwatching, jogging, and leisure time amidst nature. Many lakes have walking paths, parks, and gardens built around them, attracting locals and tourists alike. The Bengaluru lakes are essential to the city's heritage, environment, and community. Efforts to conserve and restore these water bodies are crucial for their ecological benefits and for preserving Bengaluru's identity as the "City of Lakes."

The study explores the involvement of the community in environmental activism and their efforts to restore and conserve lakes in Bengaluru city. The chapter discusses community engagement activities, mobilization, network building, and the challenges faced during the conservation initiative. However, it is important to note that this study only focuses on the community-based conservation efforts in Bengaluru city, and the context and evolution of such initiatives in other cities may differ significantly.

Hydrosystem Restoration Handbook. DOI: https://doi.org/10.1016/B978-0-443-29802-8.00019-4

19.1.1   Report of Lakshmana Rau Committee, 1986

In 1985, the Government of Karnataka (GoK) formed an expert Committee led by Sri N. Lakshmana Rau to address preserving and restoring city lakes in Bengaluru. The committee thoroughly examined the challenges of maintaining the lake environment and drafted a comprehensive report within three months. The report focused on preserving existing water bodies, preventing pollution, desilting, evicting encroachments, and promoting aquatic life. It also recommended the creation of tree parks in breached tanks, transferring certain areas to the Forest Department for tree planting, and forming an implementation agency to oversee the process. Measures to control mosquitoes and improve the groundwater table were also suggested (Rau, 1986). Despite the government accepting the report on February 11, 1988, the implementation agency was not constituted to review and enforce the recommendations periodically, leading to delays in executing the proposed measures (Patil, 2011).

19.2   Status of lakes in Bengaluru

In 1963, Bengaluru's area was only 112 km^2, and the city's population was 1.29 million (Sudhira, 2008). However, both the area and population rapidly increased. In 2011, the city population was 8.5 million, and the area was 800 km^2. If the current population growth remains unchanged, the population will increase to 20.3 million by 2031 (Bangalore Development Authority, 2017). The city is the fifth-largest urban agglomeration in India and has also achieved the country's highest population growth rate (42%) from 2001 to 2011. As per the *Bangalore Development Authority (BDA) Revised Master Plan, 2031*, the city comprises around 14.60% of the state's total population, with a land share of 0.64%.

The rapid pace of urbanization has profoundly impacted the city's environment. A study conducted by the faculty of the Indian Institute of Science (IISc), Bengaluru, has revealed that over the past four decades, the urban area has expanded by a staggering 925%, leading to a reduction in green cover by 78% and water bodies by 79%. This underscores the haphazard urban growth and planning that have contributed to the degradation of the city's ecology and environment (Ramachandra et al., 2017). The encroachment by construction companies, private enterprises, and the government is a primary factor in the city's diminishing number of lakes and water bodies. Once renowned for its many lakes, Bengaluru was deeply interconnected with them for its drinking water, irrigation, and fishing requirements.

The city lakes have a significant positive impact on the ecology and microclimate. However, the lakes' present condition is deplorable. Most lakes are polluted, encroached, or filled with garbage, and the water storage volume in the lake has significantly decreased. Many lakes are converted to bus stands, government apartments, playgrounds, golf courses, residential apartments, and dumping yards. Only about 17 healthy lakes exist today in the city (Murali, 2016).

In Bengaluru, there are no accurate statistics on lakes and water bodies, so the number of lakes varies department-wise (Thippaiah, 2009). Also, managing and protecting lakes is a significant problem. A study by Thippaiah on lakes in Bengaluru mentioned that the number of lakes varies because of the boundary limits among the different agencies like BDA, Bruhat Bengaluru Mahanagara Palike (BBMP), and Bangalore Metropolitan Regional Authority jurisdictions. According to the Minor Irrigation Department, 608 tanks (all sizes of classes) existed the city in the year 1986–87. At the same time, 652 tanks existed, according to the Directorate of Economics and Statistics. In 1961, the city had 261 lakes as per the information provided by the Government of Karnataka, 2002–03, Lake Development Authority (LDA), 2002 (Thippaiah, 2009). According to the Rao Report, 389 lakes and tanks existed in the Bengaluru Metropolitan Area in 1984. Of these, 262 and 127 lakes and tanks were in the green belt and con-urban areas (Rau, 1986). However, according to the BBMP Data, 2020, the total number of lakes was 210.

Fig. 19.1 shows the zone-wise lake distribution in the city, where the Mahadevapura zone (53 lakes) and Bommanahalli zone (49 Lakes) have a higher number of lakes, and the East zone (12 lakes) and Dasarahalli zone (6 lakes) have fewer numbers of the lakes. Suppose we increase the city boundary and include both Bengaluru Urban and Bengaluru Rural. In that case, the total number will account for 1545 (including lakes, ponds, and tanks), and out of this, Bengaluru Urban has 835 water bodies and Bengaluru Rural has 710 water bodies (Fig. 19.2).

As discussed earlier, the encroachment of lakes by different parties is one of the significant problems in the city. Dumping construction debris and garbage in the lakebed makes it easy to encroach on the lake area after being covered by debris. Private parties are not always responsible for encroachment activity; the government also plays a significant role in using the lake for different purposes. The total lake encroachment area is 10,472 acres in Bengaluru's urban and rural areas (Fig. 19.3). The total extent of the encroachment area in Bengaluru Urban is 4277 acres and that in Bengaluru Urban is 6195 acres.

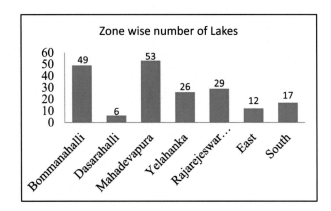

FIGURE 19.1 Zone-wise number of lakes. *From Nishanth, C. (2016). BBMP, Data on lakes in each of Bengaluru's wards. Open City-Urban Data Portal.*

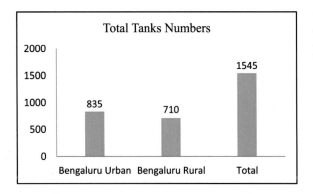

FIGURE 19.2 Total number of tanks (Sengupta, 2016). *From Sengupta, S. (2016, January 12). Karnataka government reveals sad state of Bengaluru lakes. Retrieved from DownToEarth: https://www.downtoearth.org.in/news/ water/bengaluru-not-yet-learnt-a-lesson-from-chennai-52407.*

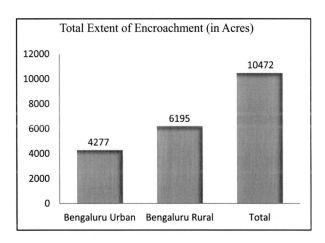

FIGURE 19.3 Total extent of encroachment (in acres). *From Sengupta, S. (2016, January 12). Karnataka government reveals sad state of Bengaluru lakes. Retrieved from DownToEarth: https://www.downtoearth.org.in/news/ water/bengaluru-not-yet-learnt-a-lesson-from-chennai-52407.*

Irrespective of being the government or private parties, encroachment is a matter of concern. We can observe the government and private encroachment, as shown in Fig. 19.4. In the Bengaluru Urban area, the government encroachment area is 2254 acres, higher than the private encroachment area, which is 2023 acres. However, in the Bengaluru Rural area, the private encroachment area is higher than the government encroachment area. The reason is that in the core city area, the government authority has proper land information and survey maps; the land price is also very high compared to periphery areas. For development or any other purpose, the government has a specific power to use this land, but private parties do not have this power. However, they seem to compete significantly with the government in land encroachment. However, in Bengaluru's rural areas, private parties have dominated compared to the government encroachment area; that is, 5162 acres and 1032 acres, respectively. The private encroachment area is significantly more than the government encroachment area.

Several industries, apartments, and government departments occupy much of the area surrounding the lake. These are the significant sources of waste dumping, untreated sewage, and encroachment. In the city, *Bangalore Water Supply*

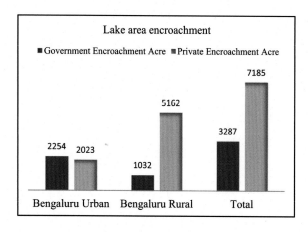

FIGURE 19.4 Lake area encroachment. *From Sengupta, S. (2016, January 12). Karnataka government reveals sad state of Bengaluru lakes. Retrieved from DownToEarth: https://www.downtoearth.org.in/news/water/bengaluru-not-yet-learnt-a-lesson-from-chennai-52407.*

and Sewerage Board (BWSSB), a parastatal body of the government created to manage the sewerage, has installed three major STPs of 248 million liters per day (MLD) at Vrishabhavathi, Kolar-Challagatta Valley and Hebbal Valleys. According to the *Comptroller and Auditor General of India Report*, 2021 (Principal Account General, 2021), 1440 MLD sewerage are generated in the BBMP area, and out of this, 660 MLD sewerage is treated by STPs. Therefore the lakes receive significant untreated sewage from different places in the city. This condition has increased numerous lake problems, that is, foul smell, frothing, fire, and reduction in flora and fauna. Lately, the lakes are gaining a significant place in daily newspapers and the media's headlines. Further, the neighboring apartments and industries built around these lakes have enhanced the problems.

In the city, numerous citizen groups, including residential welfare associations (RWAs), community groups, environmental activists, and nongovernmental organizations (NGOs), are trying to protect the lakes. Several experts have provided their views and suggestions for improving the lake situation. T. V. Ramachandra, Professor at the IISc Bengaluru, indicated that the construction builders, BBMP, and industrialists have encroached on 98% of the city lakes; the rest are sewerage-fed lakes. As a result, their water quality is highly polluted and contaminated, damaging the habitat of flora and fauna, for instance, birds and fishes. Moreover, this condition has reduced the quality of groundwater. In addition, he also mentioned that if the total number of lakes is managed and proper rainwater harvesting is done, it is sufficient to fulfil the city's water needs. Therefore there is no need to depend on the Cauvery River (Ramachandra & Kumar, 2008).

However, the current government initiatives need to be revised to protect the lakes and environment; they need a stronger push and an integrated approach to address the concerns related to lakes. Further, Sridhar Pabbisetty (SEO of Namma Bengaluru Foundation) mentioned the same issue and said that the government and its agencies are the primary ones responsible for protecting and managing the lake environment. One senior official of BBMP said that the wastewater treatment process needs to be improved compared to the present status. The city produces more than 1400 MLD of wastewater, but BBMP can treat only 720 MLD of sewage, and only 520 MLD is treated on average (Urban Water Bengaluru, 2018). On April 11, 2018, the National Green Tribunal (NGT) took action against the report the state government submitted for an action plan on Bellandur Lake. NGT stated that the report is "incorrect and misleading" and is unfit to take action on the ground. Therefore NGT formed a team to inspect Bellandur, Varthur, and Agara Lakes on April 14 and 15, 2018. The structure of this team consisted of one senior advocate, a professor from IISc, a scientist from CPCB (Central Pollution Control Board), the commissioner of BBMP, a secretary of *Bangalore Development Authority* (BDA), and an experienced lawyer. The team inspected these three lakes, prepared a detailed report based on the actual condition of the lakes, and submitted it directly to the tribunal (Gopalakrishnan, 2018).

The city boasts the largest concentration of lakes along its outskirts. The southern region features the most significant number of lakes, closely trailed by the northern, eastern, and western zones (see Fig. 19.5). The highest density of lakes can be found in the southeastern and northern regions within the city's boundaries, while the southwestern and northwestern areas exhibit a more moderate lake density. The lake density in the city's core area is notably lower than in other parts owing to key transportation hubs such as bus stations and railway stations and extensive industrial and commercial activities. This underscores the consequences of haphazard urban expansion and planning, which have degraded the city's ecology and environment (Ramachandra et al., 2017).

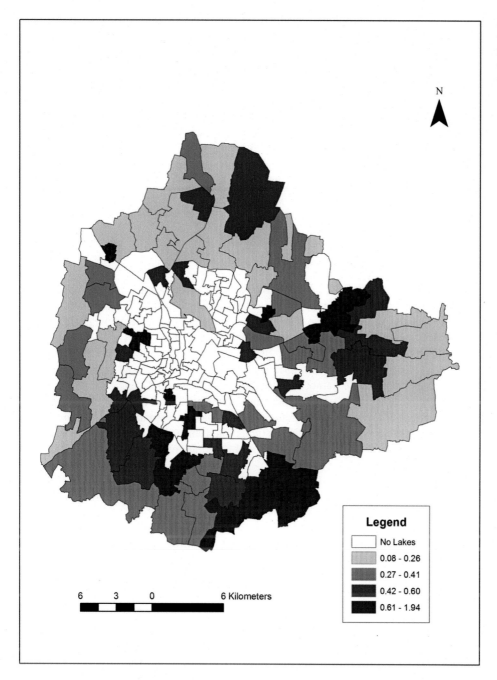

FIGURE 19.5 Map showing the ward wise lake area in Bengaluru in km² in 2011. *From Author's compilation based on BBMP's Lake Department Data, 2011.*

19.3   Issues with lakes in Bengaluru city

Several critical issues have been observed in many of the city's lakes. The primary problems include the improper disposal of garbage, the discharge of sewage effluents into the lakes, and encroachments upon the lake areas. Encroachment is a significant factor contributing to the city's lake population decline. The institutional framework has proven inefficient, lacking a robust system for ensuring accountability and transparency. This has given rise to connections among various land encroachers. The city currently produces an average of 0.5 kg of waste per person per day, given its population of approximately 13.5 million. This results in a daily waste generation of roughly 6100 metric tons (Directorate of Economics and Statistics; GoK, March 2022). Due to ineffective waste management practices, parts of the lake areas have been transformed into dumping grounds. Furthermore, the lakes and water bodies are primary receptacles for untreated sewage from residents and apartment complexes. Introducing this untreated sewage has led to a decline in water quality and increased the Biological Oxygen Demand within the lake ecosystem. In such conditions, aquatic organisms, including fish and snails, find it difficult to thrive and often face mortality, presenting a common challenge in many of the city's lakes.

In the city of Bengaluru, 90% of its lakes have become repositories for sewerage owing to the unchecked inflow of untreated sewage and industrial effluents, alongside the improper disposal of waste and construction debris onto the lakebeds. To identify the challenges of the city's lakes, we studied 40 lakes, analyzing them based on primary and secondary data sources (refer to Fig. 19.6). Our findings reveal that a substantial 85% of the lakes contend with the issue of solid waste dumping, affecting 34 of these water bodies. Sewerage effluents impact 80% of the lakes (32 in total), while encroachment concerns a staggering 92.5% (affecting 37 lakes). Additionally, 10% of lakes face issues related to weeds (4 lakes), and 27.5% grapple with various other problems (11 lakes). Among the other problems encountered, this study encompasses challenges such as idol immersion, government decisions (on occasion, government decisions have led to land use violations, where lakes are repurposed for residential layouts, road construction, playgrounds, and parks), and maintenance issues.

19.4 Number of citizen groups in lake rejuvenation

In the city, various civic groups are engaged in improving the lake environment. The study has selected 43 civic groups based on available secondary data and discussions with experts, lake activists, and field visits. These groups work at various levels and adopt different approaches to rejuvenate the lake that need attention. The rejuvenated lakes may have 1−13 groups working on the same lake, or one can engage in several lake restoration activities.

Lakes are pivotal in the collective efforts of various civic groups, NGOs, resident associations, academicians, and environmental enthusiasts. These groups typically coalesce around residents living near the lakes and collaborating with other associations, NGOs, and volunteer organizations. To garner the attention of residents and government authorities, these dedicated groups orchestrate activities such as photo sessions, advocacy campaigns, and art competitions for students. Similarly, they coordinate lake cleanup initiatives, partake in cultural and religious festivities, conduct awareness campaigns, host community gatherings, celebrate Kere Habba (lake festivals), and engage in tree-planting programs. The civic groups also engage in dialogues with local government bodies and corporations to explore viable solutions for the rejuvenation of lakes in the city.

Some citizen groups secure lake ownership from the BBMP, like *United Way of Bengaluru*, *Mahadevapura Parisara Samrakshane Mattu Abhivrudhi Samiti* (MAPSAS), and *Puttenahalli Neighborhood Lake Improvement Trust* (PNLIT). They maintain seven, five, and one lake(s), respectively.

The study has selected 40 lakes to examine the role of stakeholder involvement in the effort to revive the lakes. However, it was discovered that a total of 227 individuals are actively involved in the revival efforts across 40 lakes. This inclusive group comprises 25 political leaders, 71 resource persons (academicians, theatre figures, and social activists), 96 civic organizations, and 35 academic institutions (see Fig. 19.7). These stakeholders participate at different levels and collaborate to enhance the lake's surroundings, resulting in notable improvements and progress.

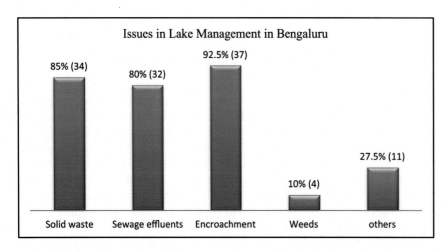

FIGURE 19.6 Issues in lake management in Bengaluru. *Compiled by author from different sources (Field survey, LDA Data, and newspapers).*

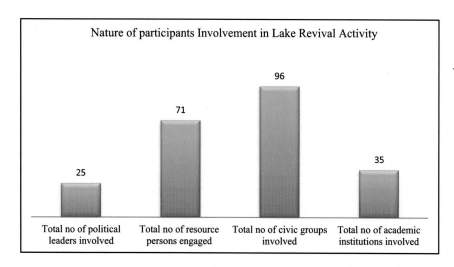

19.4.1 The formation process of lake rejuvenation civic groups

The haphazard expansion and growth of the city have had a severe impact on the city's environment. In order to preserve and safeguard the city's ambience, the residents have come together to form a range of "Human Chain Networks" with the assistance of NGOs, RWAs and volunteer groups. They aim to capture the public's attention and pressure local governing bodies.

The movement for the conservation of lakes in Bengaluru is not a recent development; it took root in 1980 through the efforts of a civic group called "Save Bengaluru's Lakes." This group established a committee to conduct an in-depth survey and offer recommendations for reviving Bengaluru's lakes. By 1990, the committee had compiled a report advocating that "we must protect our lakes and water bodies to meet the increasing demand for water in the city" (Thippaiah, 2009). The committee also recognized that the Cauvery River provided only a temporary solution, making it essential to safeguard the city's lakes and water bodies and harness rainwater to meet the city's water requirements.

In 2000, the government introduced the Public–Private–Participation (PPP) program to enhance the development of lakes within the city. As part of this policy, the government enlisted private companies' services to oversee the management and development of four lakes, Nagawara, Vengaiyana Kere, Hebbal, and Agara Lake, for a specified duration (Wirth, 2017). Additionally, the government opted to lease out three lakes to private entities. Nagavara Lake was entrusted to Lumbini Gardens Limited in 2004, Hebbal Lake was handed over to East India Hotels Limited in 2006, and Venkayana Kere was leased to Biota Natural System Private Limited in 2007. This project encompassed the creation of artificial beaches, upscale hotels, floating restaurants, and tourist attractions. Regrettably, these developments resulted in the degradation of the lake's ecosystem and the displacement of diverse forms of flora and fauna.

Moreover, the city possesses a significant population of traditional users, such as farmers, fishermen, and laundry workers, who are closely connected to the lakes from both socioeconomic and environmental standpoints. This strategy, however, limited the public's access to these shared resources due to private entities intervening in their management. Differing perspectives emerged as well. According to Fernando's research, this policy was perceived unjust to society's lower- and middle-income segments (Fernando, 2008). The state-of-the-art amenities and services provided by the private sector remained within reach only for high-income groups, leaving the economically disadvantaged with limited access. In the past, people could readily enjoy the lake environment for leisure, drinking, and irrigation purposes, but now they have to pay for these facilities.

To overturn the privatization of lake management, the Environmental Support Group (ESG), an NGO, initiated a Public Interest Litigation (PIL) at the High Court of Karnataka in April 2008. This PIL was grounded in a Supreme Court judgment that emphasized lakes as communal assets meant to serve traditional users. The High Court of Karnataka issued a directive underscoring the lakes' status as shared resources, emphasizing the responsibility of citizens and governing bodies to preserve them. Consequently, this order excluded Agara Lake from the privatization plan, while Hebbal Lake continued to operate under the program. Furthermore, the High Court established the N. K. Patil Committee to oversee the rejuvenation and restoration of urban lakes in collaboration with local stakeholders. In 2011, the Chief Minister of Karnataka allocated 150 crore rupees for the lake rejuvenation and revival initiative.

These collective efforts are receiving attention in newspapers, social media, and governing bodies. The following section highlights some of these collaborative efforts.

- On October 7, 2022, residents of Hosakerehalli organized a clean-up drive at Hosakerehalli Lake to remove debris and improve the lake bed. They received support from the Indian Plogger Army, Unicorn Art Clubs, and students. Additionally, they conducted an awareness program and sent a letter to BBMP urging them to protect the lake environment.
- In October 2018, residents of Sarjapur Road and nearby areas formed a human chain to protest poor infrastructure and government negligence, particularly regarding Kaikondrahalli Lake. They demanded better civic amenities and submitted a petition to the Chief Secretary. Despite previous appeals for neighborhood development, the BBMP took no action, leading to the demonstration involving various citizen forums and RWAs (TNN, 2018).
- In May 2018, residents of Pattandur Agrahara Lake formed a human chain to protect the lake from debris dumping and encroachment. They demanded a High-Power Committee or judicial investigation, but the BBMP failed to act, prompting protests against the government's inaction (The Indian Express, 2018).

19.4.2 Time line: formation of civic groups

Most environmental NGOs, civic groups, and volunteer organizations emerged following the economic reforms. This transformation can be attributed to the fact that, under this policy, many multinational corporations (MNCs), IT companies, and service sector industries flocked to the city, ushering in a robust job market, improved infrastructure, and elevated living standards. However, it also led to population growth and urban expansion within the city.

As previously mentioned, diverse citizen groups, volunteer organizations, NGOs, and activists have united to address the challenges and issues confronting preserving and conserving the city's environment. These groups raise awareness, mobilize citizens, establish program schedules, and accomplish targeted objectives. They frequently engage in dialogues with neighbors, politicians, academic institutions, and other groups to generate public pressure, initiate PILs to garner attention, or seek government and other agencies' funding.

Digital media, particularly social media, is pivotal in urban environmental movements. Platforms like Facebook and WhatsApp provide a convenient means of disseminating information about issues, agendas, activities, events, and contact details to the general public and other groups. These civic groups maintain Facebook pages and WhatsApp groups to deliberate on issues and propose solutions related to their concerns. Various stakeholders, including BBMP officials, urban planners, academicians, and engineers, actively engage through these social media groups.

Fig. 19.8 shows the growth of environmental groups focusing on preserving the city's lakes. Over 90% of these groups were formed after 2010, mirroring the city's population and urban expansion trends. These groups are actively addressing environmental concerns, reaching beyond local boundaries and forming collaborations at regional, national, and international levels. They mobilize the community, advocate for environmental causes, and work with the government, contributing to successful lake rejuvenation projects. Their strength lies in leadership, organization, mobilization, and media outreach, ensuring the protection and enhancement of Bengaluru's environment.

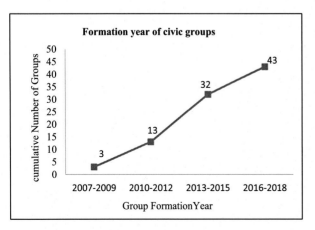

FIGURE 19.8 Formation year of civic groups. *Compiled by author from different sources (Newspaper, field survey, and LDA Data).*

19.5   Lakes condition in the city under BBMP's custody

In the 1970s, Bengaluru had 285 lakes, but there are currently 210 lakes. As per BBMP data 2020, out of the total number of 210 lakes, BBMP has custody of 167 lakes, followed by 33 lakes to BDA, five lakes to *Karnataka Forest Department*, four lakes to LDA, and four lakes to *Bengaluru Metro Rail Corporation* (BMRCL). According to the BBMP data, in January 2023, BBMP had a custody of 202 lakes. The rest of the lakes come under the BDA and *Karnataka Forest Department* out of 210 lakes in Bengaluru. In the section, the study gives information on 167 lakes under the BBMP as per BBMP data for 2020. Under the custody of BBMP of 167 lakes, 104 lakes have been surveyed, and encroachment areas occupied by various entities have been removed (Fig. 19.9). Further, to prevent encroachment and protect the lake ecosystem, 85 lakes have already been fenced, and the rest of the lakes will be fenced after completing the survey and removing the encroachment.

BBMP follows environmental planning and mapping using a geographical information system and an integrated development strategy to rejuvenate and restore the lake. Also, the body organizes various workshops and community meetings with the citizens, environmental activists, citizen groups, NGOs, and corporations to discuss the detailed project plan of the lake for their active participation. This collective action by various stakeholders helped BBMP develop 65 lakes, and 24 lakes are in the development in progress stage out of 167 lakes (Fig. 19.10).

19.6   Mobilization methods and strategies used in the conservation of lakes in the city

Lakes serve essential ecological functions in the city, such as groundwater recharge, drinking water, flood control, socioeconomic functions, and environmental sustainability. However, as mentioned earlier, there are no accurate statistics on the city's total number of lakes and survey maps. Similarly, no comprehensive lakes-related databases are available because multiple groups (eight government agencies and citizen groups) are responsible for managing and

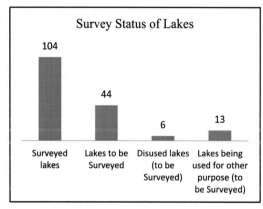 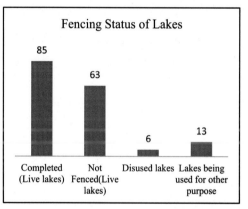

FIGURE 19.9   Survey and fencing status of lakes under BBMP. *From BBMP's Data, 2020.*

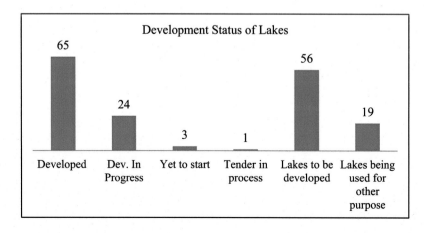

FIGURE 19.10   Development status of lakes under BBMP. *From BBMP Data, 2020.*

protecting the lakes. These eight groups are *Karnataka Forest Department, BDA, Karnataka Tank Conservation and Development Authority* (KTCDA), BBMP, *Minor Irrigation Department*, KSPCB, *Karnataka Fisheries Department*, and BWSSB (Kulranjan & Palur, 2022). As aforementioned, it is badly affected due to encroachment, garbage dumping, sewage effluents, and construction activities in the lake buffer zone. Several civic groups are engaged in protecting lakes and adopting various strategies for lake conservation and protection, but there are contradictory issues.

19.6.1   Steps followed by citizen groups in lakes rejuvenation activities

Citizen-led lake groups are engaged in various actions to save urban water bodies, such as activism, monitoring, maintenance, and restoration. The citizen groups and government agencies work together to prepare detailed project reports (DPRs) to rejuvenate and restore lakes by following specific steps (Fig. 19.11).

In the lake restoration process, many citizen groups consider fixing the boundary line as a first step, which is challenging. Removing encroachment and garbage and identifying the lake area requires support from various stakeholders, such as community groups, tehsildar, survey experts from the revenue department, and BBMP officials. Multiple players, such as construction companies, promoters, and sometimes BBMP, encroach on the lake area. Lake revival groups often face constant battles with encroached members who strongly oppose lake revival activities. For instance, a similar issue came up during the revival of Shikaripalya and Puttenahalli Lake. Ecological development is a secondary stage in the lake restoration program. Proper desilting is required to bring back the lake's water-holding capacity, and it is essential to maintain the lake's ecological balance.

In biodiversity development, a specific area is selected to fulfil a specific objective, such as an island for birds and other species, wetlands, gardens, and the installation of park amenities. The groups then organize a plantation and lake clean-up drive to increase the green cover and cleanliness in the lake area. Infrastructure development in lakes includes the installation of sewerage treatment plants, CCTV cameras, the appointment of security guards, regular monitoring, water quality checks, a drainage system, and a community meeting hall. Many people use the lake for jogging, walking, and recreation purposes; to maintain it, a proper plan is created to beautify the lake. Making of walking paths, cycle tracks, parks, open gyms, and gazebos are considered to beautify the lake.

Community engagement is crucial in lake restoration activities, preparing DPR, fencing, forming a pressure group, and negotiating with other groups and agencies. In favor of community engagement, painting, photographic sessions, plan (Dalton et al., 2010; Dutton et al., 2013; Economic Survey of Karnataka 2021−22. Bengaluru: Planning, Programme Monitoring & Statistics Department, Government of Karnataka, 2022; Gopalakrishnan, 2018; Howard &

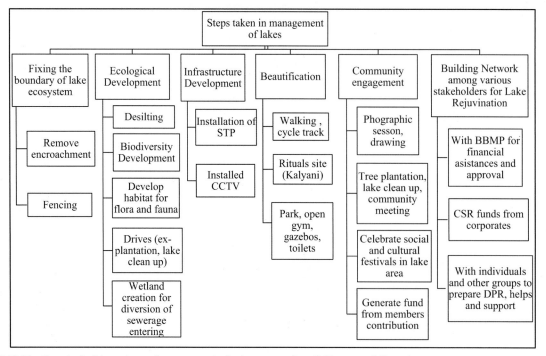

FIGURE 19.11   Steps in the lakes rejuvenation program. *Author's construct from field notes and discussions.*

Parks, 2012; Khondker, 2001; Kothari, 2014; Nishanth, 2016; Patil, 2011; Ramachandra & Kumar, 2008; Ramachandra et al., 2017; Sudhira, 2008; Thippaiah, 2009) plantation, clean-up drives, and celebration of sociocultural festivals are carried out. Lake revival groups celebrate festivals and occasions such as "Kere Habba," "Earth Day," "Environmental Day," and specific lake festival days to engage people, students, and other groups. They are building a network of lake revival groups, government agencies, NGOs, RWAs, and citizen help to acquire new ideas, innovation, technical advice, financial assistance, and support. It also adds strength to the lake restoration and management activities. Many stakeholder engagements help sustain lake management activities for a longer duration, fund generation, and strengthening pressure groups for negotiations and creating pressure on the government, polluters, and encroachment agencies.

19.6.2 Lake maintenance and community engagement

Lake revival groups follow various methods and strategies for saving Bengaluru lakes, and they try to form a network among various stakeholders and connect them with the lake environment (Fig. 19.12). Since various stakeholders are actively engaged in saving lakes, the city has seen many successful lake restoration stories, such as Hoodi Lake, Sarakki Lake, Jakkur Lake, Kaikondrahalli Lake, and many others. The success stories of these lake-saving activities motivate other people and citizen groups to rejuvenate and restore other lakes in the city.

MAPSAS started the rejuvenation activity to restore Kaikondrahalli Lake with the help of the lake's neighborhood residents. The group had campaigned and discussed with local people and created an expert group with the help of architects, ornithologists, and others who could perform well in their field of interest. Similarly, they would meet BBMP officials and philanthropists frequently for their participation in the lake revival activity. After many discussions, BBMP agreed to consider Kaikondrahalli Lake to be on the lake rejuvenation list and provide financial assistance. The adjoining school students access the lake for play, and people use the lake for walking, jogging, and recreation. These small groups are supportive and concerned about the management and protection of Kaikondrahalli Lake. In addition, various individuals, such as filmmaker ecologists, strategists, and ornithologists, play an essential role in restoring Kaikondrahalli Lake.

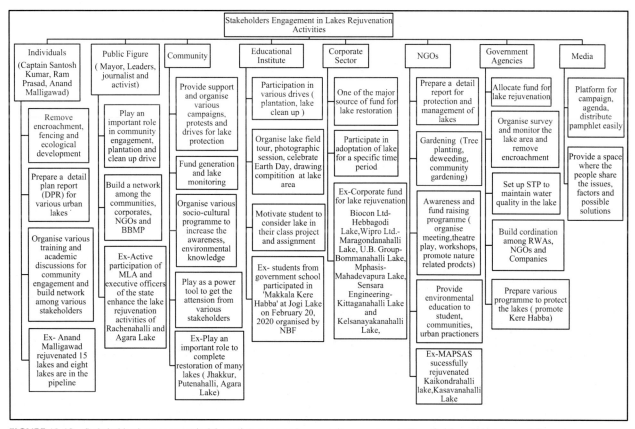

FIGURE 19.12 Stakeholders' engagement in lake maintenance and community engagement. *Compiled from field notes and discussion.*

In the city, various individual lake activists engage in lake restoration activities, like lake activists A. Malligavad, R. Prasad, S. Kumar, and others. A. Malligavad, a mechanical engineer who became a lake conservationist, has restored over 15 lakes and is restoring eight more. Kyalasanhalli Lake (36 acres) and Margondanhalli Lake (19 acres) were rejuvenated within 45 days and 75 days, respectively. It led him to become a famous lake conservationist in the city. He also began engaging with Odisha, Uttar Pradesh, and Meghalaya for similar activities. He started work in 2016 to revive lakes as he believed the city would face a hazardous situation if work to protect the lakes had not begun, as the city's water needs depend on water bodies. In the initial stage, Malligavad discussed these ideas with people to create awareness but was not met with positive responses.

After that, he studied lake ecosystems and their technical and scientific aspects, visiting nearly 180 lakes and preparing a restoration model based on his mechanical engineering knowledge. This systematic way helps him to understand the lake catchment area, topography, drainage system, and selected ecological perspective for lake rejuvenation, not only beautification of the lake. As a follow-up, he frequently met with government bodies, corporate, and philanthropists for their participation. In 2017, Sunshine Engineering Limited came forward and took responsibility for restoring Kyalasanhalli Lake, and it was ready to invest 1 crore rupees. Later, he rejuvenated many lakes one by one with the help of communities, corporations, and government agencies. By 2025, he plans to restore 45 different lakes in the city. Additionally, to educate the people, he organizes discussions and webinars with students, corporations, and the government to spread awareness and gain financial assistance. Similarly, R. Prasad from "Friends of Lakes," a volunteer group, is pioneering in saving Bengaluru lakes and coordinating among various stakeholders.

In lake restoration activities, public figures play an important role in plantation and clean-up drives, community engagement, and public attention. For instance, Meena, an editor in a local news magazine who used to write frequently about Kaikondrahalli Lake, helped the lake revival group gain attention and increased the rate of people's participation. Community engagement plays a crucial role in the integrated lake restoration program, removing encroachments and handling legal cases in court. Community participation has helped the legal case for revival activity at Sarraki, Doddakallasandra, and Chunchaghatta Lake and created a strong pressure group to negotiate with other stakeholders.

Engagement of academic institutions with lake revival activities helps students learn about the lake environment's features and components and its importance to the city environment. On the other hand, they participated in various programs like tree plantation, clean-up, and celebrations of other socio-cultural events, which helped increase stakeholder participation. The Jalmitra group organized various study tours in the lake area, where students collected data from field surveys to analyze different aspects of the lake's importance in the city's environment.

Adequate financial assistance is required to execute any plan for lake restoration activity. BBMP, the corporate sector, and members' contributions fill this gap. Many corporations have adopted lakes under CSR in the city for a specific period. During this period, they can provide financial assistance to manage and revive lakes. For example, Bommanahalli Lake was taken by U.B. Mahadevapura Lake by Mphasis, and Sansera Engineering Private Limited adopted Kelsanayakanahalli Lake. NGOs also help the group with fund generation through campaigns and discussions with the government and philanthropists.

Additionally, they select activists to promote their brand and product name, which builds a strong connection between corporate and lake groups. For example, a well-known jewellery brand, Tanishq, selected Odette Kateak (cofounder of Beautiful Bengaluru and an expert in strategy for zero waste), a sustainability changemaker and activist, as their brand ambassador to recognize her work in improving the city environment. Sensing Local, an NGO, provides an integrated urban lake management plan, data services, and techniques to prepare a DPR, which helps to get financial help from stakeholders.

Another NGO called "Save Green" is engaged with lake clean-up drives, mobilizing the people for their active engagement in protecting the lake environment. Tree plantation is one of the essential tasks in lake revival activity. Concerning this, "Say Tree," an NGO, has organized numerous plantation drives with various lake revival groups in the city and planted many trees in lake areas. An NGO called Go Green has done similar work. They engage in various activities to mobilize people by providing tree saplings on different occasions and celebrations where organizers can gift or donate them to the participants. To save the city's lakes, Clean Bengaluru (NGO) adopted the "Clean Lake" initiative to protect the lake from garbage dumping and encroachment, as these two issues are expected for most lakes in the city.

The organization is well connected with various environmental experts, legal advisors, urban planners, and citizen groups, which help to create environmental awareness and take legal action against the encroachment parties. As mentioned earlier in the governance of the city lake, eight government bodies are engaged in custodianship, rejuvenation, planning, management, regulation and policies, monitoring, and removing encroachment from the lake buffer zone. Also, the governing bodies have a good network among RWAs, communities, and NGOs for integrated lake management and program organizations (Kere Habba) for community engagement toward saving lakes in the city.

Social media, such as Facebook, Twitter, and Gmail, are powerful tools for youth participation in social movements (Ellison & Boyd, 2013). Utilizing online and offline networks facilitates gatherings and critical thinking, aiding mass mobilization and offering access to protest-related information. Multiple communication channels and peer acceptance through online platforms enhance group cohesion and standards (Howard & Parks, 2012). Specialized groups and blogs on social media provide a platform for participants to join and access relevant information, news, and event schedules that might be challenging to obtain from other sources. Additionally, social media promotes social interaction, fostering the exchange of information, knowledge, and organizational activities, thereby strengthening networks and group engagement (Dalton et al., 2010).

Earlier discussions have shown that various NGOs and civil society are engaged in different urban settings in Bengaluru. Social media is a helpful tool for them to mobilize and get attention from citizens. The study has listed 55 registered and unregistered groups related to urban waste and lakes, focusing on urban movements related to urban waste and lakes. Facebook is one of the dominating social media tools, and most people have a Facebook account. Some groups are very active and regularly update their profile about group activities (tree plantation, garbage cleaning drive, spot-fixing date, pamphlet, discussion of government policy, and regulation) and collect public opinions.

WhatsApp is another vital tool in social media communication; it is easy to communicate with participants and voluntary groups. Some groups have category-wise groups like core committees, participants, and voluntary groups (Agara Lake Protection and Managing Society). Twitter is another essential part of social media as a mobilization tool, and issues of fire and frothing at Varthur Lake and Bellandur Lake have drawn massive attention from the people. Additionally, regular discourse is carried out on Twitter, with most activists sharing lake-related activities regularly on the platform.

19.6.2.1 Puttenahalli Lake rejuvenation activity

Puttenahalli Lake, situated in BBMP ward no. 187, Jaya Prakash Nagar, Bengaluru South, covers an area of 13 acres and 25 guntas. Initially facing neglect, the lake deteriorated due to lack of maintenance, waste dumping, untreated sewage, and encroachment. Concerned citizens initiated campaigns, seminars, and discussions to save the lake. They formed the PNLIT in 2007, seeking to restore and protect the lake.

The lake had degraded to a marshy dump yard, and its ecological significance, including groundwater recharge and habitat for flora and fauna, was compromised. The restoration strategy involved two phases: first, identifying and fencing the lake boundary, followed by desilting, infrastructure development, and beautification. Activities included creating a walking track, gazebo, exercise bars, and rainwater harvesting.

The restoration journey began with a signature campaign and evolved into broader community engagement. PNLIT fostered public awareness through events like photography sessions, clean-up drives, and tree planting. Financial support initially came from BBMP and later involved public donations and corporate CSR funds. PNLIT's efforts were recognized with awards and honors from various organizations.

However, challenges persisted. The lake area's encroachment by hut dwellers led to conflicts, with some resisting restoration efforts due to political considerations. PNLIT emphasized changing citizens' attitudes toward lake protection, advocating for community participation. Currently, PNLIT faces financial constraints for lake maintenance. Despite hurdles, their commitment to lake revival and community involvement is a success story, inspiring other civic groups to prioritize grassroots engagement and collaborate with local stakeholders for sustainable lake restoration.

19.6.2.2 Jakkur Lake rejuvenation activity

The Jakkur Lake rejuvenation project is a successful example of various stakeholders cooperating to protect urban water bodies and lakes. Initially plagued by garbage dumping and sewage, the lake was restored through the efforts of citizens, community groups, NGOs, and government agencies. They collaborated to fence the lake, desilt it, and install a sewage treatment plant, thus improving water quality and biodiversity.

Despite initial opposition, the fishing community played a crucial role in the restoration, contributing their traditional knowledge and expertise in managing the lake ecosystem. Conflicts among different stakeholders were addressed through the efforts of an urban planner and water conservationist who facilitated discussions and encouraged collaboration.

Industrial effluents and raw sewage were significant issues affecting water quality, but the group's efforts led to installing a sewage treatment plant to address the problem. Their successful rejuvenation efforts earned them recognition and awards from the National Water Mission.

The Jakkur Lake project's success has inspired other citizen-led organizations to rejuvenate and manage other lakes and water bodies in the city, fostering a positive impact on the urban environment.

TABLE 19.1 Status of water bodies in Bengaluru City limits and Greater Bengaluru.

	Bengaluru city		Greater Bengaluru	
	No. of water bodies	Area (ha)	No. of water bodies	Area (ha)
SOI	58	406	207	2342
1973	51	321	159	2003
1992	38	207	147	1582
2002	25	135	107	1083
2007	17	87	93	918

Source: From Ramachandra, T. V., Kumar, U. (2008). Wetlands of Greater Bangalore, India: Automatic delineation through pattern classifiers. Electronic Green Journal, 26. http://egj.lib.uidaho.edu/index.php/egj/article/view/3171/3227.

19.7 Conclusion

Stakeholder engagement is a critical aspect of successful lake conservation efforts in Bengaluru. The involvement and collaboration of various stakeholders, specifically government agencies, environmental organizations, local communities, and citizens, are essential for addressing complex challenges and achieving meaningful outcomes. A collective understanding of the value and significance of Bengaluru's lakes can be fostered through stakeholder engagement. This shared awareness can lead to coordinated efforts and the allocation of resources for lake conservation initiatives. Stakeholders can collaborate on planning and implementing strategies such as pollution control measures, restoration projects, and sustainable management practices.

Government agencies play a crucial role as critical stakeholders in lake conservation. Their commitment to enforcing regulations, developing effective policies, and providing necessary funding is vital for the success of conservation efforts. Additionally, active involvement and support from environmental organizations and research institutions can contribute valuable expertise, scientific knowledge, and technical assistance to lake conservation projects. Engaging local communities and citizens is equally essential for the long-term sustainability of lake conservation efforts. Empowering communities through education, awareness campaigns, and capacity-building initiatives can foster a sense of ownership, responsibility, and pride in the lakes. Encouraging citizen participation in lake clean-up drives, tree plantation programs, and monitoring activities can create a sense of stewardship and promote a culture of environmental consciousness.

Successful stakeholder engagement also involves effective communication channels and platforms for sharing information, feedback, and collaboration (Fig. 19.1 and Table 19.1). Regular consultations, public hearings, and dialogue sessions can facilitate transparent decision-making and foster stakeholder trust. In conclusion, engaging diverse stakeholders in lake conservation efforts in Bengaluru is crucial for achieving tangible and sustainable outcomes. By actively involving government agencies, environmental organizations, local communities, and citizens, a collective approach can be fostered, harnessing the expertise, resources, and commitment needed to address the challenges Bengaluru's lakes face. Through collaborative efforts, stakeholder engagement can pave the way for effective lake conservation practices, ensuring the preservation and revitalization of these critical natural assets for present and future generations.

Declaration

The author has no point of conflict of interest.

References

Bangalore Development Authority (2017). Revised Master Plan 2031, Bangalore Development Authority, Bangalore. https://data-opencity.sgp1.cdn.digitaloceanspaces.com/Documents/Recent/RMP2031-English-Final-2017-01-04-R1.pdf.

Directorate of Economics and Statistics. (March, 2022). Economic Survey of Karnataka 2021−22. Bengaluru: Planning, Programme Monitoring & Statistics Department, Government of Karnataka, India.

Ellison, N., & Boyd, D. M. (2013). *Sociality through social network sites*. UK: Oxford University Press.

Fernando, V. (2008). *November). Disappearance and privatisation of lakes in Bangalore. Retrieved October 30, 2018.* Available from http://base.d-p-h.info/en/fiches/dph/fiche-dph-7689.html.

Gopalakrishnan, S. (2018, April 17). NGT Appointed Team Inspects Bengaluru Lakes. Retrieved from India Water Portal: https://www.indiawaterpor-tal.org/articles/ngt-appointed-team-inspects-bengaluru-lakes.

Howard, P. N., & Parks, M. R. (2012). Social media and political change: Capacity, constraint, and consequence. *Journal of Communication, 62*(2), 359−362. Available from https://doi.org/10.1111/j.1460-2466.2012.01626.x.

Khondker, H. H. (2001). Environment and the global civil society. *Asian Journal of Social Science, 29*(1), 53−71. Available from https://doi.org/10.1163/156853101X00325.

Kothari, R. (1986). NGOs, the state and world capitalism, Economic and Political Weekly, undefined*Economic and Political Weekly, 2014*(28), 2177−2182. Available from http://www.epw.in/system/files/special_articles_ngos_the_state_and_world_capitalism.pdf.

Kulranjan, R., Palur, S. (2022, March 12). How Many Lakes Does Bengaluru Really Have? A Crowdsourcing Initiative Is Finding Out. Retrieved August 10, 2022, from Scroll.in: https://scroll.in/article/1018934/how-many-lakes-does-bengaluru-really-have-a-crowdsourcing-initiative-is-find-ing-out.

Murali, J. (2016, August 9). Bengaluru Demolition Drive: Govt Should Assess the Damage Caused to Families. Retrieved October 19, 2018, from Firstpost: https://www.firstpost.com/india/bengaluru-demolition-drive-govt-should-assess-damage-caused-to-families-2944508.html.

Nishanth, C. (2016). *BBMP, data on lakes in each of Bengaluru's Wards.* India: Open City-Urban Data Portal.

Patil, N. K. (2011). *Preservation of lakes in the city of Bangalore: Report of the committee constituted by the honorable high court of Karnataka.* Bangalore, India: Bangalore, Honorable High Court of Karnataka.

Principal Accountant General (2021). Performance Audit on Storm Water Management in Bengaluru Urban Area. Principal Accountant General (Audit-I), Bengaluru, Karnataka. https://cag.gov.in/uploads/download_audit_report/2021/7.%20Chapter%205-0614304649467e1.80708377.pdf.

Ramachandra, T. V., & Kumar, U. (2008). *Wetlands of greater Bangalore, India: automatic delineation through pattern classifiers, . Electronic Green Journal* (26). 2008 ISSN: 1076-7975. Available from http://egj.lib.uidaho.edu/index.php/egj/article/view/3171/3227.

Ramachandra, T. V., Bharath, H. A., Gouri, K., & Vinay, S. (2017). Green spaces in Bengaluru: Quantification through geospatial techniques. *Indian Forester, 143*(4), 307−320.

Rau, L. (1986). Restoration of the Existing Tanks in Bangalore Metropolitan Area. Report of the Expert Committee, Government of Karnataka, India. https://admin.indiawaterportal.org/sites/default/files/iwp2/report_of_the_expert_committee_for_preservation_restoration_or_otherwise_of_the_ex-isting_tanks_in_bangalore_metropolitan_area_laxman_rau_1986.pdf.

Sengupta, S. (2016, January 12). Karnataka Government Reveals Sad State of Bengaluru Lakes. Retrieved from DownToEarth: https://www.downto-earth.org.in/news/water/bengaluru-not-yet-learnt-a-lesson-from-chennai-52407.

Sudhira, H. S. (2008). Studies on Urban Sprawl and Spatial Planning Support System for Bangalore, India. Centre for Sustainable Technologies and Department of Management Studies Indian Institute of Science Bangalore-560012, India.

The Indian Express (2018 May 27). Bengaluru Residents Plan Human Chain to Save Pattandur Agrahara Lake. https://www.newindianexpress.com/cit-ies/bengaluru/2018/May/27/bengaluru-residents-plan-human-chain-to-save-pattandur-agrahara-lake-1819961.html.

Thippaiah, P. (2009). Vanishing Lakes: A study of Bangalore City. In D. Rajasekhar (Ed.), *ISEC Working Paper Series.* Bangalore, India: Institute of Social and Economic Change.

TNN (2018, October 28). Residents form human chain to protest poor infrastructure, Palike's apathy. The Times of India. https://timesofindia.india-times.com/city/bengaluru/residents-form-human-chain-to-protest-poor-infrastructure-palikes-apathy/articleshow/66397706.cms.

Urban Water Bengaluru (2018). Wastewater Treatment and Reuse. Bengaluru, Karnataka. http://bengaluru.urbanwaters.in/wastewater-treatment-and-reuse-49/.

Ahmadi, M., & Eslamian, S. (2022). Hydrological resilience of large lakes. In S. Eslamian, & F. Eslamian (Eds.), Flood Handbook. In: *Flood Impact and Management.* (3). USA: Taylor and Francis, CRC Group.

Dalton, R., Van Sickle, A., & Weldon, S. (2010). The individual−institutional nexus of protest behaviour. *British Journal of Political Science, 40*(1), 51−73. Available from https://doi.org/10.1017/S000712340999038X, http://uk.cambridge.org/journals/jps.

Dutton, W. H., Ellison, N. B., Boyd, D. M. (2013). Sociality through Social Network Sites. Oxford University Press, Vol. 1. Available from https://doi.org/10.1093/oxfordhb/9780199589074.013.0008.

Economic Survey of Karnataka (2022). 2021−22. Bengaluru: Planning, Programme Monitoring & Statistics Department, Government of Karnataka.

Chapter 20

Rejuvenation of streams and rivers through decentralized and community-driven rainwater capture for dignified livelihoods, climate resilience, and peace: living examples from Rajasthan, India

Indira Khurana

Indian Himalayan River Basins Council, New Delhi, India

20.1 Introduction

The world is staring at a serious water crisis (Barthakur & Khurana, 2014). Groundwater resources are depleting faster than they get replenished (Konikow & Kendy, 2005; Wada et al., 2010). Rivers are drying up and water pollution is increasing (Damania et al., 2019), thus posing additional challenges to equitable, sustainable economic and social development. Much of this has to do with the current economic development paradigm, which is based on unsustainable natural resource extraction (Correspondents, 2017; Ennis et al., 2017). Climate change is adding to the problem with increasing temperatures, unpredictable rainfall, and increasing extreme weather events such as droughts, floods, storms, and fires (http://www.worldweatherattribution.org). It is getting difficult to manage with little or too much rain, also because natural water holding sources and drainage systems have been heavily tampered with. In an inequal world, the implications of these disasters manifest themselves in various forms—social, economic, political, uprooting, displacement, and migration (Migration Policy Institute, 2022)—perpetuating inequality and conflict.

Instead of continuing to breach nature's boundaries, there is a need for nature rejuvenation. The world is wounded and in need of healing for which water is critical. Well-managed water resources—its conservation and judicious and efficient use by involved communities rejuvenate society and the planet.

Rivers are an integral part of water resources. Unfortunately, globally rivers are under threat. Efforts to restore seasonal or dry streams that often feed into large rivers can lead to revival, rejuvenation, and perennial flow not only of these streams but also of the large rivers.

Rain is a primary source of water: Its conservation leads to replenishment of ground and surface water, reestablishment of the symbiotic relationships between surface and groundwater, and drought and flood mitigation and resilience. Rain brings livelihoods and restores biodiversity and soils. If not managed well, rain also brings pain in the form of flooding and soil erosion. Subsequent deposition of this soil elsewhere compounds the problems of reduced water-holding capacities (Eslamian & Eslamian, 2021a,2021b).

Decentralized capture of rainwater by communities using their local knowledge leads to decentralized water and food security and economic prosperity. It helps restore the water cycle and rejuvenate rivers.

This chapter informs about successful and decades-old examples of this approach and the use of indigenous/local knowledge and wisdom in climate action and resilience.

Hydrosystem Restoration Handbook. DOI: https://doi.org/10.1016/B978-0-443-29802-8.00020-0

20.2 Is the world running out of water? Climate change and water

The world is running out of water and climate change is adding to the problem. In its Global Risks Report, 2019, the World Economic Forum listed water scarcity as one of the largest global risks (World Economic Report, 2019).

The infographic below gives an account of water stress scenarios in 2050 (Fig. 20.1).

This infographic shows that 51 of the 164 countries analyzed are expected to suffer from high to extremely high water stress by 2050, an equivalent of 31% of the population. In addition to the entire Arabian Peninsula, Iran, and India, most North African countries are expected to consume at least 80% of the available water by 2050 (Armstrong, 2024). The phenomenon of water scarcity is not limited to emerging countries, as European countries such as Portugal, Spain, and Italy are reportedly already under high water stress, and the situation in Spain is set to worsen significantly by 2050. For France and Poland, Armstrong states that Water Resources Institute (WRI) experts assume medium to high water stress, which corresponds to a consumption rate of 20%−40% of available resources.

Water and climate are very closely linked to each other, where change in one affects the other (Kundzewicz, 2008). A recent report by the Germany-based United Nations University (UNU) categorically states that human actions are behind the rapid and catastrophic changes taking place on the planet. The world is quickly reaching multiple risk "tipping points" with respect to climate change, which will hamper disaster coping attempts and mechanisms. A risk tipping point is reached when the systems that humans rely on for our lives and societies cannot buffer risks and stop functioning like we expect them to do (UNU EHS, 2023a, 2023b). According to the Interconnectedness Disaster Risk Report (https://interconnectedrisks.org/summaries/2023-executive-summary), one tipping point is groundwater. Twenty-one of 37 of the world's largest aquifers are being depleted faster than they can be replenished. The water stored in these aquifers accumulated over thousands of years and it would take thousands of years to replenish these. With 70% of the global food production dependent on groundwater, food security is at great risk. India is the world's largest groundwater extractor, and when the groundwater depletes, so will the livelihoods of farmers (https://s3.eu-central-1.amazonaws.com/interconnectedrisks/reports/2023/TR_231115_ Groundwater.pdf).

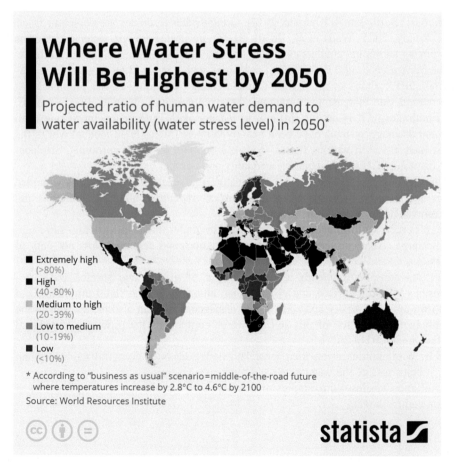

FIGURE 20.1 Map lines delineate study areas and do not necessarily depict accepted national boundaries. Map lines delineate study areas and do not necessarily depict accepted national boundaries. *Courtsey Armstrong M. (March 2024) Where water stress will be highest in 2050. https://www.statista.com/chart/26140/water-stress-projections-global/.*

Other tipping points being approached are accelerating extinctions, mountain glaciers melting, space debris, unbearable heat, and uninsurable future.

Climate change is increasing the scale, duration, and intensity of water-related disasters such as drought, flood, storms, and cyclones. Droughts are among the greatest threats to sustainable development. Since 2000, the number of droughts has increased by 29%. When more than 2.3 billion people already face water stress, this is a huge problem. While more than 1.8 million people worldwide are at risk of severe floods, several of these have been attributed to climate change (Singh & Khurana, 2023a, 2023b).

Most of the disaster due to climate change are water-related and yet, it was only in COP 27 (Conference of Parties of the United Nations Framework Convention on Climate Change) in 2022 that the importance of water in climate adaptation and ecosystem restoration was recognized (Singh & Khurana, 2022a, 2022b, 2022c).

There is a direct influence of global warming on precipitation. Increased heating leads to greater evaporation and surface drying, thereby increasing intensity and duration of drought. The dry and fragile soil blows and settles downstream reducing water-holding capacities of rivers and other waterbodies. Moreover, with a 1°C rise in temperature, the water-holding capacity of the atmosphere increases by 7%, leading to increased water vapor in the atmosphere. Hence, intense and more precipitation events occur. According to Richard Seager of the Centre for Climate and Energy Solutions, University of Columbia, "When the atmosphere holds more moisture, it also transports more moisture from one place to another. Winds can blow away these moisture-laden clouds resulting in drought" (Banerjee, 2023). These droughts and floods affect food security, lives, and livelihoods, leading to conflicts and mass exodus.

Already, insurance companies are hesitating to pay insurance for infrastructure damaged by floods. Homeowners in the United States are struggling to get insurance money for flood-destroyed homes. Some insurance companies have gone bankrupt, others have either hiked up the insurance fee or are not willing to insure homes in high-risk flood areas. These global problems can be addressed through local and decentralized solutions to conserve, manage, and use water efficiently and equitably.

20.3   The importance of rivers

A complex of small streams and rivers often combine to produce a main river, akin to the human circulatory system, wherein arteries veins and capillaries that extend to every part of the human provide nourishment and remove waste, even from extremities. Streams and small rivers add to the base flow of larger rivers and often empty themselves into larger rivers, adding to the latter's flow. They help decentralized groundwater recharge, serving as water sources to local communities and other life forms close by. The health of larger rivers is thus also dependent on the health of these streams, and if a river must be saved, the springs, streams, and rivulets that feed into it must be healthy and flourishing, since their drying will impact the overall flow of the river (Singh & Khurana, 2022a, 2022b, 2022c).

Without rivers, there can be no life: As they flow, rivers contribute to the water cycle, biodiversity, livelihoods, food and nutrition security, and transportation, for instance. Civilizations thrived alongside rivers.

With respect to the water cycle, one critical function that rivers perform is maintaining the balancing act with groundwater. Rivers and groundwater mutually feed each other. As one dries, so does the other; if one is replete, it recharges and replenishes the other (Singh & Khurana, 2022a,b,c). Rivers thus help maintain a balance with groundwater aquifers, enabling water availability during water scarcity and drought. Replete groundwater aquifers provide base flows to rivers and river flows recharge groundwater aquifers. During the monsoon, the river replenishes the aquifer. A healthy relationship between the two enables water security, and if this relationship is disrupted, then rivers dry up and groundwater tables decline, overall impacting water availability and leading to water scarcity. Rivers with their functional floodplains provide space for the spreading of waters from heavy rainfall, thereby reducing flooding risk and intensity. This cushioning is an important role for climate change resilience.

Rivers provide a sense of well-being. Rivers are linked to our culture, spirituality, and sense of well-being and security, providing psychological support and peace of mind (Khurana, 2021a).

Globally, rivers are under threat due to mismanagement and pollution, overextraction, encroachment, intensive agriculture, catchment disruption, deforestation, sand mining, and dam building. As a result, several perennial rivers have turned seasonal. Some have even disappeared. River ecology and the habitat of river species have been affected. Sand mining has altered riverbeds, resulting in a river changing course, causing flood and reducing recharge. Domestic, sewer and industrial pollution have affected water quality and turned rivers to receptacles of human-generated waste. The pollution of rivers does not remain confined to the rivers themselves, but spills over into groundwater, lakes, floodplains, and seas and oceans. Also, from there, this pollution enters the human body, causing debilitating disease and even death.

Thirty of the 47 largest river systems, which together discharge half of the global runoff to the oceans, are at risk: Only 37% of the planet's longest rivers remain free flowing. Nearly 60,000 dams have been built worldwide, with more than 3700 under construction or planned (Grill et al., 2019).

20.4   River and climate change

Rivers help in climate mitigation, adaptation, and resilience. River systems are already under threat and are themselves affected by climate change. How is climate change affecting our rivers and, through rivers, us?

1. As temperatures rise, rapid melting of glaciers will increase flow in rivers, causing floods, and as glaciers melt away, glacier-fed rivers will gradually run dry. According to Srinivasan (2019), small glaciers of less than 1 km^2 have been retreating quickly in the Indian Himalaya. In the Chenab river basin in Himachal Pradesh, the area of small glaciers has decreased by 38% between 1962 and 2004, while the area of large glaciers (greater than 10 km^2) has decreased by 12% in the same period (Khurana, 2021b).
2. A recent report by the Indian Space Research Organization (ISRO) found that 89% of the 2431 Himalayan glacial lakes identified in 2016−17 have significantly expanded—more than doubled their size (ISRO, 2024). This poses risk of these lakes breaching and causing sudden flooding and devastation in lower lying areas.
3. Droughts, floods, and extreme weather events are increasing across continents and ecologies. Droughts can lead to drying up of rivers: In the United States, drought has led to decline in water levels in rivers such as the Mississippi, clogging shipping lanes and affecting drinking water supplies (*New York Times*, 2022). This type of drought earlier expected once in 400 years can now be expected 30 years according to researchers (Singh & Khurana, 2023a, 2023b). In the first half of 2021 itself, drought was prevalent in most continents according to an August 2021 paper published in *Nature* (Langenbrunner, 2021). The number of people affected by drought was nearly 62.5 million (Erick Burguneo Salas, 2023). Groundwater overextraction compounds the problem and the effect of climate-related drought.
4. As the monsoon pattern changes, increased precipitation within short intense spells will lead to floods in rivers and flooding of adjacent areas.
5. As climate-related disasters increase, so will displacement, distress, and migration, with incalculable health, psychological, social, and economic burdens.

Healthy rivers cushion against climate change impacts, acting like sponges to soak excess water and redirecting it into groundwater "banks," which are then available in the times of drought. When the rivers are allowed to flow uninterrupted, water security is possible. For this, the connection and symbiotic relationship between groundwater and surface water needs to be functional: Rainwater needs to recharge groundwater and rivers need to be allowed to flow.

Below are time-tested examples of decentralized and community-centered rainwater conservation efforts that led to rejuvenation of rivers, revival of agricultural livelihoods and economies, and resilience. The nongovernmental organization Tarun Bharat Sangh (TBS) facilitated these village communities to take charge of their own lives and water security.

20.5   How decentralized community-driven water conservation measures are implemented

The communities are facilitated by the TBS team. Several meetings are held in the village to understand their problems and build confidence and trust between each other and within themselves. At the onset, three different pad yatras (walks within the village are undertaken) around the campaigns of planting and saving trees, saving water and making johads, and village independence. Discussions in the village are held around the importance of water conservation. The villagers are informed that they will need to make one-third of the contribution and be responsible for the O and M.

Site selection of the water conservation works supported by TBS take several perspectives into account. The villagers are informed about the facilitation and the financial support they will get from TBS and the contributions that the villagers will need to make. If during the discussions, the villagers suggest a site for the structure, immediately the TBS team visits the site along with the villagers, or alternately invites the villagers to the TBS office for discussions.

Discussions are held with the village elders including agriculturists and livestock herders and owners to understand the primary purpose of the water conservation structure, whether it is for agricultural irrigation, feeding livestock, or groundwater recharge. Potential sites and then visited and finalized. Discussions are held to understand the water sources available, the demand, the catchment area, the flow and amount of rainfall, the soil, and geology; then, collectively, decisions around the type of water conservation structure and site selection are made. Based on the above information, the size and cost of the structures are finalized and engineering drawings prepared accordingly. As far as possible, locally available materials are used. The villagers are encouraged to cultivate crops that have low water requirements and is suited to the ecology and to reduce this demand using microirrigation.

20.6 Life with and without water: Sherni–Parvati river rejuvenation in Rajasthan

The following are excerpts taken from the author's publication *Decentralized and equitable climate resilience through Community knowledge, Rainwater capture*, authored by Singh and Khurana (2023a, 2023b).

The Parvati–Sherni rivers were perennial rivers in the past, emerging from the deep forests nearby. The Sherni river is 70 km long and flows through 108 villages. It is a tributary of the Chambal, which in turn flows into the Yamuna, which originates from the Himalaya.

Mining started in these forests name of development. First, the wells dried up, then the tube wells, and by the year 2010, the bore wells also became defunct. Both these perennial rivers became rain-fed rivers in 1980. In 1990, although some of these mines were closed on the orders of the Supreme Court due to the efforts of TBS, some illegal mining continued on gun power.

Communities in several villages faced extreme marginalization by society and, coupled with a lack of water, legitimate livelihood opportunities were hard to come by in this challenging terrain. Meeting basic needs of family and putting food on the table was a daily challenge and led to despair, desperation, forced migration, family fragmentation, and resorting to illegal occupations, even dacoity, for survival. Conflicts were many and an atmosphere of fear and uncertainty prevailed.

When the Sherni river dried, income sources in villages situated around the source of the river—Maharajpura, Bhudkheda, Daudpura, Ballapura, Arodara, Khunda, Koripura, and others also dried up. These were perhaps the most affected and desperate villages, and as dignified and honest options to livelihood vanished, so did these people, either migrating in search of labor or taking to a life of crime and absconding. It was in these very villages that TBS began working with the villagers. They gained the trust and respect of the villagers, who, in turn, got trained in water conservation measures. Water conservation efforts began on different streams of the Sherni and Parvati. This relationship and water conservation work brought in water prosperity and helped revive the Sherni river.

20.6.1 Winds of change

Both the Sherni and Parvati rivers together form a total catchment area of 841 km^2. In this, area, 300-plus water conservation structures have been built by the villagers with the support of TBS. While the number of structures is not enough to protect the entire watershed, and more work is needed in this area, change is visible.

Due to water conservation, the ground water level increased, and as the water table came up, wells recharged and small water streams that fed into the river revived. Stretches of the river began to flow for longer periods of time and 2019 onwards, the Sherni rejuvenated back into a perennial river. When the area started getting water, the gun toting men turned to agriculture, halting Illegal mining and theft. As agricultural activities picked up, dignified livelihoods replaced violence and crime. Local prosperity improved and fear vanished. It was water that brought in this change.

As agricultural and livestock activities picked up, the men stopped evading the law and began appearing in court seeking their release. They began staying in the village the year round for their agricultural activities. Crop production is good because of the soil moisture. Fish rearing is thriving, and water chestnuts are also being cultivated. Green fodder is available for the livestock. Fuelwood is available within the village. All this has brought comfort in the lives of women.

Water brings in livelihood. Life. Dignity. The conscience buried deep under desperation and debt then has the space to awaken. As honest livelihoods opportunities improved, the men laid down their arms and embarked on a journey of nonviolence and rejuvenation. They realised that it was better to accept the charges and surrender, be free once and for all, rather than continue to live trapped in fear.

Entire villages would live in fear even when police cases would be filed against only one person: The entire village would be considered as criminal and face consequences. With the surrender and acceptance of one person, the entire village became free. When one criminal of Budhkheda village, who had 40 cases registered against him with the police surrendered in the courts, the entire village was able to breathe and live in freedom. This is the power and potential of water.

20.6.2 Sherni and Parvati Catchment areas, rainfall harvested with the structures created

I. Available rainfall for harvesting in Sherni and Parvati Catchments
Total catchment area = 840 sq km = 840 × 1000,000 sq m
Average rainfall = 616 mm, approximately 0.6 m
Runoff = 40% or 0.4 of the rainfall
Rainfall available for harvesting = 840 × 1000000 × 0.6 × 0.4 = 201.6 × 1000,000 cum = 201.6 million cum = 20.16 crore
cum = 201,600 million litres = 20,160 crore litres
II. Water harvesting and storage capacity created

Size of one lake behind one of the large structures	= 200 m × 150 m × 12 m
average depth of water	= 360 million litres
No of macro (9, with middle and micro structures)	= 160
Average water storage behind one structure	= 95 million litres
Total capacity of 160 structures	= 15,200 million litres = 1,520 crore litres
Groundwater recharge is about twice the capacity, which is also making the river flow perennially	= 3,040 crore litres
Total harvesting done	= **4,560 crore litres, or 22.62% of the rainfall which is available for harvesting.**

20.6.2.1 Catchment area

Smaller than 500 ha or 5 sq km = Micro. 100 ha = 1 sq km
Greater than 500 ha or 5 sq km = Macro
Conclusion — what the Sherni river tells us

If approximately 25 per cent of the rainfall in the catchment area is harvested, and the groundwater storage is recharged twice, even in the most challenging areas, revival of stretches of rivers is possible within a couple of years. The entire river can become perennial within ten years and at low cost.

In the Sherni catchment area, cost of storage created was between Rs 4—6/kiloliter.

20.7   Peace in water because of water: community chorus

Intensive interaction with village communities in 29 villages in the river basin by the author revealed the many layers of change that water brought about. Below is a glimpse of this change:

1. *The villagers now have dignity.* They live in peace and security, without fear and heads held high. The shroud of shame that enveloped the entire village, even when only one person had taken to crime, has lifted.
2. The men who had previously engaged in criminal activities are no longer on the run: They have surrendered on their own, served sentences if required and now stay with their families. *This has provided much relief and succor to the family members and the restless men themselves.*
3. Agricultural activities are expanding. Livestock is thriving and milk production has increased. The able-bodied youth who earlier migrated as labor are now engaging laborers from other villages. *The employment generating potential has thus increased.*
4. Agricultural activities have actualized dreams and brought in much needed money. The villagers are busy repairing or upgrading their homes, educating their children and improving their quality of life. *Aspirations are being met.*
5. Kitchen vegetable gardens and fruit trees are being planted. *This has led to decentralized availability of food and nutrition security, fodder, and fuelwood.*
6. Women, who rarely handled cash, now have some at their disposal and tied to the end of their saris (*pallu*), as is the practice in rural settings, *leading to their financial empowerment.*
7. *Rainwater conservation has led to the revival of ecosystems.* Grass cover is increasing, and forests are recovering as their biodiversity is increasing. Wildlife has adequate water and food. All this is improving the local ecosystem. Because of the increase in greenery, carbon sequestration is taking place. *This is climate mitigation.*
8. With the flowing Sherni river and recharged groundwater aquifers, the villagers are confident that they will be able to withstand erratic monsoon for at least 3—4 years. A good monsoon year will provide them a safety net for between 8 and 10 years. *The villages are moving towards climate resilience.*
9. *Only big expenditures/investments lead to big returns is a myth. Small is beautiful and gives manifold social, ecological, and economic returns, as these water conservation structures have proven. It leads to the creation of wealth in an affordable, nondestructive, rejuvenative, decentralized, and equitable manner, as opposed to the concentration of wealth in the hands of a few.*

20.8   Community validation of water conservation efforts made between 1980s and 1990s

When relationships between groundwater and surface water are maintained, and these feed into each other, water security is achieved. This is possible through decentralized, community-driven and ecologically sound rainwater conservation. As the soil health improves, the greenery increases and forests revive and thrive. With adequate water and green nutritious fodder available, livestock health and productivity improve. Dignified livelihoods become reality.

Over time, these efforts lead to cushioning against drought and flood. When communities realize what is possible and are willing to work toward a better and shared common future, they adapt themselves; through their water conservation measures, nature rejuvenation begins to take shape. Coupled with a respect for local ecology and shaping livelihoods around it, there is peace and security. Carbon is sequestered. The earth is healed. Shocks are withstood. Climate mitigation, adaptation, and resilience becomes possible, all through water conservation.

A visit to villages in three different river basins in Alwar district of Rajasthan (January to early March 2023) by the author (Indira Khurana), where water conservation works were undertaken in the mid-1080s and early 1990s by TBS and where the rains have been erratic and insufficient in the past 4 years reaffirmed this.

20.8.1   Immediate impact, long-term impact: Rada ka Baas village in Jahazwali river basin

Like other villages in Alwar district of Rajasthan, Rada ka Baas suffered from water scarcity and consequent hardships. Water was not available for livestock, domestic use, and agriculture.

Drought-stricken and facing hardships, in the mid-1980s, villagers of Rada ka Baas in Alwar district of Rajasthan, India, approached TBS seeking support for water conservation measures. The villagers were willing to meet 25% of the costs.

Water conservation through the construction of new *johads* and repair of existing ones began in 1986−88. A *johad* is a water percolation pond, a community-owned traditionally harvested rainwater storage waterbody. Together, there are 10 such water conservation structures within the village boundary. The community decided the location of the new structures, and with support of TBS, these structures were made, in a serial manner.

The ponds created by these structures filled up after the first rains itself, and within three days the wells began to have water. Where even a blade of grass as not visible, agricultural activities began. Livestock began to thrive, and forests began to rejuvenate.

March 1, 2023: A visit was made to two water conservation structures (*bandhs* that created a pond or *talaab*): school ka bandh and Sona ka bandh. The former talab contained water from which livestock and wild animals were seen quenching their thirst.

The Sona ka bandh *talaab* was full behind the *bandh*, where downstream plush wheat fields were visible. The monsoon was deficient for the past 4 years, (2019−22), but in this village, there continues to be water in the wells, buffaloes and goats have adequate water to drink, and agriculture is continuing. Because of vegetation, siltation in the *bandh* was less. The population of buffalos has increased, and milk production has shot up between 5 and 10 times.

The elders in the village explained, "As the flow of rainwater slows down and accumulates because of these structures, groundwater recharge takes place. There are some 10−15 dug wells in the village which are downstream of the Sona ka *bandh* structure. These wells fill up. Soil health has improved. Because of water availability, agricultural production has gone up. With the millet, wheat, and mustard cultivation, we meet our food requirements and there is fodder availability for the livestock. With ample water being available for livestock, milk production is good, with zero cost and with good selling price. Wildlife also has adequate water to drink."

The catchment of Sona ka bandh is large, and the rainwater flows with speed from the surrounding hills. The *bandh* acts like a "speed breaker," while groundwater recharge occurs at the same time. The excessive water that flows from the weirs flows with lesser speed and since there are natural nallahs, the water flows without destroying the farms, and all along the way, groundwater recharge continues. This helps in drought proofing and mitigating floods.

Drought and flood mitigation is possible by the "speed breaker effect" that is created by numerous ponds as part of decentralized rainwater conservation measures. These provide space for holding the water. The overflowing water then flows at a slower rate to the next structure and so on, thus slowing the flow, recharging along the way. The flow is also slowed due to the grown forests and the soil quality is maintained. Since the soil contains moisture and is no longer fragile, it is not blown away in the dry season. All these factors provide climate resilience.

During our visit, we observed that despite inadequate rainfall for past 4 years, water was being withdrawn from the wells downstream for cultivation and the village had hundreds of livestock whose thirst was being quenched, the pond was replete with water, and at about 14 ft deep at the center.

Over the years, the rainwater conservation had replenished groundwater reserves, given moisture to the soil and greened the forests. A livestock herder informed us, "In March we still have water available on the surface, we are confident that last year's monsoon water will be present to meet and intermingle with this year's rains in June—July. One good rainfall and we will be equipped to withstand erratic rainfall for 4—5 years."

The rainwater conservation through structures built with local wisdom has helped in drought and flood proofing. Over the years, the water conservation structures have provided space for holding water, leading to *climate mitigation*. An increase in the greenery has helped in sequestering carbon and reducing soil erosion, which is *climate mitigation*. The ability to withstand climate shocks is *climate resilience*. The cultivation of crops such as mustard and millets, which are not water intensive is *climate adaptation*.

20.8.2 Mandalvaas village in Sariska buffer zone, Bhagani river basin

This village where water conservation structures were made is inside the Sariska forest reserve, and was visited on January 10, 2023. Like other villages in the region, this village too had received scanty rainfall for last 3—4 years. Before reaching the main *johad*, the agricultural fields on both sides of the road were lush with crops. Six engines were drawing water from the dug well to agricultural fields at the entrance of the village. Moving further to the *johad*, long stretches of water in the river were observed. Speaking to the villagers and from physical observations, it was evident that the thriving agriculture was a result of sufficiency of water reserves. There was no sense of panic. The soil was moist enough to support greenery and not fragile and dry for a wind would blow away and cause siltation downstream or desertification, the villagers informed that with one good monsoon the structures would all be replenished, and they would be secure for the next 10 years.

The history of this village is that there was drought for four consecutive and long years in 1985 and most of the population had shifted to other villages and nearby towns.

TBS got involved in 1987, and after surveying the area along with the villagers, TBS saw the possibility of constructing several johads, leading to the formation of ponds for rainwater retention and groundwater recharge.

To build confidence, TBS held discussions with the villagers and organized exposure visits to villages where the people were reaping benefits of their water conservation work. All decisions were taken collectively by representatives of all households in the village. The villagers contributed 25% of the costs incurred.

This collective community participation had a direct impact on the production (agriculture, horticulture, animal husbandry, fisheries), employment, and economic prosperity of the families. Except for two families, all other families returned to the village, back to their roots. Expenses for *johad* maintenance and other community activities are being met by giving out fishing contracts to other parties.

This village is yet another example of climate mitigation, adaptation, and resilience.

20.8.3 Gadbasi village, Sabi river basin

In 1984—85 the situation was alarming because of the drought. Wells were dry and agriculture was minimal. People began moving out in search of work.

By 1992 the drought intensified. TBS decided to begin water conservation work here. However, there was a problem: People were not coming forward to work together for community conservation. The good thing was that work began on three individual lands. Meanwhile, dialogue with the villagers for community work continued.

In 1998 community afforestation measures began in a sacred grove. Alongside, water conservation discussions continued. In 1999 persistence paid, and the village community began to work for construction of the Mansa Sagar *bandh* on a *nallah*/stream.

The results were soon visible. With the first monsoon, water was available and visible in some stretches of the *nallah*/small stream. The clear water was considered sacred, and several rituals were performed here.

In a public meeting in 2004, the villagers put a proposal before TBS to build a large water conservation structure, with a commitment to meet 33% of the cost through their labor. Although the work began soon, it was completed only in 2008 because of differences within the village.

For the first time the *bandh* was full in 2008. Water began to flow through springs and for more than 3 km in the river. The people were overjoyed. Some 40—50 wells had water. Subsequently, water levels rose in downstream villages.

In January 2023, even though the monsoon was below average, there was greenery all around, and cultivation of wheat and mustard was thriving.

20.8.4 Villages in Gogunda valley: 35 years later

Gogunda block is a tribal block in Udaipur district of Rajasthan with around 230 villages. The geography is beautiful with undulating hills and forests. Following the completion of water conservation work of the civil society organization, Alert (ALERT) with support of TBS, the local population was able to initiate irrigation and ensure a regular supply of water, creating employment opportunities in agriculture. Girls began attending school, and the community actively participated in forest preservation efforts. This was the scenario, when ALERT withdrew from these villages 30 years ago.

On December 30, 2022, as the year came to an end, we visited this work in Gogunda valley along with members from ALERT.

This work was old, the structure was in the hills, in difficult terrain. The group had withdrawn from the village. Deforestation in the upper catchment had increased. The structure was built in the forest area decades ago. Would water be available? As we alighted from the vehicle and began climbing up the hills, we saw a small stream of water flowing downstream from the structure through a channel made of stones by the villagers. These were feeding into agricultural fields downstream.

The anicut was made in the upper regions of the forest area. After all these years, while there was some siltation in the structure, stretches of water were visible. The small streams of water flowing through the water channels was proof enough that the structure continued to serve its purpose and the tribals of the village. Discussing with the villagers, we found that this water continued to cater to around 100 villages.

Some of the men came forward and identified themselves. They were among the youth who had, 30 plus years ago, carried material up the hill slopes, on their backs and on mules, to the anicut construction site. Nostalgically, they recalled the challenges and the joy when the structure was complete.

This again was evidence enough that with the involvement of the locals, and suitable identification of site for conserving rainwater, the benefits extend to long term and include social impact, little envisaged while working on water conservation. This work led to a life of nonviolence against nature.

20.9 Scientific validation of community efforts

20.9.1 Revival of the Maheshwara river, Rajasthan

A detailed scientific evaluation of the change brought about in Khajura village revealed a 275.03% rise in income between 2009 and 2013 due to the water conservation efforts (Singh, 2013). A visit to the Morewala bandh site in 2024 revealed water security and flourishing agriculture (Table 20.1).

The Maheshwara river is 27 km long and has a catchment area of 102 km^2. The origins of Maheshwara river lies in Khejura village of Sapotara tehsil of Karauli district and in Bandhan ka pura village. The two streams join together in Khejura and then flow onward. Work began in 1998 and continued in 1999 and 2000. Up to March 2013, TBS had supported communities to make 107 water conservation structures.

A comparative analysis was undertaken in village Khejura to understand the change between 2009 and 2013 because of water availability.

As can be seen form the table above, between 2009 and 2013, there was a 275.03% increase in total village income.

As part of the People's Water and Peace Summit held in Karauli, on March 20−21, 2024, we revisited the village and the Morewala talab. Despite low rainfall in the past few years, the water in the center of the pond was approximately 25 ft deep. Downstream of the bandh, fields with wheat cultivation were blooming awaiting harvest in a few weeks. These fields were being irrigated through siphon irrigation from the water in the pond.

20.9.2 Reaffirmation of centuries-old traditional wisdom by modern science and engineering principles

This evaluation was undertaken in 1996 by Prof. G. D. Agarwal, Former Professor and Head of Civil Engineering, IIT Kanpur. The following are quoted from different sections of the evaluation report.

20.9.2.1 Objectives

- How well the works would fare if tested on well-accepted engineering principles and practices?
- What have been the significant quantifiable impacts of the works?

A basic tenet of TBS philosophy has been that rural communities have traditional knowledge of appropriate technologies for their welfare. All that is needed is encouragement and regeneration of their self-confidence, coupled with

TABLE 20.1 Comarative analysis of different parametres (in numbers and as percentage) in Khajura village.

No.	Parameter	2009	2013	Percentage (%) change between 2009 and 2013
1	Number of families	37	40	8.11
2	Population	206	236	14.56
3	Male population	115	129	12.17
4	Female population	91	107	17.58
5	Male literacy	63	97	53.97
6	Female literacy	29	47	62.07
7	Cow population	209	229	9.57
8	Buffalo population	192	316	64.58
9	Goat population	325	641	97.23
10	Milk production in kg	156	320	105.13
11	Land under agriculture (in bigha)	201	240	19.40
12	Rice production in kg	500	507	1.40
13	Wheat production in kg	470	804	71.06
14	**Total village income in INR**	**11,23,900**	**42,15,000**	**275.03**

Source: Modified from Singh R. (2013) Maheshwara Nadi (Maheshwara River). Tarun Bharat Sangh [in Hindi].

facilitation and financial support. All decisions are taken in a transparent manner by the villagers themselves through a *gram sabha*.

20.9.2.2 Findings

- As much as 36% of the works have the right capacity. Only 10% and 3% fell into the superfluous and excessive category.
- Small capacity is not a disadvantage, as the surplus flowing downstream is arrested by a downstream storage, thus dividing a large capacity requirement into several small storages.
- The best proof of adequacy and safety is that these works stood the test of time and the ravages of intense rainfalls of 1995 and 1996 when several engineer-designed structures maintained by the government failed, creating tragic floods. The crux lies not in the design or construction but in maintenance and protection. Also, villagers maintain and protect their possessions with utmost care and manage them so as to cause minimum damage when emergencies arise.
- An interesting aspect is the extremely low costs at which these storages were created. No engineering organization would have been able to create storage at such low costs.

20.9.2.3 Conclusions

- The water conservation structures built with the involvement of TBS at extremely low costs of Rs 1.00−2.00 per cubic meter storage costs have not only stood the test of time but are, by and large, engineering-wise sound and appropriate.
- A high coefficient of correlation establishes that the rise in groundwater table was an impact of the *johad* building effort. The optimal *johad* storage for these areas would be 1000−1500 cubic meters per hectare, which would raise the annual average groundwater table by 20 ft.
- The correlation between the increase in economic production and level of investment on water conservation work is must stronger than between *the johad* work and the rise in groundwater table. This is because the rise in groundwater table does not include the gains in soil conservation and the improvement in forest and grassland vegetation.
- There can be no other type of investment in the conditions of these villages that can yield quicker or better returns. An extremely high correlation (over 90%) shows the immediate and definite impact of investments on water conservation on village incomes. There can be no better investments than on *johads*.

As per the author, all investment in these areas should be put into *johads* (water conservation structures). Everything else can be taken care by the villagers themselves form their rising incomes that results from this investment (Agarwal, 1996).

20.10 The sense behind people's science

People's science or local wisdom is steeped in centuries of collective knowledge and wisdom, as a response, tried and tested in different scenarios. The examples covered in this publication prove the relevance of this knowledge even today.

India can be divided into 15 different ecological zones, which include high altitude dry and cold deserts, hot and dry deserts, coastal plains, plateaus, and mountain ranges. Based on these ecological zones, the rainfall, and the geology, over centuries, Indians developed a range of traditions and decentralized technologies to harvest water, be it rain, groundwater, or surface water to meet their water needs. From the northern most point of India to the southernmost tip, from the eastern boundary to the western boundary, there were water conservation structures that provided water for life.

Local/indigenous knowledge or "*lok vigyan*" is based on the collective wisdom of communities—not an individual—and is based on centuries of experience. A comprehensive understanding of this knowledge and its practice of all this helped communities conserve water, engage in dignified livelihoods, and live in peace and security. One significant lesson that can be drawn from this local knowledge was the importance of rainwater conservation, even in present times.

There was people's science behind this local wisdom, which included (a) an understanding the ecology, (b) the importance of rainwater conservation and managing flow of water: what are the structures to be built and where, and for what purpose—storage, groundwater recharge, or direct use, (c) lay of the land: where and how water flows, where it can be stored, and soil characteristics, (d) local biodiversity behavior: forest biodiversity, how plants are growing, behavior of animals, birds, insects, aquatic flora, and fauna to understand rainfall pattern, and surface water and groundwater availability, (e) cloud pattern, wind flow, and direction, (f) importance of localized food and cultural practices, water behavior, and (g) recognition of the importance of community action and common good.

Traditional harvesting systems have passed the test of time and are suited to the specific environments in which they evolved. These worked differently in different contexts, for example, drought and flood. The case studies presented here are contemporary validation of a local knowledge system that continues to provide livelihood and life needs.

Some of the learnings that can be drawn from this people's science as in the case studies discussed and field research are given below.

20.10.1 Drought and floods leads to degraded landscapes and desperation

Water scarcity degrades ecosystems, which then lose their capacity to support livelihoods and life. Soils become fragile and eroded and are blown away. They no longer have capacities to support biodiversity and agriculture. Drinking water becomes a challenge. In the absence of food and presence of hunger, residents are forced to migrate.

Floods sweep away lives, livelihoods, and soil. Ecosystems are degraded. Public health challenges mount as waterborne disease spread. Poverty levels rise and distress migration increases. Principles are sometimes sacrificed as desperation leads to violence and a life of crime. Droughts and floods have intergenerational impacts.

20.10.2 There is life in water: water is transformative, and this water can be made available through rainwater conservation

Rain is a major primary source of water. Barring Calama in the Atacama Desert in Chile where no rain has ever been recorded, there is no other place where it does not rain. Efficient capture and management of rain reduces, even prevents, disasters such as drought and flood.

Clouds store water. For centuries, India has had a rich tradition of rainwater harvesting and use through indigenous technologies that were ecologically sound and decentralized. Slowing the flow of rain and directing it into the groundwater aquifers and surface waterbodies provides cushioning against drought and flood. Recharge of groundwater aquifers is possible through water conservation structures whose designs are based on local ecology, rainfall, and terrain and made with community involvement. There are several examples of the revival of these indigenous, decentralized, and people-managed technologies and structures in different parts of India. These small structures when made in a contiguous manner can replenish groundwater and even revive rivers. These structures are restorative and rejuvenate and not based on displacement or destruction.

20.10.3 Small means more water and efficient capture of rainwater

a) Water can be captured either in large reservoirs that cover large catchments by building large dams; as rain runoff from small catchments in small tanks and ponds; or, by storing it in such a way that it recharges groundwaters and is then stored as groundwater.

According to Evanari et al. (1971), who conducted research on rainwater harvesting in the Negev desert, Israel, the volume of water harvested from small watersheds per hectare (ha) of watershed area is much more than that collected over large watersheds. This is because in large watersheds, the runoff travels longer distances before it is collected and there are water losses along the way. This loss can be significantly high as shown in Table 20.2.

As the above table indicates, 3000 microcatchments of 0.1 ha each will give 5 times more water together than one catchment of 300 ha, even though the total area over which the rain is harvested remains the same.

b) *During drought year, the loss of water due to evaporation and other causes can be significantly high.* According to a former Director of Central Soil and Water Conservation Research and Training Institute, and quoted by Agarwal et al. (2001a, 2001b), small catchments give more water. In semiarid regions, every 10-fold increase in catchment area reduces average annual runoff by about 36% (Boughton & Stone, 1985).

c) *National water security is possible through decentralized rainwater capture*: For water security, storage is essential, either on the surface or in underground reserves. Meeting the water needs of all life forms, of "everyone everywhere,"—and the Earth—is not possible through the creation of large storages. In any case, these are not the most efficient as indicated above. There is now a growing movement for destroying dams (Habel et al., 2020).

Water needs are decentralized and so the approach to meet these needs should also decentralized. It is possible to make entire countries water secure through rainwater capture and storage in surface waterbodies, traditional tanks, and other such structures and through groundwater recharge or for direct use. A good combination of rainwater harvesting and cultivation of water-saving crops such as millets and maize will lead to water and local food and livelihood security.

20.10.4 Decentralized small structures are cost-effective and powerful

Small/microrainwater conservation structures are powerful. They lead to immediate benefits in surrounding areas and if constructed over a large area, lead to macronature rejuvenation and livelihood benefits, without displacement or destruction. Over time, with judicious use of water, these structures contribute to river flows and to cushioning against extreme weather events.

Small structures to capture rain can be built with locally available knowledge and material. With community contribution, there is ownership and hence better maintenance. This was also the finding of Prof. G.D. Agarwal in his evaluation (Agarwal, 1996).

The costing of some of the rainwater conservation structures that led to the revival of the Sherni river is another example of how water availability can be enhanced through small and decentralized structures.

20.10.5 Rainwater conservation rejuvenates groundwater and surface water relationships

Groundwater and surface water have dynamic and symbiotic relationships that support each other. When the relationship between the two is snapped (groundwater or river water overextraction), the water sources begin to dry.

TABLE 20.2 Quantity of water harvested as percentage of rainfall in different size catchment areas.

No.	Size of catchment (ha)	Quantity of water harvested (cubic meters per ha)	Percentage of annual rainfall collected
1	Microcatchment (up to 0.1 ha)	160	15.21
2	20 ha	100	9.52
3	300 ha	50	3.33

Note: Microcatchment size is up to 1000 m^2 or 0.1 ha.
Source: Evanari, M., et al. (1971). The Negev: The Challenge of a Desert. Oxford University Press, UK. Agarwal, A., Sunita, N., Khurana, I. (Eds) (2001) *Making water everybody's business: Practice and policy of water harvesting.* CSE, New Delhi.

The reverse is also true. When water is captured and directed underground, it provides base flows to rivers and over time rejuvenates them. Rivers become perennial. During the monsoon, the flows in the river replenish groundwater resources. Rejuvenated rivers help groundwater recharge. Relationships between groundwater and surface water are restored and water is available year-round. This is what rejuvenation of Sherni and other rivers have shown. Healthy and rejuvenated rivers are important for climate resilience. Flowing rivers act as a cushion against drought and flood, soaking up excess water during heavy rains and releasing water during lean season.

20.10.6   Local food security depends on water conservation and availability

In India, irrigated agriculture occupies about 52% of the India's net sown area (Patel et al., 2020). Thus, for the remaining 48% of rain-fed sown area, from the agricultural livelihood and food security perspective, decentralized rainwater conservation cannot be ignored, since rain-fed agriculture contributes to a large portion of India's food production. This should not be confused with irrigation projects that provide for large-scale agriculture and food security (Agarwal et al., 2001a, 2001b).

20.10.7   Investing in community-centered decentralized rainwater conservation paves the way for socio-economic prosperity

The living examples mentioned in this chapter prove that once water became available, dignified livelihoods increased. The villagers became busier as agricultural activities increased, and drudgery reduced. Health improved as there was adequate food available. Education levels increased and more children began to go to school and pursue higher education. Living spaces improved.

20.10.8   Rainwater conservation leads to ecological regeneration

The earth needs healing through ecological regeneration. This healing is not possible without water. Rainwater therefore ecological regeneration that takes place in successive stages comes with its own economic impact (Agarwal et al., 2001a, 2001b). This was also visible in the work of TBS, in the ecological regeneration visible after the rejuvenation of river Sherni in Karauli district of Rajasthan (Singh & Khurana, 2022a, 2022b, 2022c) and in villages of Alwar and Udaipur districts of Rajasthan, revisited in 2023.

20.10.9   Rainwater conservation helps in climate mitigation, adaptation, and resilience

Structures created for rainwater conservation replenish water resources. Capacities to hold water are increased in a decentralized manner. These ponds, *johads*, and rejuvenated rivers, provide cushioning against erratic rainfall, thus mitigating both drought and flood and providing resilience. As soil moisture increases, soil erosion is less. Riverbeds are not silted up. This is another example of mitigation. Coupled with water-saving agricultural practices, adaption is possible.

20.10.10   Decentralized water conservation ushers in peace, security, and well-being

Decentralized and community-driven rainwater conservation brings water to the people, in the shortest time possible. If the water conservation structures are ecologically appropriate, costs are low.

When there is water, there is peace. Many conflicts at local, national, and international levels have their roots in the tension created due to water scarcity. Water flows do not recognize physical boundaries created by humans and simply flows. Lack of water leads to desperation. With water, the happiness index increases, there is peace and security and a feeling of well-being (Singh and Khurana, 2022a, 2022b, 2022c).

20.11   Conclusions

Solutions to global problems are rooted in local decentralized and community-driven actions that are just, equitable, and in harmony with the local environment. Such action recharges groundwater aquifers and provides flow to rivers, reviving them and making them perennial. This is what the work of TBS with village communities has proved. This combination is what brings in ownership and respect for nature and each other and the planet. As humankind and nature rejuvenate, climate resilience is achieved.

Water ushers in equality, peace, and security. It enhances a feeling of well-being and enables dignity. Water underpins survival, fulfilment of human rights, nature regeneration, and ecosystem restoration. Water thus supports life.

The examples cited in this chapter are living and time-tested testimonies of dramatic and sustainable turnarounds that were possible through decentralized community-driven water conservation in villages in some of India's most challenging geographies. *Availability of water led to a dramatic shift from a violence to a nonviolence way of life. Migrants returned to their villages as honest livelihoods became possible.*

Over time, water conservation led to revival of groundwater resources, rivers, and local economies. As ecologies regenerated, dignified livelihoods led to economic and social prosperity and an environment of peace and security. Local food security was possible. *Decentralized water conservation bridged the inequality gap and led to decentralized wealth generation.* These villages were able to withstand climate shocks of erratic monsoon. Thus, *sustained water conservation led to climate mitigation, adaptation, and resilience.*

The world is facing unprecedented environmental crisis deepened by climate change. Droughts and floods have increased in the world, and there is increasing water and food insecurity, loss of livelihood and dignity, and distress migration. Conflicts are on the rise. *The earth is wounded and in need of healing, which is possible through nature restoration and nature-based solutions. For this, water is critical.*

Solutions that are financially, culturally, socially, and ecologically equitable and sustainable are required. Water needs are decentralized. Rainfall is also decentralized. *Conservation of rain through decentralized ecologically appropriate structures can resolve the crisis of drought and flood and provide inequitable access to water. These structures are of low cost and easy to manage. When community-driven and based on local wisdom, these efforts are sustained.*

Rivers play a critical role in water resources availability and management. Rivers revive ecosystems and help cushion against drought and flood. *Rivers can be rejuvenated and must be allowed to flow.*

The examples from the above villages across three different river basins have stood the test of time, erratic monsoon behavior, and technical evaluation. The Sherni river rejuvenation example validates the prosperity, dignity, and nonviolence that is possible with water. *As we move toward an increasingly uncertain climate, the lessons from these villages offer validated ray of hope as solutions for resilient livelihoods and economic prosperity that is restorative, healing, nonviolent, decentralized, and equitable.*

References

Agarwal, A., Narain, S., & Khurana, I. (Eds.). (2001a). Making Water Everybody's Business: Practice and Policy of Water Harvesting (3rd reprint, pp. xvi–xx). CSE, New Delhi.

Agarwal, A., Narain, S., & Khurana, I. (Eds.). (2001b). Making Water Everybody's Business: Practice and Policy of Water Harvesting (3rd reprint, pp. xvi–xxvii). CSE, New Delhi.

Agarwal, G. D. (1996). An Engineer's Evaluation of Water Conservation Efforts of Tarun Bharat Sangh in 36 Villages of Alwar District.

Armstrong, M. (2024). Where Water Stress will be Highest in 2050. https://www.statista.com/chart/26140/water-stress-projections-global/.

Banerjee, K. (2023). Climate Change: Impact and Implications. *Millennium Post*, India.

Barthakur, R., & Khurana, I. (Eds.). (2014). Reflections on Managing Water: Earth's Greatest Natural Resource. Balipara Foundation, Assam, India.

Boughton, W. C., & Stone, J. J. (1985). Variation of runoff with watershed area in a semiarid location. *Journal of Arid Environment, 9*, 13–15.

Correspondents (January 12, 2017) Economic Inequality Top Global Risk: WEF. https://www.thenews.com.pk/print/178394-Economic-inequality-top-global-risk.

Damania, R., Desbureaus, S., Rodella, A.-S., Russ, J., & Zaveri, E. (2019). Quality Unknown. The Invisible Water Crisis. World Bank.

Ennis, S., et al. (2017). Inequality: A Hidden Cost of Market Power. http://www.oecd.org/daf/competition/inequality-a-hidden-cost-of-market-power.htm.

Erick Burguneo Salas. (June 26, 2023). Number of People Affected by Drought Worldwide from 1990–2021. https://www.statista.com/statistics/1293329/global-number-of-people-affected-drought/.

Eslamian, S., & Eslamian, F. (2021a). Handbook of Water Harvesting and Conservation. In *Basic Concepts and Fundamentals*. (1, p. 495). New Jersey, USA: John Wiley & Sons, Inc.

Eslamian, S., & Eslamian, F. (2021b). Handbook of Water Harvesting and Conservation. In *Case Studies and Applications Examples*. (2, p. 496). New Jersey, USA: John Wiley & Sons, Inc.

Evanari, M., et al. (1971). *The Negev: The Challenge of a Desert.* UK: Oxford University Press.

Grill, G., Lehner, B., Thieme, M., et al. (2019). Mapping the world's free-flowing rivers. *Nature, 569*, 215–221. Available from https://doi.org/10.1038/s41586-019-1111-9.

Habel, M., Mechkin, K., Podgorska, K., et al. (2020). Dam and reservoir removal projects: A mix of social-ecological trends and cost-cutting attitudes. *Scientific Reports, 10*, 19210. Available from https://doi.org/10.1038/s41598-020-76158-3.

ISRO. (April 22, 2024). Satellite Insights: Expanding Glacial Lakes in the Indian Himalayas. https://www.isro.gov.in/Satellite_Insights_Expanding_Glacial_Lakes_Indian_Himalayas.html.

Khurana, I. (November 3, 2021a). Why Rue a Dead River? https://globalbihari.com/why-rue-a-dead-river/.

Khurana, I. (September 2021b). Rivers for Climate Resilience. https://timesofindia.indiatimes.com/blogs/eco-sensitive/rivers-for-climate-resilience/.

Konikow, L. F., & Kendy, E. (2005). Groundwater depletion: A global problem. *Hydrogeology Journal, 13*, 317–320.

Kundzewicz, Z. W. (2008). Climate change impacts on the hydrological cycle. *Ecohydrology & Hydrobiology, 8*(2–4), 195–120.

Langenbrunner, B. (2021). Water, water not everywhere. *Nature Climate Change, 11*, 650. Available from https://doi.org/10.1038/s41558-021-01111-9.

Migration Policy Institute. (2022). Top 10 Migration Issues of 2022. https://www.migrationpolicy.org/programs/migration-information-source/top-10-migration-issues-2022.

Patel, N., Bruno, D., & Nagaich, R. (2020). A New Paradigm for Indian Agriculture: From Agroindustry to Agroecology. Niti Ayog, Delhi, India.

Singh, R., & Khurana, I. (2022a). Rejuvenation of Rivers. Climate Change. Resilience. Dignity: Living Examples. https://www.researchgate.net/publication/366633507_Rejuvenation_of_Rivers.

Singh, R., & Khurana, I. (2022b). Rejuvenation of Rivers. Climate Resilience. Livelihoods. Dignity. Living Examples. https://norden.iofc.org/sites/default/files/2022-09/river-rejuvenation2-1_0.pdf.

Singh, R., & Khurana, I. (2022c). Rejuvenation of Rivers: Climate Resilience, Livelihoods, Dignity (Epilogue).

Singh, R., & Khurana, I. (March 2023a). Drought, Flood and Climate Change: Global challenge, local solution. Water Conservation for mitigation, Adaptation and Resilience: A Reaffirmation.

Singh, R., & Khurana, I. (November 2023b). Decentralised and Equitable Climate Resilience through Community knowledge, Rainwater Capture, Experience, Evidence. ISBN 978-93-6039-792-0.

Singh, R. (2013). Maheshwara Nadi (Maheshwara River). Tarun Bharat Sangh [in Hindi].

Srinivasan, J. (2019). Impact of Climate Change on India. In N. K. Dubash (Ed.), *India in a Warming World: Integrating Climate Change and Development*. Oxford University Press. Available from https://doi.org/10.1093/oso/9780199498734.001.0001, Online ISBN 9780199098408.

UNU EHS. (2023a). The Interconnectedness Disaster Risk Report. https://interconnectedrisks.org/summaries/2023-executive-summary.

UNU EHS. (2023b). The Interconnectedness Disaster Risk Report. https://s3.eu-central1.amazonaws.com/interconnectedrisks/reports/2023/TR_231115_Groundwater.pdf.

Wada, Y., et al. (2010). Global depletion of groundwater resources. *Geophysical Research Letters, 37*, L20402.

World Economic Report. (2019). The Global Risks Report 2019. https://web.archive.org/web/20190325101510/https://www.weforum.org/reports/the-global-risks-report-2019.

Further reading

https://www.nytimes.com/2022/11/03/us/mississippi-river-drought.html?action = click&pgtype = Article&state = default&module = styln-extreme-weather&variant = show&region = MAIN_CONTENT_1&block = storyline_top_links_recirc.

Index

Note: Page numbers followed by "*f*" and "*t*" refer to figures and tables, respectively.